Invasion Genomics

_____

Invasion Genomics

Edited by

Dan G. Bock and Marc Rius

CABI

CABI is a trading name of CAB International

CABI
Nosworthy Way
Wallingford
Oxfordshire OX10 8DE
UK

CABI
200 Portland Street
Boston
MA 02114
USA

Tel: +44 (0)1491 832111
E-mail: info@cabi.org
Website: www.cabi.org

Tel: +1 (617)682-9015
E-mail: cabi-nao@cabi.org

A catalogue record for this book is available from the British Library, London, UK.

ISBN-13: 9781800626249 (hardback)
9781800626256 (ePDF)
9781800626263 (ePub)

DOI: 10.1079/9781800626263.0000

Commissioning Editor: David Hemming
Editorial Assistant: Theresa Regueira
Production Editor: James Bishop

Typeset by Straive, Pondicherry, India
Printed in the USA

Contents

Contributors

_____

Malika L. Ainouche, University of Rennes, Rennes, France. ORCID iD: https://orcid.org/0009-0004-6979-3405. E-mail: malika.ainouche@univ-rennes1.fr

Elena Barni, University of Turin, Turin, Italy. ORCID iD: https://orcid.org/0000-0001-7256-0064. E-mail: elena.barni@unito.it

Paul Battlay, School of Biological Sciences, Monash University, 25 Rainforest Walk, Clayton 3800, Victoria, Australia. ORCID iD: https://orcid.org/0000-0001-6050-1868. E-mail: Paul.Battlay@monash.edu

Jingwen Bi, Fudan University, Shanghai, China. ORCID iD: https://orcid.org/0009-0005-5161-3952. E-mail: jwbi20@fudan.edu.cn

Vanessa C. Bieker, Department of Natural History, NTNU University Museum, Norwegian University of Science and Technology (NTNU), Trondheim, Norway. ORCID iD: https://orcid.org/0000-0002-2061-9041. E-mail: vanessa.bieker@ntnu.no

Bernd Blossey, Department of Natural Resources, Cornell University, Ithaca, New York, USA. ORCID iD: https://orcid.org/0000-0001-7043-4435. E-mail: bb22@cornell.edu

Dan G. Bock, School of Environment and Science, Griffith University, Nathan, QLD 4111, Australia. ORCID iD: https://orcid.org/0000-0001-7788-1705. Email: dan.bock@griffith.edu.au

Oliver Bossdorf, University of Tübingen, Tübingen, Germany. ORCID iD: https://orcid.org/0000-0001-7504-6511. E-mail: oliver.bossdorf@uni-tuebingen.de

Bethany Burns, University of South Florida, Tampa, Florida, USA. ORCID iD: https://orcid.org/0009-0006-3822-2823. E-mail: bethanyburns@usf.edu

Peipei Cao, Fudan University, Shanghai, China. ORCID iD: https://orcid.org/0009-0004-0026-9787. E-mail: ppcao20@fudan.edu.cn

Armand Cavé-Radet, University of Rennes, Rennes, France; University of Tübingen, Tübingen, Germany. ORCID iD: https://orcid.org/0000-0001-9254-4993. E-mail: armand.cave-radet@hotmail.fr

Bing Chen, Department of Ecology and Evolutionary Biology, School of Life Sciences, Fudan University, Shanghai, China. ORCID iD: https://orcid.org/0000-0002-5259-7086. E-mail: chenbing23@outlook.com

Yiyong Chen, Research Center for Eco-Environmental Sciences, Chinese Academy of Sciences, Beijing, China. ORCID iD: https://orcid.org/0000-0002-8285-6472. E-mail: yychen@rcees.ac.cn

Manpreet K. Dhami, Biocontrol and Molecular Ecology, Manaaki Whenua Landcare Research, Lincoln, New Zealand. ORCID iD: https://orcid.org/0000-0002-8956-0674. E-mail: dhamim@landcareresearch.co.nz

Stacy B. Endriss, University of North Carolina Wilmington, Wilmington, NC, USA; Virginia Tech, Blacksburg, VA, USA. ORCID iD: https://orcid.org/0000-0001-9688-4741. E-mail: endriss@vt. edu

Elisa Giaccone, University of Turin, Turin, Italy. ORCID iD: https://orcid.org/0000-0001-6440-2533. E-mail: elisa.giaccone@gmail.com

Uta Grünert, University of Tübingen, Tübingen, Germany. ORCID iD: https://orcid.org/0009-0002-0297-7328. E-mail: uta.gruenert@uni-tuebingen.de

Yaolin Guo, Fudan University, Shanghai, China. ORCID iD: https://orcid.org/0000-0002-2203-1970. E-mail: ylguo19@fudan.edu.cn

Kathryn A. Hodgins, School of Biological Sciences, Monash University, 25 Rainforest Walk, Clayton 3800, Victoria, Australia. ORCID iD: https://orcid.org/0000-0003-2795-5213. E-mail: kathryn. hodgins@monash.edu

Juntao Hu, Department of Ecology and Evolutionary Biology, School of Life Sciences, Fudan University, Shanghai, China. ORCID iD: https://orcid.org/0000-0003-1857-8700. E-mail: juntao_hu@ fudan.edu.cn

Xuena Huang, Research Center for Eco-Environmental Sciences, Chinese Academy of Sciences, Beijing, China. ORCID iD: https://orcid.org/0000-0002-1517-7496. E-mail: xnhuang@rcees. ac.cn

Ruth A. Hufbauer, Department of Agricultural Biology, Colorado State University, Fort Collins, CO 80523 USA; Graduate Degree Program in Ecology, Colorado State University, Fort Collins, CO 80523 USA. ORCID iD: https://orcid.org/0000-0002-8270-0638 E-mail: ruth.hufbauer@ colostate.edu

Ramona E. Irimia, University of Tübingen, Tübingen, Germany. ORCID iD: https://orcid.org/0000-0002-5304-8660. E-mail: ramona-elena.irimia@uni-tuebingen.de

Ruiting Ju, University of Rennes, Rennes, France. ORCID iD: https://orcid.org/0000-0001-9265-8245. E-mail: jurt@fudan.edu.cn

Sophie Karrenberg, Uppsala University, Uppsala, Sweden. ORCID iD: https://orcid.org/0000-0002-7146-588X. E-mail: sophie.karrenberg@ebc.uu.se

Kyle Keefer, University of South Florida, Tampa, FL, USA. ORCID iD: https://orcid.org/0009-0003-4156-0610. E-mail: kylekeefer@usf.edu

Olga Kozhar, Department of Agricultural Biology, Colorado State University, Fort Collins, CO 80523 USA; ORCID iD: https://orcid.org/0000-0002-0976-7334. E-mail: olga.kozhar@gmail. com

Carol Eunmi Lee, Department of Integrative Biology, University of Wisconsin-Madison, Madison, WI 53703, USA. ORCID iD: https://orcid.org/0000-0001-6355-0542. E-mail: carollee@wisc. edu

Katie Lee, Department of Natural Resources, Cornell University, Ithaca, NY, USA. ORCID iD: https://orcid.org/0000-0002-8930-7505. E-mail: kl528@cornell.edu

Bo Li, Department of Natural Resources, Cornell University, Ithaca, NY, USA; Yunnan University, Kunming, China. ORCID iD: https://orcid.org/0000-0002-0439-5666. E-mail: bool@fudan.edu. cn

Zhiyong Liao, Xishuangbanna Tropical Botanical Garden, Xishuangbanna, China. ORCID iD: https://orcid.org/0000-0001-6411-2042. E-mail: liaozy@xtbg.org.cn

Man Luo, Department of Ecology and Evolutionary Biology, School of Life Sciences, Fudan University, Shanghai, China. ORCID iD: https://orcid.org/0009-0009-8676-1938. E-mail: mlman7@ 163.com

Michael D. Martin, Department of Natural History, NTNU University Museum, Norwegian University of Science and Technology (NTNU), Trondheim, Norway. ORCID iD: https://orcid.org/0000-0002-2010-5139. Email: mike.martin@ntnu.no

Angela McGaughran, Te Aka Mātuatua/School of Science, University of Waikato, Hamilton, New Zealand. ORCID iD: https://orcid.org/0000-0002-3429-8699. E-mail: angela.mcgaughran@ waikato.ac.nz

Jane Molofsky, Department of Plant Biology, University of Vermont, Burlington, VT 05405, USA; ORCID iD: https://orcid.org/0000-0001-7927-516X. E-mail: Jane.Molofsky@uvm.edu

Kattia Palacio-Lopez, Natural Science Department, University of Houston-Downtown, Houston, TX 77002, USA; ORCID iD: 0000-0002-9575-270X. E-mail: kattiapalacio@gmail.com

Madalin Parepa, University of Tübingen, Tübingen, Germany. ORCID iD: https://orcid.org/0000-0002-3567-3884. E-mail: madalin.parepa@uni-tuebingen.de

Elahe Parvizi, Te Aka Mātuatua/School of Science, University of Waikato, Hamilton, New Zealand. ORCID iD: https://orcid.org/0000-0002-1695-8817. E-mail: ellie.parvizi@waikato.ac.nz

Marta Pascual, Departament de Genètica, Microbiologia i Estadística, Facultat de Biologia, Universitat de Barcelona, 08028 Barcelona, Spain; Institut de Recerca de la Biodiversitat (IRBio), Universitat de Barcelona, Barcelona, Spain. ORCID iD: https://orcid.org/0000-0002-6189-0612. E-mail: martapascual@ub.edu

Christina L. Richards, University of South Florida, Tampa, Florida, USA; University of Tübingen, Tübingen, Germany; Xishuangbanna Tropical Botanical Garden, Xishuangbanna, China. ORCID iD: https://orcid.org/0000-0001-7948-5165. E-mail: clr@usf.edu

Marc Rius, Department of Marine Ecology, Centre for Advanced Studies of Blanes (CEAB), Spanish Research Council (CSIC), 17300 Blanes, Spain; Department of Zoology, Centre for Ecological Genomics and Wildlife Conservation, University of Johannesburg, Auckland Park Johannesburg 2006, South Africa. ORCID iD: https://orcid.org/0000-0002-2195-6605. E-mail: mrius@ceab.csic.es

Lee A. Rollins, Evolution & Ecology Research Centre, University of New South Wales, Sydney, NSW 2052, Australia. ORCID iD: https://orcid.org/0000-0002-3279-7005. E-mail: l.rollins@unsw.edu.au

Armel Salmon, University of Rennes, Rennes, France. ORCID iD: https://orcid.org/0000-0003-4479-0801. E-mail: armel.salmon@univ-rennes1.fr

Anna W. Santure, School of Biological Sciences, University of Auckland, Auckland, New Zealand. ORCID iD: https://orcid.org/0000-0001-8965-1042. E-mail: a.santure@auckland.ac.nz

Marc W. Schmid, MWSchmid GmbH, Glarus, Switzerland. ORCID iD: https://orcid.org/0000-0001-9554-5318. E-mail: contact@mwschmid.ch

Nicole Sebesta, University of Turin, Turin, Italy. ORCID iD: https://orcid.org/0000-0002-0976-3560. E-mail: nicole.sebesta@unito.it

Jane E. Stewart, Department of Agricultural Biology, Colorado State University, Fort Collins, CO 80523, USA; Graduate Degree Program in Ecology, Colorado State University, Fort Collins, CO 80523 USA. ORCID iD: https://orcid.org/0000-0001-9496-6540. E-mail: Jane.stewart@colostate.edu

Katarina C. Stuart, School of Biological Sciences, University of Auckland, Auckland, New Zealand; Evolution and Ecology Research Centre, University of New South Wales, Sydney, NSW 2052, Australia. ORCID iD: https://orcid.org/0000-0002-0386-4600. E-mail: Katc.stuart@gmail.com

Carolyn Tepolt, Department of Biology, Woods Hole Oceanographic Institution, Woods Hole, MA 02543-1050, USA. ORCID iD: https://orcid.org/0000-0002-7062-3452. Email: ctepolt@whoi.edu

Isolde van Riemsdijk, University of Tübingen, Tübingen, Germany; Lund University, Lund, Sweden. ORCID iD: https://orcid.org/0000-0001-9739-6512. E-mail: isolde.van-riemsdijk@uni-tuebingen.de

Amy L. Vaughan, Biocontrol and Molecular Ecology, Manaaki Whenua Landcare Research Group, Bioeconomy Science Institute, Lincoln, New Zealand and Kura Ngahere School of Forestry, University of Canterbury, Christchurch, New Zealand. ORCID iD: https://orcid.org/0000-0003-0309-8851. E-mail: amy.vaughan@canterbury.ac.nz

Frédérique Viard, ISEM, University of Montpellier, CNRS, EPHE, IRD, Montpellier, France. ORCID iD: https://orcid.org/0000-0001-5603-9527. E-mail: frederique.viard@umontpellier.fr

Shengyu Wang, Fudan University, Shanghai, China; Université Paris-Saclay, Palaiseau, France. ORCID iD: https://orcid.org/0000-0002-2161-5317. E-mail: shengyuwang20@fudan.edu.cn

Jihua Wu, Fudan University, Shanghai, China. ORCID iD: https://orcid.org/0000-0001-8623-8519. E-mail: jihuawu@fudan.edu.cn

Tianchun Wu, School of Biological Sciences, Monash University, 25 Rainforest Walk, Clayton 3800, Victoria, Australia.

Wei Yuan, Max Planck Institute, Tübingen, Germany. ORCID iD: https://orcid.org/0000-0001-5117-3639. E-mail: wyuan@tuebingen.mpg.de

Aibin Zhan, Research Center for Eco-Environmental Sciences, Chinese Academy of Sciences, Beijing, China; University of Chinese Academy of Sciences, Chinese Academy of Sciences, Beijing, China. ORCID iD: https://orcid.org/0000-0003-1416-1238. E-mail: azhan@rcees.ac.cn

Lei Zhang, Fudan University, Shanghai, China. ORCID iD: https://orcid.org/0009-0006-0479-8058. E-mail: zhang_l20@fudan.edu.cn

Weihan Zhao, Fudan University, Shanghai, China; University of Konstanz, Konstanz, Germany. ORCID iD: https://orcid.org/0009-0005-6575-7623. E-mail: weihan.zhao@uni-konstanz.de

Yujie Zhao, Fudan University, Shanghai, China. ORCID iD: https://orcid.org/0009-0003-6226-524X E-mail: 22110850041@m.fudan.edu.cn

Xin Zhuang, Fudan University, Shanghai, China University of Helsinki, Helsinki, Finland. ORCID iD: https://orcid.org/0000-0002-1626-2043. E-mail: xin.zhuang@helsinki.fi

1 From the Ecology to the Genetics and Genomics of Invasive Species

Dan G. Bock[1]* and Marc Rius[2,3]

[1]*School of Environment and Science, Griffith University, Nathan, Australia;*
[2]*Department of Marine Ecology, Centre for Advanced Studies of Blanes (CEAB),
Spanish Research Council (CSIC), Blanes, Spain;* [3]*Department of Zoology, Centre
for Ecological Genomics and Wildlife Conservation, University of Johannesburg,
Auckland Park Johannesburg, South Africa*

Abstract

Invasion science has evolved from a narrowly focused subdiscipline of ecology into a dynamic, multi-disciplinary field at the forefront of ecological and evolutionary research. Genomic tools have played a pivotal role in expanding the scope of this field, contributing to both fundamental and applied research. This chapter traces the development of invasion science, from its ecological roots to the integration of genetic and genomic technologies, culminating in the transformative impact of high-throughput sequencing over the past two decades. Within this framework, it introduces the chapters of this volume, which showcase the breadth of research in invasion genomics and identify key unresolved questions about biological invasions that genomics is beginning to answer.

1.1 Introduction

Since at least the mid-1800s, researchers have used invasive species as model systems to study and understand the natural world (Huey *et al.*, 2005; Cadotte, 2006; Sax *et al.*, 2007; Simberloff, 2013). Early naturalists, evolutionary biologists and ecologists recognized that biological invasions represent a unique asset, because they occur on spatial and temporal scales that far exceed the scales of most planned field experiments. Charles Darwin, for example, relied on naturalized invasive species to illustrate key concepts in his magnum opus, *On the Origin of Species* (Darwin, 1859). In Chapter 3 of his book, Darwin points the reader to 'introduced plants which have become common throughout whole islands', to illustrate that, when released from the constraints of competition, populations can rapidly expand in size (Darwin, 1859). Throughout much of the 19th century, and the early 20th century, invasive species were viewed mainly as rare curiosities, or as fortuitous opportunities to understand nature (Cadotte, 2006; Richardson and Pysek, 2007). Although prominent ecologists like Charles Elton brought attention to the 'ecological explosions' of invasive species, and the potential that 'instead of six continental realms of life... there will only be one world' (Elton, 1958; Ricciardi and MacIsaac, 2008), biological invasions were not yet considered a global burden.

The view that invasive species are an important driver of extinctions and economic

*Corresponding author: dan.bock@griffith.edu.au

© CAB International 2025. *Invasion Genomics* (Eds D. Bock and M. Rius)
DOI: 10.1079/9781800626263.0001

loss while also representing a hazard to human health started to permeate the scientific literature in the mid-1980s. This occurred on the heels of an international SCOPE (Scientific Committee on Problems of the Environment) programme on biological invasions (Drake *et al.*, 1989). Initiated in 1982, this programme sponsored scientific meetings in the UK, South Africa, the Netherlands, the USA and Australia, covering topics ranging from attributes of invasive species and invaded habitats to the management of biological invasions (Huenneke *et al.*, 1988; Richardson and Pysek, 2007; Simberloff, 2011, 2013). Researchers at that time noted the potential magnitude of bioinvasions in both marine and terrestrial ecosystems (Carlton, 1987; Carlton and Geller, 1993; Lodge, 1993). Over the following decades, SCOPE and other programmes had a prominent role in inspiring research on bioinvasions, establishing modern invasion science as a rapidly expanding scientific discipline (Simberloff, 2011, 2013; Ricciardi *et al.*, 2017).

Over the last decades, technological and methodological advances have dramatically increased the accessibility of genetic tools to researchers studying biological invasions. The cost of DNA sequencing has plunged in the last 20 years and high-throughput sequencing has become a routine method used in biodiversity research. This has had revolutionary implications for several scientific disciplines including invasion science. Since the seminal work of Baker and Stebbins (1965), there has been a sharp increase in studies exploring the genetics and genomics of invasive populations (Bock *et al.*, 2015; Rius *et al.*, 2015a,b; North *et al.*, 2021; McGaughran *et al.*, 2024; Hodgins *et al.*, 2025), particularly following the introduction and widespread adoption of high-throughput sequencing. Researchers are now able to explore genomic regions that can facilitate rapid adaptation during colonization, or unravel the composition and dynamics of entire biological communities affected by invasive species via metabarcoding and metagenomics (Darling and Mahon, 2011; Bell *et al.*, 2024).

This chapter first introduces foundational ecological research (section 1.2) and then considers the context in which evolutionary and genetic information were initially incorporated in the study of invasive species (section 1.3). Subsequently, it reviews how new technological advancements that now allow us to sequence nearly all regions of the genome are revolutionizing invasion science (section 1.4). This includes advancing our understanding of how invasive species originate, how they interact with the biotic and abiotic environment in their new ranges, and what their evolutionary trajectory might be in a rapidly changing environment. Taken together, this chapter aims to provide the context for the rest of the volume, as well as to highlight how high-throughput sequencing, particularly when combined with careful experimentation and detailed ecological information, can help us tackle key questions about invasive species that are still largely unresolved.

1.2   The Rise of Invasion Biology and Ecological Research

Early invasion biology research (1980s–1990s) focused predominantly on community ecology. This included, for instance, the role that predation by invasive species has in precipitating species extinctions. A textbook example is the invasion of the brown tree snake on the island of Guam, which decimated most of the forest bird fauna of the island and caused global extinctions of endemics such as the Guam flycatcher (Savidge, 1987; Lockwood *et al.*, 2013; Roy *et al.*, 2024). An equally striking example is represented by the collapse of endemic species flocks of haplochromine cichlids in Lake Victoria of East Africa, which began in the early 1980s, following the introduction of the invasive Nile perch (Ogutu-Ohwayo, 1990; Witte *et al.*, 1992; Pringle, 2011). The cascading impacts of bioinvasions on multiple trophic levels were also highlighted during this time (Vitousek *et al.*, 1987; Spencer *et al.*, 1991; reviewed by Ehrenfeld, 2010). A case in point is the 1981 opossum shrimp invasion of Flathead Lake (Montana, USA), which led to the collapse of kokanee salmon, which in turn contributed to the precipitous decline of local populations of bald eagles and grizzly bears (Spencer *et al.*, 1991).

Research that followed in the 1990s and early 2000s expanded to also include predictive ecology and the identification of characteristics that can be used to forecast bioinvasions (Lodge, 1993; Kolar and Lodge, 2001). This work focused on attributes of the release events, on characteristics of the invasive species and on properties of the recipient communities. Regarding release events, the probability of establishment success was found to increase with the number and size of introduced propagules (i.e. propagule pressure; Lockwood *et al.*, 2005, 2009; Colautti *et al.*, 2006; Simberloff, 2009). With regard to the characteristics of invasive species, building on earlier descriptions of invasiveness traits (Baker, 1965; reviewed by van Kleunen *et al.*, 2015), characteristics such as rapid dispersal, high phenotypic plasticity and a propensity for associations with humans were highlighted as representative of invaders that are likely to become ecological 'winners' in an increasingly human-dominated world (Lodge, 1993; McKinney and Lockwood, 1999; Colautti *et al.*, 2006). Finally, at the level of recipient communities, interactions among species were proposed to either facilitate invasions, when they occur among invasive taxa (i.e. invasional meltdown; Simberloff and von Holle, 1999), or to prevent invasions, when invasive species interact with species-rich native communities (i.e. biotic resistance; Case, 1991; Stachowicz *et al.*, 1999).

Far from having noteworthy contributions only during the early stages of invasion science, ecological research continues to drive important breakthroughs in our understanding of invasive species and our ability to manage them. Some of the most consequential problems currently being tackled by invasion ecologists relate to the synergistic effects between biological invasions and other aspects of global change (reviewed by Ricciardi *et al.*, 2017, 2021). This includes, for example, the effects of ice loss and associated globalization of the Arctic and the Antarctic on local invasion risks (e.g. Duffy *et al.*, 2017; Chan *et al.*, 2019), or the extent to which global warming can change the direction and intensity of species interactions, affecting the spread of invasive species (e.g. Duell *et al.*, 2019; Ricciardi *et al.*, 2021). From a geopolitical perspective, it includes evaluating the extent to which intercontinental trade agreements or military conflicts can, as an unintended consequence, magnify the invasive species problem by expanding transportation networks (Ricciardi *et al.*, 2017), or by reducing the resistance of local biological communities to invasions (e.g. Guyton *et al.*, 2020).

1.3 Invasion Science and Early Contributions of Genetics

While ecology had a prominent role in the development of the fledgling field of invasion science, evolution and genetics had, by comparison, more modest contributions at first (Barrett, 2015). This may seem surprising, given that *The Genetics of Colonizing Species*, the first volume focused on the genetics and evolution of weedy and invasive taxa, was published in 1965 (Baker and Stebbins, 1965). In addition, genetic information had started to be used for the study of invasive species since the 1930s. For example, chromosome counts were studied to clarify the origin of introduced taxa and their hybrids (e.g. Huskins, 1931). The major impediment to the broad-scale incorporation of evolutionary and genetic thinking in invasion biology was a paradigm engrained in other biological fields as well, that viewed evolution as a slowly unfolding process. Evolutionary transitions that could be observed in the geological record were generally assumed to happen over thousands and millions of years, far too slowly to influence the course of invasions (reviewed by Reznick *et al.*, 2019; but see Pimentel, 1968, as an early example of a study that departed from this status quo).

By the late 1990s, the tide had started to turn, and contemporary evolution, which occurs over the course of months or years, was no longer regarded as a curiosity (Reznick *et al.*, 2019). This change in perception was facilitated by evidence that natural selection can be quite strong in wild populations (Endler, 1986), by observational data that pointed to rapid evolution in response to natural selection in the wild (e.g. Johnston and Selander, 1964; McNeilly and Antonovics, 1967; Boag and Grant, 1981), and by field experiments that demonstrated evolution in real time (e.g. Reznick *et al.*, 1990; Losos *et al.*, 1997). As a result, research in disciplines such as medicine, conservation biology and fisheries biology started to increasingly incorporate a genetic focus, with the goal of understanding how evolution shapes population responses to conditions such as antibiotic use, contemporary climate change and harvesting (Palumbi, 2002).

Invasion science was similarly influenced by this paradigm shift (Reznick *et al.*, 2019). The broadening of invasion science to include genetic and evolutionary perspectives was also facilitated by concurrent advancements in DNA sequencing, which allowed allozyme-based surveys of genetic variation (Holland, 2000) to be replaced with methods such as microsatellite genotyping (reviewed by Allendorf, 2017). These molecular markers provided increased resolution for resolving population structure and facilitated landscape/seascape-scale surveys of genetic composition and diversity of invasive species (e.g. Pascual et al., 2007; Rius et al., 2012; reviewed by Allendorf, 2017). The stage was set, and genetic and evolutionary explanations for invasion success started to be evaluated increasingly often (Sakai et al., 2001; Lee, 2002; Prentis et al., 2008; Bock et al., 2015; Hodgins et al., 2018). For example, building on rapidly accumulating evidence of invasive populations that originated from intra- or interspecific mating (e.g. Burdon and Brown, 1986; Kolbe et al., 2004; Lavergne and Molofsky, 2007), hybridization was proposed as a stimulus for the evolution of invasiveness (Ellstrand and Schierenbeck, 2000; Schierenbeck and Ellstrand, 2009; Rius and Darling, 2014). An additional possibility advanced at this time, focusing on evolutionary mechanisms of invasion success, was that invasions are enabled by the reallocation of resources from defence to growth and reproduction across generations (Blossey and Notzold, 1995). Such reallocation would be favoured by selection when natural enemies such as herbivores are lost during the colonization event. Lastly, some of the first analyses of genes and loci that may drive invasion success under natural conditions were published during this time (e.g. Paterson et al., 1995; Krieger and Ross, 2002), expanding on earlier work on the genetics of pesticide resistance (e.g. Georghiou and Pasteur, 1978; Plapp and Tripathi, 1978).

Aside from investigating genetic causes of invasions, researchers in the early 2000s focused on the genetic consequences of human-mediated range expansions. For example, contrasts in genetic diversity between invasive and native populations were increasingly used to evaluate the severity of genetic bottlenecks (Dlugosch and Parker, 2008) and to unravel invasion routes (Estoup and Guillemaud, 2010). The goal in this case was to understand how invasive populations can thrive in new environments, given that population genetic theory predicted that they should retain only a small fraction of the diversity that segregates in the ancestral range (Allendorf and Lundquist, 2003). This so-called 'genetic paradox' of invasions was resolved gradually, as accumulating genetic data revealed that reductions in genetic diversity in the introduced range are not as common and do not have the severity that might have been predicted at face value (reviewed by Roman and Darling, 2007). This occurs because bioinvasions are often derived from repeated introductions and/or large propagule pools (Roman and Darling, 2007), and human-mediated intraspecific hybridization may enhance genetic variation. Moreover, even when molecular diversity does decline, variation in quantitative traits, which include most traits relevant for invasion success, is rarely impacted (Dlugosch and Parker, 2008).

Consequences of invasions may also involve interspecific interactions. Rhymer and Simberloff (1996), for example, considered whether rare native species that hybridize with invasive species might be driven to extinction. The detrimental effects of interspecific mating could be manifested, in this case, as either demographic swamping or as genetic swamping (Wolf et al., 2001; Todesco et al., 2016). Demographic swamping can occur when hybrid fitness is low, ultimately driving population growth rates below levels that are required for population replacement (Wolf et al., 2001). Genetic swamping, by contrast, can occur when hybrid fitness is high, leading to the replacement of pure parental species with hybrids (Wolf et al., 2001). Rhymer and Simberloff (1996) described several examples of genetic exchange between native and invasive taxa, particularly from disturbed and heterogenous habitats, which are predicted to favour hybridization (Anderson, 1948). Recent literature surveys provide a broader view of this process. Todesco et al. (2016), for example, summarized information from 143 cases of hybridization linked to extinction. Genetic exchange with an invasive species was indeed revealed as a significant predictor of extinction risk and involved rare as well as common native species (Todesco et al., 2016). This result underscores the importance of management actions that limit the introduction of closely related taxa, or that restore degraded habitats, thereby increasing habitat resistance to invasions and hybrid establishment (Todesco *et al.*, 2016).

Other applications of genetic data in invasion science were inspired from forensics and aimed to reconstruct the invasion histories and dynamics of post-establishment spread (reviewed by Estoup and Guillemaud, 2010; Cristescu, 2015). These studies typically relied on dense landscape-level sampling, and on inferences of ancestor–descendant relationships. When data from both the native and introduced ranges were available, information could be obtained on the identity of the invasive species (e.g. by revealing cryptic species complexes; Zhan *et al.*, 2010; Bock *et al.*, 2012) or on the origin of introduced populations and the likely vectors of spread (e.g. Jousson *et al.*, 1998; Durka *et al.*, 2005; Rius *et al.*, 2012). In invasive species such as the green alga *Caulerpa taxifolia*, this information was critical for mobilizing funding and for activating a successful rapid eradication programme (Simberloff *et al.*, 2013). When sampling focused exclusively on the introduced range, information could be extracted on mechanisms of post-establishment spread (e.g. Darling and Folino-Rorem, 2009; Lacoursière-Roussel *et al.*, 2012). In addition, genetically isolated invasive populations could be discerned and targeted for management (e.g. Rollins *et al.*, 2009, 2011). Even beyond applied uses, these analyses were critical for revealing multiple independent colonization events and post-introduction admixture among divergent lineages (e.g. Lee, 1999; Kolbe *et al.*, 2004). Often, this motivated subsequent genomic analyses of the effects of hybridization and mechanisms of repeated adaptation to novel environments (Rius *et al.*, 2015a,b).

1.4 The Genomic Study of Invasive Species

Genomic technologies, which started to be incorporated in the study of non-model species after 2010 (Andrew *et al.*, 2013), have triggered another more recent inflection point in invasion science (reviewed by Bock *et al.*, 2015; Rius *et al.*, 2015a,b; Viard *et al.*, 2016; North *et al.*, 2021; McGaughran *et al.*, 2024; Hodgins *et al.*, 2025). This section highlights how these major developments are enabling long-standing questions on biological invasions to be answered. In doing

so, it introduces important topics covered in this volume and directs the reader to the relevant chapters.

The sharp reduction in the cost of high-throughput sequencing has democratized access to resources that were previously available for only a small number of genome-enabled species. Over two decades after the human genome was first released (International Human Genome Sequencing Consortium, 2001), we are now able to scan genomes from end to end in almost any organism at a fraction of the cost. Therefore, genetic studies of evolution and adaptation can be pursued outside model species, and outside confined laboratory environments (Barrett and Hoekstra, 2011; Stenseth *et al.*, 2022). This has had a transformative impact on the study of invasive species (see Chapter 2, this volume, for an introduction to genomic technologies followed by an in-depth look at their capabilities in Chapter 3). For example, analyses on genetic drivers of invasions are now possible even in organisms for which genetic mapping populations, which have classically been used to study the genetic and molecular underpinnings of trait variation, cannot easily be obtained. Chapter 5 of this volume illustrates this point using a discussion of genomics-enabled studies of adaptation in marine invasive species. Second- and third-generation sequencing technologies also allow a broader range of specimens to be studied for any given species. Most notably, this includes historical specimens maintained in museums or herbaria for hundreds of years, which can provide a critical perspective on lineages that first seeded an invasion, or on invasions that have died out before they could be studied (reviewed by Kim *et al.*, 2023). Chapter 8 of this volume describes how genomic analyses of historical DNA obtained from herbarium specimens have bolstered evolutionary studies of invasive plants, while Chapter 9 provides an example of how whole-genome sequencing of herbarium specimens can be used to resolve invasion histories.

Aside from broadening the types of specimens that can be included in evolutionary studies of invasions, recent innovations in sequencing technology have also allowed vastly more genomic regions to be interrogated in any one species (Allendorf, 2017). These data can then be leveraged to reveal aspects of the history or the biology of invasive species that were previously untraceable using a small number of genes. For example, based on high-resolution information provided by tens or hundreds of thousands of genome-wide markers, invasion histories can now be reconstructed even in species with complex dynamics, including multiple introductions followed by admixture and bridgehead events (e.g. Lombaert *et al.*, 2010; Vallejo-Marín *et al.*, 2021). Similarly, comprehensive genome-wide scans have helped identify the traces of past hybridization even in invasive species that were previously assumed to have evolved without interspecific genetic exchange (e.g. Le Moan *et al.*, 2021; Rosinger *et al.*, 2021; see also the discussion on the genomics of hybridization in Chapter 5, this volume). Invasive species known to have experienced one or more bouts of hybridization can then be used to disentangle the beneficial or deleterious effects of this process during post-introduction evolution. In this context, Chapter 6 of this volume includes a discussion of transgressive segregation in invasive populations of animals and plants.

Dense marker sets that thoroughly cover the genome have also allowed invasion biologists to identify narrow chromosomal segments that depart from genome-wide trends due to the action of natural selection (e.g. Bieker *et al.*, 2022; Tepolt *et al.*, 2022; Stuart *et al.*, 2023; North *et al.*, 2024), which co-vary with invasiveness traits (e.g. Bock *et al.*, 2018; Battlay *et al.*, 2023), or which are associated with environmental variables important for invasive spread (e.g. Stern and Lee, 2020). Knowledge of these candidate invasiveness loci then sets the stage for exploring other important questions on the origin of invasions. For example, does the success of invasive species depend on the introduction of standing genetic variants that natural selection can immediately act on, or on the fortuitous occurrence of new adaptive mutations? On the one hand, current empirical evidence (e.g. Stern and Lee, 2020; Tepolt *et al.*, 2022; Battlay *et al.*, 2023), as well as theoretical predictions (Barrett and Schluter, 2008), favour standing genetic variants. On the other hand, conditions encountered in the novel range can also promote genome reorganization and *de novo* mutations, some of which may prove to be advantageous in the new range (Stapley *et al.*, 2015; see also Chapter 4, this volume, for a discussion on the role of transposable elements as a source of novel genetic variation in invasive populations). Chapter 10 of this volume illustrates how genomic data sets have been used to illuminate the role of standing genetic variants during invasions, using as a model the calanoid copepod *Eurytemora affinis*.

Even beyond the age and origin of mutations, genomic data sets now allow researchers to evaluate whether polymorphisms such as structural genetic variants (e.g. chromosomal inversions) or epigenetic variants have a particularly important role during invasions. These analyses can now be performed in a range of invasive species and environmental conditions. For instance, analyses of chromosomal inversions and their contribution to range expansion no longer need to rely on privileged systems such as *Drosophila* fruit flies (e.g. Carson, 1965; Dobzhansky, 1965). Instead, we can evaluate whether inversions or other recombination cold spots tend to facilitate or impede rapid adaptation depending on whether an invasive species colonizes an analogous environment (e.g. Battlay *et al.*, 2023; see also Chapter 8, this volume) or an entirely different environment (e.g. Bock *et al.*, 2024). In this way, predictions made by theoretical models on the constructive and destructive consequences of linking multiple adaptive alleles together (e.g. Roesti *et al.*, 2022) can be evaluated in a range of conditions and study systems. Chapter 4 of this volume provides an in-depth look at how genomics can be used to disentangle the contribution of structural variants and transposable elements during invasions. Chapter 7 of this volume further discusses how genomic technologies allow epigenomic variation to be measured and incorporated into studies of invasion success. Collectively, studies of adaptive genetic and epigenetic variants can help us understand the ecological, evolutionary and molecular mechanisms that drive biological invasions (reviewed by Bock *et al.*, 2015; Rius *et al.*, 2015a,b; North *et al.*, 2021; McGaughran *et al.*, 2024; Hodgins *et al.*, 2025). They may also provide information that is critical for the management of invasions. Chapters 11 and 12 of this volume cover some of these applied uses of genomics, including for anticipating the future distribution of invasive species or for deploying genetic biocontrol.

Research has repeatedly demonstrated that the management of biological invasions should prioritize prevention (e.g. by constricting intro-duction pathways), early detection of new in-cursions, and decisive action on invaders that have not yet achieved ecological dominance (Rejmánek and Pitcairn, 2002; Simberloff *et al.*, 2013). In this*context, genomic technologies have revolutionized our early detection capabil-ities. Environmental DNA, for example, can enable the rapid and high-resolution recon-struction of microbial and macrobial communi-ties (Kuczynski *et al.*, 2012; Zhan *et al.*, 2013; Lacoursière-Roussel *et al.*, 2018). Thus, while there are still important obstacles to overcome before these methods can be widely implemented for invasion control (Darling and Mahon, 2011; Sepulveda *et al.*, 2020), they hold tremendous promise for improving detection of known and unknown invaders via active and passive biosur-veillance. Chapter 12 of this volume highlights some of these capabilities.

This volume consists of 13 chapters that explore how researchers working at the leading edge of invasion science are using genomic tech-nologies to fast-track our understanding of – and our ability to contain – invasive species. The first section (Chapters 2 and 3) provides a background on sequencing and analytical methodologies that are commonly used to study biological inva-sions. The second section (Chapters 4–7) relates to mechanisms of bioinvasion and how these can be studied using a genomics lens. The third section (Chapters 8–10) consists of case studies of well-known model systems in the field. The goal in this case is to provide examples of what genomics can teach us about bioinvasions. The fourth section (Chapters 11 and 12) departs from basic science and instead attempts to emphasize the applied uses of genomics in the con-text of invasion science. Finally, we end the vol-ume by summarizing the current state of the field and identifying promising directions for future study. Chapters have been contributed by leading researchers in invasion science from 13 countries. While we attempted to broadly canvas the field, we emphasize that this volume does not provide a comprehensive summary of research that is being done on invasion genomics. Rather, we hope to highlight some of the opportunities as well as challenges of this exciting and rapidly developing branch of invasion science.

References

Allendorf, F.W. (2017) Genetics and the conservation of natural populations: allozymes to genomes. *Molecular Ecology* 26, 420–430. DOI: 10.1111/mec.13948

Allendorf, F.W. and Lundquist, L.L. (2003) Introduction: population biology, evolution, and control of invasive species. *Conservation Biology* 17, 24–30.

Anderson, E. (1948) Hybridization of the habitat. *Evolution* 2, 1–9. DOI: 10.2307/2405610

Andrew, R.L., Bernatchez, L., Bonin, A., Buerkle, C.A., Carstens, B.C. *et al.* (2013) A road map for molecu-lar ecology. *Molecular Ecology* 22, 2605–2626. DOI: 10.1111/mec.12319

Baker, H.G. (1965) Characteristics and modes of origin of weeds. In: Baker, H.G. and Stebbins, G.L. (eds) *The Genetics of Colonizing Species*. Academic Press, New York, pp. 147–172.

Baker, H.G. and Stebbins, G.L. (eds) (1965) *The Genetics of Colonizing Species*. Academic Press, New York.

Barrett, R.D. and Hoekstra, H.E. (2011) Molecular spandrels: tests of adaptation at the genetic level. *Nature Reviews Genetics* 12, 767–780. DOI: 10.1038/nrg3015

Barrett, R.D. and Schluter, D. (2008) Adaptation from standing genetic variation. *Trends in Ecology and Evolution* 23, 38–44. DOI: 10.1016/j.tree.2007.09.008

Barrett, S.C. (2015) Foundations of invasion genetics: the Baker and Stebbins legacy. *Molecular Ecology* 24, 1927–1941. DOI: 10.1111/mec.13014

Battlay, P., Wilson, J., Bieker, V.C., Lee, C., Prapas, D. *et al.* (2023) Large haploblocks underlie rapid adap-tation in the invasive weed *Ambrosia artemisiifolia*. *Nature Communications* 14: 1717. DOI: 10.1038/s41467-023-37303-4

Bell, K.L., Campos, M., Hoffmann, B.D., Encinas-Viso, F., Hunter, G.C. and Webber, B.L. (2024) Environ-mental DNA methods for biosecurity and invasion biology in terrestrial ecosystems: progress, pitfalls, and prospects. *Science of the Total Environment* 926: 171810. DOI: 10.1016/j.scitotenv.2024.171810

Bieker, V.C., Battlay, P., Petersen, B., Sun, X., Wilson, J. *et al.* (2022) Uncovering the genomic basis of an extraordinary plant invasion. *Science Advances* 8: eabo5115. DOI: 10.1126/sciadv.abo5115

Blossey, B. and Notzold, R. (1995) Evolution of increased competitive ability in invasive nonindigenous plants: a hypothesis. *Journal of Ecology* 83: 887889. DOI: 10.2307/2261425

Boag, P.T. and Grant, P.R. (1981) Intense natural selection in a population of Darwin Finches (Geospizinae) in the Galapagos. *Science* 214, 82–85. DOI: 10.1126/science.214.4516.8

Bock, D.G., MacIsaac, H.J. and Cristescu, M.E. (2012) Multilocus genetic analyses differentiate between widespread and spatially restricted cryptic species in a model ascidian. *Proceedings of the Royal Society B: Biological Sciences* 279, 2377–2385. DOI: 10.1098/rspb.2011.2610

Bock, D.G., Caseys, C., Cousens, R.D., Hahn, M.A., Heredia, S.M. *et al.* (2015) What we still don't know about invasion genetics. *Molecular Ecology* 24, 2277–2297. DOI: 10.1111/mec.13032

Bock, D.G., Kantar, M.B., Caseys, C., Matthey-Doret, R. and Rieseberg, L.H. (2018) Evolution of invasiveness by genetic accommodation. *Nature Ecology and Evolution* 2, 991–999. DOI: 10.1038/s41559-018-0553-z

Bock, D.G., Baeckens, S., Kolbe, J.J. and Losos, J.B. (2024) When adaptation is slowed down: genomic analysis of evolutionary stasis in thermal tolerance during biological invasion in a novel climate. *Molecular Ecology* 33: e17075. DOI: 10.1111/mec.17075

Burdon, J.J. and Brown, A.H. (1986) Population genetics of *Echium plantagineum* L. target weed for biological control. *Australian Journal of Biological Sciences* 39, 369–378. DOI: 10.1071/BI9860369

Cadotte, M.W. (2006) Darwin to Elton: early ecology and the problem of invasive species. In: Cadotte, M.W., McMahaon, S.M. and Fukami, T. (eds) *Conceptual Ecology and Invasion Biology: Reciprocal Approaches to Nature*. Springer, Dordrecht, Netherlands, pp. 15–33.

Carlton, J.T. (1987) Patterns of transoceanic marine biological invasions in the Pacific Ocean. *Bulletin of Marine Science* 41, 452–465.

Carlton, J.T. and Geller, J.B. (1993) Ecological roulette: the global transport of nonindigenous marine organisms. *Science* 261, 78–82. DOI: 10.1126/science.261.5117.78

Carson, H.L. (1965) Chromosomal morphism in geographically widespread species of Drosophila. In: Baker, H.G. and Stebbins, G.L. (eds) *The Genetics of Colonizing Species*. Academic Press, New York, pp. 503–531.

Case, T.J. (1991) Invasion resistance, species build-up and community collapse in metapopulation models with interspecies competition. *Biological Journal of the Linnean Society* 42, 239–266. DOI: 10.1111/j.1095-8312.1991.tb00562.x

Chan, F.T., Stanislawczyk, K., Sneekes, A.C., Dvoretsky, A., Gollasch, S. *et al.* (2019) Climate change opens new frontiers for marine species in the Arctic: current trends and future invasion risks. *Global Change Biology* 25, 25–38. DOI: 10.1111/gcb.14469

Colautti, R.I., Grigorovich, I.A. and MacIsaac, H.J. (2006) Propagule pressure: a null model for biological invasions. *Biological Invasions* 8, 1023–1037. DOI: 10.1007/s10530-005-3735-y

Cristescu, M.E. (2015) Genetic reconstructions of invasion history. *Molecular Ecology* 24, 2212–2225. DOI: 10.1111/mec.13117

Darling, J.A. and Folino-Rorem, N.C. (2009) Genetic analysis across different spatial scales reveals multiple dispersal mechanisms for the invasive hydrozoan *Cordylophora* in the Great Lakes. *Molecular Ecology* 18, 4827–4840. DOI: 10.1111/j.1365-294X.2009.04405.x

Darling, J.A. and Mahon, A.R. (2011) From molecules to management: adopting DNA-based methods for monitoring biological invasions in aquatic environments. *Environmental Research* 111, 978–988. DOI: 10.1016/j.envres.2011.02.001

Darwin, C. (1859) *On the Origin of Species*. John Murray, London.

Dlugosch, K.M. and Parker, I.M. (2008) Founding events in species invasions: genetic variation, adaptive evolution, and the role of multiple introductions. *Molecular Ecology* 17, 431–449. DOI: 10.1111/j.1365-294X.2007.03538.x

Dobzhansky, T. (1965) "Wild" and "domestic" species of Drosophila. In: Baker, H.G. and Stebbins, G.L. (eds) *The Genetics of Colonizing Species*. Academic Press, New York, pp. 533–546.

Drake, J.A., Mooney, H.A., di Castri, F., Groves, R.H., Kruger, F.J. and Williamson, M. (eds))1989) *Biological Invasions. A Global Perspective*. Wiley, Chichester, UK.

Duell, E.B., Zaiger, K., Bever, J.D. and Wilson, G.W. (2019) Climate affects plant–soil feedback of native and invasive grasses: negative feedbacks in stable but not in variable environments. *Frontiers in Ecology and Evolution* 7: 419. DOI: 10.3389/fevo.2019.00419

Duffy, G.A., Coetzee, B.W., Latombe, G., Akerman, A.H., McGeoch, M.A. and Chown, S.L. (2017) Barriers to globally invasive species are weakening across the Antarctic. *Diversity and Distributions* 23, 982–996. DOI: 10.1111/ddi.12593

Durka, W., Bossdorf, O., Prati, D. and Auge, H. (2005) Molecular evidence for multiple introductions of garlic mustard (*Alliaria petiolata*, Brassicaceae) to North America. *Molecular Ecology* 14, 1697–1706. DOI: 10.1111/j.1365-294X.2005.02521.x

Ehrenfeld, J.G. (2010) Ecosystem consequences of biological invasions. *Annual Review of Ecology, Evolution, and Systematics* 41, 59–80. DOI: 10.1146/annurev-ecolsys-102209-144650

Ellstrand, N.C. and Schierenbeck, K.A. (2000) Hybridization as a stimulus for the evolution of invasiveness in plants? *Proceedings of the National Academy of Sciences USA* 97, 7043–7050. DOI: 10.1073/pnas.97.13.7043

Elton, C.S. (1958) *The Ecology of Invasions by Animals and Plants.* Methuen, London.

Endler, J.A. (1986) *Natural Selection in the Wild.* Princeton University Press, Princeton, New Jersey.

Estoup, A. and Guillemaud, T. (2010) Reconstructing routes of invasion using genetic data: why, how and so what? *Molecular Ecology* 19, 4113–4130. DOI: 10.1111/j.1365-294X.2010.04773.x

Georghiou, G.P. and Pasteur, N. (1978) Electrophoretic esterase patterns in insecticide-resistant and susceptible mosquitoes. *Journal of Economic Entomology* 71, 201–205. DOI: 10.1093/jee/71.2.201

Guyton, J.A., Pansu, J., Hutchinson, M.C., Kartzinel, T.R., Potter, A.B. *et al.* (2020) Trophic rewilding revives biotic resistance to shrub invasion. *Nature Ecology and Evolution* 4, 712–724. DOI: 10.1038/s41559-019-1068-y

Hodgins, K.A., Bock, D.G. and Rieseberg, L.H. (2018) Trait evolution in invasive species. *Annual Plant Reviews Online* 1, 1–37. DOI: 10.1002/9781119312994.apr0643

Hodgins, K.A., Battlay, P. and Bock, D.G. (2025) The genomic secrets of invasive plants. *New Phytologist* 245, 1846–1863. DOI: 10.1111/nph.20368

Holland, B.S. (2000) Genetics of marine bioinvasions. *Hydrobiologia* 420, 63–71. DOI: 10.1023/A:1003929519809

Huenneke, L., Glick, D., Waweru, F.W., Brownell, R.L. and Goodland, R. (1988) SCOPE program on biological invasions: a status report. *Conservation Biology* 2, 8–10.

Huey, R.B., Gilchrist, G.W. and Hendry, A.P. (2005) Using invasive species to study evolution. In: Sax, D.F., Stachowicz, J.J. and Gaines, S.D. (eds) *Species Invasions: Insights to Ecology, Evolution and Biogeography.* Sinauer Associates, Sunderland, Massachusetts, pp. 139–164.

Huskins, C.L. (1931) Origin of *Spartina townsendii. Nature* 127, 781–781. DOI: 10.1038/127781b0

International Human Genome Sequencing Consortium (2001) Initial sequencing and analysis of the human genome. *Nature* 409, 860–921. DOI: 10.1038/35057062

Johnston, R.F. and Selander, R.K. (1964) House sparrows: rapid evolution of races in North America. *Science* 144, 548–550. DOI: 10.1126/science.144.3618.548

Jousson, O., Pawlowski, J., Zaninetti, L., Meinesz, A. and Boudouresque, C.F. (1998) Molecular evidence for the aquarium origin of the green alga *Caulerpa taxifolia* introduced to the Mediterranean Sea. *Marine Ecology Progress Series* 172, 275–280. DOI: 10.3354/meps172275

Kim, A.S., Kreiner, J.M., Hernández, F., Bock, D.G., Hodgins, K.A. and Rieseberg, L.H. (2023) Temporal collections to study invasion biology. *Molecular Ecology* 32, 6729–6742. DOI: 10.1111/mec.17176

Kolar, C.S. and Lodge, D.M. (2001) Progress in invasion biology: predicting invaders. *Trends in Ecology and Evolution* 16, 199–204. DOI: 10.1016/S0169-5347(01)02101-2

Kolbe, J.J., Glor, R.E., Rodríguez Schettino, L., Lara, A.C., Larson, A. and Losos, J.B. (2004) Genetic variation increases during biological invasion by a Cuban lizard. *Nature* 431, 177–181. DOI: 10.1038/nature02807

Krieger, M.J. and Ross, K.G. (2002) Identification of a major gene regulating complex social behavior. *Science* 295, 328–332. DOI: 10.1126/science.1065247

Kuczynski, J., Lauber, C.L., Walters, W.A., Parfrey, L.W., Clemente, J.C. *et al.* (2012) Experimental and analytical tools for studying the human microbiome. *Nature Reviews Genetics* 13, 47–58. DOI: 10.1038/nrg3129

Lacoursiere-Roussel, A., Bock, D.G., Cristescu, M.E., Guichard, F., Girard, P. *et al.* (2012) Disentangling invasion processes in a dynamic shipping–boating network. *Molecular Ecology* 21, 4227–4241. DOI: 10.1111/j.1365-294X.2012.05702.x

Lacoursière-Roussel, A., Howland, K., Normandeau, E., Grey, E.K., Archambault, P. *et al.* (2018) eDNA metabarcoding as a new surveillance approach for coastal Arctic biodiversity. *Ecology and Evolution* 8, 7763–7777. DOI: 10.1002/ece3.4213

Lavergne, S. and Molofsky, J. (2007) Increased genetic variation and evolutionary potential drive the success of an invasive grass. *Proceedings of the National Academy of Sciences USA* 104, 3883–3888. DOI: 10.1073/pnas.0607324104

Le Moan, A., Roby, C., Fraisse, C., Daguin-Thiébaut, C., Bierne, N. and Viard, F. (2021) An introgression breakthrough left by an anthropogenic contact between two ascidians. *Molecular Ecology* 30, 6718–6732. DOI: 10.1111/mec.16189

Lee, C.E. (1999) Rapid and repeated invasions of fresh water by the copepod *Eurytemora affinis*. *Evolution* 53, 1423–1434. DOI: 10.1111/j.1558-5646.1999.tb05407.x

Lee, C.E. (2002) Evolutionary genetics of invasive species. *Trends in Ecology and Evolution* 17, 386–391. DOI: 10.1016/S0169-5347(02)02554-5

Lockwood, J.L., Cassey, P. and Blackburn, T. (2005) The role of propagule pressure in explaining species invasions. *Trends in Ecology and Evolution* 20, 223–228. DOI: 10.1016/j.tree.2005.02.004

Lockwood, J.L., Cassey, P. and Blackburn, T.M. (2009) The more you introduce the more you get: the role of colonization pressure and propagule pressure in invasion ecology. *Diversity and Distributions* 15, 904–910. DOI: 10.1111/j.1472-4642.2009.00594.x

Lockwood, J.L., Hoopes, M.F. and Marchetti, M.P. (2013) *Invasion Ecology*. Wiley, Chichester, UK.

Lodge, D.M. (1993) Biological invasions: lessons for ecology. *Trends in Ecology and Evolution* 8, 133–137. DOI: 10.1016/0169-5347(93)90025-K

Lombaert, E., Guillemaud, T., Cornuet, J.M., Malausa, T., Facon, B. and Estoup, A. (2010) Bridgehead effect in the worldwide invasion of the biocontrol harlequin ladybird. *PLOS One* 5: e9743. DOI: 10.1371/journal.pone.0009743

Losos, J.B., Warheit, K.I. and Schoener, T.W. (1997) Adaptive differentiation following experimental island colonization in *Anolis* lizards. *Nature* 387, 70–73. DOI: 10.1038/387070a0

McGaughran, A., Dhami, M.K., Parvizi, E., Vaughan, A.L., Gleeson, D.M. *et al.* (2024) Genomic tools in biological invasions: current state and future frontiers. *Genome Biology and Evolution* 16: evad230. DOI: 10.1093/gbe/evad230

McKinney, M.L. and Lockwood, J.L. (1999) Biotic homogenization: a few winners replacing many losers in the next mass extinction. *Trends in Ecology and Evolution* 14, 450–453. DOI: 10.1016/S0169-5347(99)01679-1

McNeilly, T. and Antonovics, J. (1967) Evolution in closely adjacent plant populations. IV. Barriers to gene flow. *Heredity* 23, 205–218. DOI: 10.1038/hdy.1968.29

North, H.L., McGaughran, A. and Jiggins, C.D. (2021) Insights into invasive species from whole-genome resequencing. *Molecular Ecology* 30, 6289–6308. DOI: 10.1111/mec.15999

North, H.L., Fu, Z., Metz, R., Stull, M.A., Johnson, C.D. *et al.* (2024) Rapid adaptation and interspecific introgression in the North American crop pest *Helicoverpa zea*. *Molecular Biology and Evolution* 41: msae129. DOI: 10.1093/molbev/msae129

Ogutu-Ohwayo, R. (1990) The decline of the native fishes of lakes Victoria and Kyoga (East Africa) and the impact of introduced species, especially the Nile perch, *Lates niloticus*, and the Nile tilapia, *Oreochromis niloticus*. *Environmental Biology of Fishes* 27, 81–96. DOI: 10.1007/BF00001938

Palumbi, S.R. (2002) *Evolution Explosion: How Humans Cause Rapid Evolutionary Change*. W.W. Norton and Company, New York.

Pascual, M., Chapuis, M.P., Mestres, F., Balanya, J., Huey, R.B. *et al.* (2007) Introduction history of *Drosophila subobscura* in the New World: a microsatellite-based survey using ABC methods. *Molecular Ecology* 16, 3069-3083. DOI: 10.1111/j.1365-294X.2007.03336.x

Paterson, A.H., Schertz, K.F., Lin, Y.R., Liu, S.C. and Chang, Y.L. (1995) The weediness of wild plants: molecular analysis of genes influencing dispersal and persistence of johnsongrass, *Sorghum halepense* (L.) Pers. *Proceedings of the National Academy of Sciences USA* 92, 6127–6131. DOI: 10.1073/pnas.92.13.6127

Pimentel, D. (1968) Population regulation and genetic feedback: evolution provides foundation for control of herbivore, parasite, and predator numbers in nature. *Science* 159, 1432–1437. DOI: 10.1126/science.159.3822.1432

Plapp, F.W. and Tripathi, R.K. (1978) Biochemical genetics of altered acetylcholinesterase resistance to insecticides in the house fly. *Biochemical Genetics* 16, 1–11. DOI: 10.1007/BF00484380

Prentis, P.J., Wilson, J.R., Dormontt, E.E., Richardson, D.M. and Lowe, A.J. (2008) Adaptive evolution in invasive species. *Trends in Plant Science* 13, 288–294. DOI: 10.1016/j.tplants.2008.03.004

Pringle, R.M. (2011) Nile perch. In: Simberloff, D. and Rejmanek, M. (eds) *Encyclopedia of Biological Invasions*. University of California Press, Berkeley, California, p. 484.

Rejmánek, M. and Pitcairn, M.J. (2002) When is eradication of exotic pest plants a realistic goal. In: Veitch, C.R. and Clout, M.N. (eds) *Turning the Tide: The Eradication of Invasive Species*. International Union for Conservation of Nature, Gland, Switzerland, pp. 249–253.

Reznick, D.A., Bryga, H. and Endler, J.A. (1990) Experimentally induced life-history evolution in a natural population. *Nature* 346, 357–359. DOI: 10.1038/346357a0

Reznick, D.N., Losos, J. and Travis, J. (2019) From low to high gear: there has been a paradigm shift in our understanding of evolution. *Ecology Letters* 22, 233–244. DOI: 10.1111/ele.13189

Rhymer, J.M. and Simberloff, D. (1996) Extinction by hybridization and introgression. *Annual Review of Ecology and Systematics* 27, 83–109. DOI: 10.1146/annurev.ecolsys.27.1.83

Ricciardi, A. and MacIsaac, H. (2008) The book that began invasion ecology. *Nature* 452: 34. DOI: 10.1038/452034a

Ricciardi, A., Blackburn, T.M., Carlton, J.T., Dick, J.T., Hulme, P.E. *et al.* (2017) Invasion science: a horizon scan of emerging challenges and opportunities. *Trends in Ecology and Evolution* 32, 464–474. DOI: 10.1016/j.tree.2017.03.007

Ricciardi, A., Iacarella, J.C., Aldridge, D.C., Blackburn, T.M., Carlton, J.T. *et al.* (2021) Four priority areas to advance invasion science in the face of rapid environmental change. *Environmental Reviews* 29, 119–141. DOI: 10.1139/er-2020-0088

Richardson, D.M. and Pysek, P. (2007) Classics in physical geography revisited: Elton, C.S. 1958. *The ecology of invasions by animals and plants*. *Progress in Physical Geography* 31, 659–666. DOI: 10.1177/0309133307087089

Rius, M. and Darling, J.A. (2014) How important is intraspecific genetic admixture to the success of colonising populations? *Trends in Ecology and Evolution* 29, 233–242. DOI: 10.1016/j.tree.2014.02.003

Rius, M., Turon, X., Ordóñez, V. and Pascual, M. (2012) Tracking invasion histories in the sea: facing complex scenarios using multilocus data. *PLOS ONE* 7: e35815. DOI: 10.1371/journal.pone.0035815

Rius, M., Bourne, S., Hornsby, H.G. and Chapman, M.A. (2015a) Applications of next-generation sequencing to the study of biological invasions. *Current Zoology* 61, 488–504. DOI: 10.1093/czoolo/61.3.488

Rius, M., Turon, X., Bernardi, G., Volckaert, F.A. and Viard, F. (2015b) Marine invasion genetics: from spatio-temporal patterns to evolutionary outcomes. *Biological Invasions* 17, 869–885. DOI: 10.1007/s10530-014-0792-0

Roesti, M., Gilbert, K.J. and Samuk, K. (2022) Chromosomal inversions can limit adaptation to new environments. *Molecular Ecology* 31, 4435–4439. DOI: 10.1111/mec.16609

Rollins, L.A., Woolnough, A.P., Wilton, A.N., Sinclair, R.O. and Sherwin, W.B. (2009) Invasive species can't cover their tracks: using microsatellites to assist management of starling (*Sturnus vulgaris*) populations in Western Australia. *Molecular Ecology* 18, 1560–1573. DOI: 10.1111/j.1365-294X.2009.04132.x

Rollins, L.A., Woolnough, A.P., Sinclair, R., Mooney, N.J. and Sherwin, W.B. (2011) Mitochondrial DNA offers unique insights into invasion history of the common starling. *Molecular Ecology* 20, 2307–2317. DOI: 10.1111/j.1365-294X.2011.05101.x

Roman, J. and Darling, J.A. (2007) Paradox lost: genetic diversity and the success of aquatic invasions. *Trends in Ecology and Evolution* 22, 454–464. DOI: 10.1016/j.tree.2007.07.002

Rosinger, H.S., Geraldes, A., Nurkowski, K.A., Battlay, P., Cousens, R.D. *et al.* (2021) The tip of the iceberg: genome wide marker analysis reveals hidden hybridization during invasion. *Molecular Ecology* 30, 810–825. DOI: 10.1111/mec.15768

Roy, H.E., Pauchard, A., Stoett, P.J., Renard Truong, T., Meyerson, L.A. *et al.* (2024) Curbing the major and growing threats from invasive alien species is urgent and achievable. *Nature Ecology and Evolution* 8, 1216–1223. DOI: 10.1038/s41559-024-02412-w

Sakai, A.K., Allendorf, F.W., Holt, J.S., Lodge, D.M., Molofsky, J. *et al.* (2001) The population biology of invasive species. *Annual Review of Ecology and Systematics* 32, 305–332. DOI: 10.1146/annurev.ecolsys.32.081501.114037

Savidge, J.A. (1987) Extinction of an island forest avifauna by an introduced snake. *Ecology* 68, 660–668. DOI: 10.2307/1938471

Sax, D.F., Stachowicz, J.J., Brown, J.H., Bruno, J.F., Dawson, M.N. *et al.* (2007) Ecological and evolutionary insights from species invasions. *Trends in Ecology and Evolution* 22, 465–471. DOI: 10.1016/j.tree.2007.06.009

Schierenbeck, K.A. and Ellstrand, N.C. (2009) Hybridization and the evolution of invasiveness in plants and other organisms. *Biological Invasions* 11, 1093–1105. DOI: 10.1007/s10530-008-9388-x

Sepulveda, A.J., Nelson, N.M., Jerde, C.L. and Luikart, G. (2020) Are environmental DNA methods ready for aquatic invasive species management? *Trends in Ecology and Evolution* 35, 668–678. DOI: 10.1016/j.tree.2020.03.011

Simberloff, D. (2009) The role of propagule pressure in biological invasions. *Annual Review of Ecology, Evolution, and Systematics* 40, 81–102. DOI: 10.1146/annurev.ecolsys.110308.120304

Simberloff, D. (2011) Charles Elton: neither founder nor siren, but prophet. In: *Fifty Years of Invasion Ecology: The Legacy of Charles Elton*. Wiley, Chichester, UK, pp. 11–24.

Simberloff, D. (2013) Biological invasions: much progress plus several controversies. *Contributions to Science* 9, 7–16. DOI: 10.2436/20.7010.01.158

Simberloff, D. and von Holle, B. (1999) Positive interactions of nonindigenous species: invasional meltdown? *Biological Invasions* 1, 21–32. DOI: 10.1023/A:1010086329619

Simberloff, D., Martin, J.L., Genovesi, P., Maris, V., Wardle, D.A. *et al.* (2013) Impacts of biological invasions: what's what and the way forward. *Trends in Ecology and Evolution* 28, 58–66. DOI: 10.1016/j.tree.2012.07.013

Spencer, C.N., McClelland, B.R. and Stanford, J.A. (1991) Shrimp stocking, salmon collapse, and eagle displacement. *BioScience* 41, 14–21. DOI: 10.2307/1311536

Stachowicz, J.J., Whitlatch, R.B. and Osman, R.W. (1999) Species diversity and invasion resistance in a marine ecosystem. *Science* 286, 1577–1579. DOI: 10.1126/science.286.5444.1577

Stapley, J., Santure, A.W. and Dennis, S.R. (2015) Transposable elements as agents of rapid adaptation may explain the genetic paradox of invasive species. *Molecular Ecology* 24, 2241–2252. DOI: 10.1111/mec.13089

Stenseth, N.C., Andersson, L. and Hoekstra, H.E. (2022) Gregor Johann Mendel and the development of modern evolutionary biology. *Proceedings of the National Academy of Sciences USA* 119: e2201327119. DOI: 10.1073/pnas.2201327119

Stern, D.B. and Lee, C.E. (2020) Evolutionary origins of genomic adaptations in an invasive copepod. *Nature Ecology and Evolution* 4, 1084–1094. DOI: 10.1038/s41559-020-1201-y

Stuart, K.C., Edwards, R.J., Sherwin, W.B. and Rollins, L.A. (2023) Contrasting patterns of single nucleotide polymorphisms and structural variation across multiple invasions. *Molecular Biology and Evolution* 40, msad046. DOI: 10.1093/molbev/msad046

Tepolt, C.K., Grosholz, E.D., de Rivera, C.E. and Ruiz, G.M. (2022) Balanced polymorphism fuels rapid selection in an invasive crab despite high gene flow and low genetic diversity. *Molecular Ecology* 31, 55–69. DOI: 10.1111/mec.16143

Todesco, M., Pascual, M.A., Owens, G.L., Ostevik, K.L., Moyers, B.T. *et al.* (2016) Hybridization and extinction. *Evolutionary Applications* 9, 892–908. DOI: 10.1111/eva.12367

Vallejo-Marín, M., Friedman, J., Twyford, A.D., Lepais, O., Ickert-Bond, S.M. *et al.* (2021) Population genomic and historical analysis suggests a global invasion by bridgehead processes in *Mimulus guttatus*. *Communications Biology* 4: 327. DOI: 10.1038/s42003-021-01795-x

van Kleunen, M., Dawson, W. and Maurel, N. (2015) Characteristics of successful alien plants. *Molecular Ecology* 24, 1954–1968. DOI: 10.1111/mec.13013

Viard, F., David, P. and Darling, J.A. (2016) Marine invasions enter the genomic era: three lessons from the past, and the way forward. *Current Zoology* 62, 629–642. DOI: doi.org/10.1093/cz/zow053

Vitousek, P.M., Walker, L.R., Whiteaker, L.D., Mueller-Dombois, D. and Matson, P.A. (1987) Biological invasion by *Myrica faya* alters ecosystem development in Hawaii. *Science* 238, 802–804. DOI: 10.1126/science.238.4828.80

Witte, F., Goldschmidt, T., Wanink, J., van Oijen, M., Goudswaard, K. *et al.* (1992) The destruction of an endemic species flock: quantitative data on the decline of the haplochromine cichlids of Lake Victoria. *Environmental Biology of Fishes* 34, 1–28. DOI: 10.1007/BF00004782

Wolf, D.E., Takebayashi, N. and Rieseberg, L.H. (2001) Predicting the risk of extinction through hybridization. *Conservation Biology* 15, 1039–1053. DOI: 10.1046/j.1523-1739.2001.0150041039.x

Zhan, A., Macisaac, H.J. and Cristescu, M.E. (2010) Invasion genetics of the *Ciona intestinalis* species complex: from regional endemism to global homogeneity. *Molecular Ecology* 19, 4678–4694. DOI: 10.1111/j.1365-294X.2010.04837.x

Zhan, A., Hulák, M., Sylvester, F., Huang, X., Adebayo, A.A. *et al.* (2013) High sensitivity of 454 pyrosequencing for detection of rare species in aquatic communities. *Methods in Ecology and Evolution* 4, 558–565. DOI: 10.1111/2041-210X.12037

2 Population Genomics in Invasion Research: A Primer for Ecologists

Olga Kozhar[1,2], Jane E. Stewart[1,2], and Ruth A. Hufbauer[1,2]*

[1]*Department of Agricultural Biology, Colorado State University, Fort Collins, Colorado, USA;* [2]*Graduate Degree Program in Ecology, Colorado State University, Fort Collins, Colorado, USA*

Abstract

Use of genomic data to study biological invasions has exploded in recent years. Yet, many scientists with expertise in the biology and ecology of invasive species do not have expertise in genomics (and vice versa). Here, we provide background on genomics for ecologists to provide the conceptual and practical understanding necessary for productive collaborations. We discuss questions that can be addressed using genomic data, as well as approaches needed, with a particular focus on the importance of robust sampling to address questions with rigor. The integration of molecular data and metadata is a strength for invasion biology research, and our focus is discussing challenges and questions that can aid in the development of collaborations that unite phenotyping and genotyping.

2.1 Introduction

The integration of population genomic data into the field of invasion biology has provided new insights into the successful establishment and spread of invasive species (North *et al.*, 2021; Kołodziejczyk *et al.*, 2025). For example, genomic data can provide unique insights into the movement and gene flow of organisms, allowing invasion pathways to be traced, not just from the native range but also among introduced areas (e.g. Fraimout *et al.*, 2017; Blumenfeld *et al.*, 2021; Fontaine *et al.*, 2021; Boscolo Agostini *et al.*, 2025). Additionally, genomic data can reveal adaptive evolution associated with invasion (e.g. Parvizi *et al.*, 2024; Battlay *et al.*, 2025), providing insights into the genetic basis of invasiveness (Olazcuaga *et al.*, 2020), as well as exposing evolutionary stasis (e.g. Bock *et al.*,

2024). With the results from genomic data, we can set management priorities that limit the occurrence of biological invasions through biosecurity surveillance (Poland and Rassati, 2019), by blocking off invasive pathways (Estoup and Guillemaud, 2010) and by determining high-risk potential invasive species with characteristics that could lead to invasiveness (van Kleunen *et al.*, 2010; Lawson Handley *et al.*, 2011; Schulz *et al.*, 2021).

Many scientists who work on biological invasions have expertise in ecology, physiology or other fields, rather than in population genomics and bioinformatics. These scientists can gain access to genomic data generated by private companies, and bioinformatic analyses can be run with such services or through using available software packages. However, bringing meaning to the data and, indeed, collecting

---

*Corresponding author: ruth.hufbauer@colostate.edu

© CAB International 2025. *Invasion Genomics* (Eds D. Bock and M. Rius)
DOI: 10.1079/9781800626263.0002

meaningful data, is likely to be more effective through the deep expertise in genomics of a collaborator. Genomic approaches require meticulous processing of large volumes of data, with multiple filtering steps necessary to achieve robust inference (McGaughran *et al.*, 2024). Bioinformatics tools enable us to address patterns in these data, but they typically rely on assumptions about the sampled populations, such as random mating (Pearman *et al.*, 2022) or particular models of evolution (Johri *et al.*, 2022). Experts can rigorously weigh these assumptions, understand whether analytical tools are robust to or sensitive to violation of these assumptions, and consider alternative explanations for data patterns.

Here, we provide a brief tutorial of the most recent genomics tools and sampling approaches to aid ecologists and other biologists who are not experts in genomics in understanding the basic vocabulary, challenges and questions that can be useful in developing collaborations that unite phenotyping and genotyping.

Although our focus is to provide a genomic background for ecologists, we would also like to point out the crucial role that ecologists play. Ecologists provide deep expertise in the biology of the organism that makes collecting relevant populations and phenotyping relevant traits possible. Many species of interest are not common in their native range, and the boundaries of their introduced range may not be known. Thus, biological expertise and connections with others working on the taxa are crucial to find samples that are from relevant locations, and to sample in an unbiased way. Furthermore, phenotypic traits of interest can be quite complex, such as behaviors, physiology, and maximum and minimum thermal tolerances (Aguirre-Liguori *et al.*, 2021). Collecting such phenotypic data on individuals from multiple populations across a range requires deep ecological expertise. True collaborations between geneticists and ecologists or other experts in the biology of the taxa of interest can lead to the best insights into biological invasion (Zamudio *et al.*, 2016).

2.2 Questions about Biological Invasions That Can Be Addressed with Genomic Data

Prior to obtaining genetic data, researchers must first determine what the main objectives or questions are for their project. This will shape many aspects of the research, including sampling and collection design, the type of genetic sequencing required, associated phenotypic metadata (e.g. behavioral, physiological) needed to address the question, and downstream bioinformatic analyses. A broad array of questions can be asked with genomic data or with a union of genomic and phenotypic data. Some of these include:

- Where are the invaders from?
- What are the routes of invasion?
- Were there multiple introductions?
- Are individuals from distinct locations in the native range hybridizing in the introduced range?
- What genes or markers underlie the phenotypic changes observed?
- What aspects of the introduced range (e.g. temperature, precipitation) are selecting for differences that evolve within the introduced range?
- What species could be introduced, with the risk of the next invasion in a given country?
- How have invaders adapted to new locations, and what regions of the genome are under selection?

2.3 A Brief Overview of Genomic Data for Non-geneticists

Over the past 30 years, genetics and genomics have advanced rapidly, from the ability to sequence a short section of an organism's genome to the ability to sequence many genomes end to end, to compare species and populations. The tools that utilized next-generation sequencing have revolutionized our ability to capture and analyze genetic data, and have broadened the scope of questions we can now address with genetic data. We provide a summary of the approaches, uses and limitations in Table 2.1.

2.3.1 High-throughput sequencing and bioinformatics pipelines

To understand high-throughput sequencing, it is useful to initially review the first approach to sequencing DNA that became widespread: Sanger sequencing (Sanger *et al.*, 1977). Sanger

Table 2.1. Overview of different types of genomic data in invasion biology studies.

Genomic data type	Sequencing methods[a]	Amount of information generated	Key applications	Limitations
Whole-genome sequencing (WGS)	Long reads: PacBio or MinION with accuracies of 92% and 99%, respectively. Data generation is 160 Gb for PacBio and 48 Gb for MinION. Costs are US$86 for a sample of PacBio and US$72 for MinION. Short reads: Illumina with an accuracy of 99.9% and data generation of 180 Gb and cost of ~US$20 per sample.	High: provides the entire genome sequence, capturing all genetic variation.	Comprehensive studies of genome-wide diversity, adaptation and evolutionary history. Enables detection of both neutral and adaptive variation. Suitable for detailed phylogenomic studies, detecting selection, understanding complex traits and tracing invasion pathways.	Expensive and resource-intensive, especially for large genomes. Requires substantial bioinformatic expertise and computational resources. May generate more data than necessary for some questions, leading to challenges in data management.
Low-coverage WGS	Short reads: Illumina with an accuracy of 99.9% and data generation of 180 Gb and cost of ~US$20 per sample.	Moderate: captures a broader, although less detailed, overview of genome-wide variation.	Cost-effective alternative to full WGS for studying population structure, gene flow and adaptation, particularly in non-model organisms and in the context of invasion biology. Useful for identifying candidate regions under selection.	Lower resolution compared with high-coverage WGS, potentially missing rare variants. Requires robust bioinformatics pipelines to handle lower-coverage data.
Restriction-site-associated DNA sequencing (RAD-seq)	Short reads with Illumina methods.	Moderate to high: captures thousands to hundreds of thousands of SNPs across the genome.	Useful for population structure, understanding pathways of invasion and studying recent evolutionary events, especially in invasive species. Ideal for detecting population differentiation and gene flow and identifying loci under selection in invasion contexts. Suitable for organisms with little genomic data.	Less comprehensive than WGS, as it samples only a subset of the genome. Potential biases in marker distribution. May miss rare alleles or loci under selection.

Continued

Table 2.1. Continued.

Genomic data type	Sequencing methods[a]	Amount of information generated	Key applications	Limitations
Pool sequencing (pool-seq)	Only a single sample is sequenced because each consists of the pooled samples within a single library. Best completed with short-read Illumina data.	Moderate: aggregates genetic variation from pooled samples, providing allele frequency estimates.	Cost-effective approach for population genomics studies in invasion biology. Useful for estimating allele frequencies, detecting selection and identifying adaptive variants in populations, especially when working with large numbers of individuals or limited resources.	Pooling samples obscures individual-level variation, making it difficult to detect within-population diversity or rare variants. Potential for allele frequency biases due to unequal pooling.
Simple sequence repeats (SSRs)	Allele fragment length data. The cost varies per sample. Short reads New methods include next-generation sequencing of the SSRs and their flanking regions.	Low to moderate: typically generates data for 10–20 loci.	Effective for studying recent evolutionary history, population structure and genetic diversity in invasive species. Commonly used for paternity analysis, detecting clones and tracing colonization events. Relatively inexpensive and straightforward to analyze.	Limited genomic coverage. Homoplasy (genetic sequences that appear identical but are not identical by descent) can complicate interpretations. Less informative for deep evolutionary questions. May not detect subtle patterns of differentiation.

SNP, single-nucleotide polymorphism.
[a]Cost of sequencing estimates are similar to those in Espinosa *et al.* (2024) and Sarhanová *et al.* (2018).

sequencing relies on the polymerase chain reaction (PCR) to amplify a DNA segment of interest, incorporating chain-terminating dideoxynucleotide triphosphates (ddNTPs) that are radioactively labeled, with each nucleotide tagged with a different florescent dye (Hutchison, 2007). PCR results in many copies of the DNA segment of interest, producing fragments of varying lengths (Metzker, 2010), and the amplified products can be size-separated by electrophoresis, either on an agarose gel or in a capillary. Using the chain-termination method, DNA sequences are end-labeled with fluorescently-labeled dideoxy nucleotides and then size-separated through gel electrophoresis. The sequence is determined by using fluorescence detection (Heather and Chain, 2016). The output is a four-color chromatogram with the sequence being read from the fluorescent labels of the peaks that correspond to each nucleotide (Al-Shuhaib and Hashim, 2023).

High-throughput sequencing, often called next-generation sequencing or massively parallel sequencing, uses the same principles as Sanger sequencing; the main difference is that rather than sequencing a single DNA fragment, millions of fragments are sequenced in parallel (Metzker, 2010; Slatko et al., 2018). The basic process is to break up the DNA into fragments and use PCR to tag the fragments by attaching (ligating) constructed unique short sequences (adapters) to them. The set of tagged fragments from a genome (or a set of genomes) is called a "library" (Head et al., 2014). Each adapter is ligated to genome fragments that have complementary nucleotide fragments that are attached to a flow cell (a hollow glass slide with one or more channels that hold the DNA fragments to be sequenced). The constructed library can then be loaded on to a flow cell where the tags stick to their complementary fragments on the surface of the flow cell (Ambardar et al., 2016). These DNA fragments are then sequenced through the addition of complementary nucleotides in parallel, allowing the sequencing of millions of base pairs simultaneously (Ekblom and Wolf, 2014). There are many variations of high-throughput sequencing that range from short to longer reads, but they have in common the production of a tremendous amount of data that can be used for conservation and invasion biology (Fuentes-Pardo and Ruzzante, 2017).

Commonly, the analysis of large-scale genetic data is regarded as a bottleneck, and one of the most challenging parts is taking large amounts of information and reducing it in meaningful ways for addressing questions. The first step in the bioinformatics pipeline, and one of the most important, is filtering of raw sequence data to remove sequencing errors. Error rates for next-generation approaches can be much higher than for Sanger sequencing (Shendure and Ji, 2008), so removing these errors is critical. However, how filtering is done and how stringent the filtering is can affect biological understanding, such as being able to identify meaningful associations between the environment and certain genotypes (Ahrens et al., 2021). Many tools have been developed, each with strengths and weaknesses (Wee et al., 2019; Hemstrom et al., 2024). However, to use these tools effectively and conduct biologically meaningful bioinformatic analyses, one must have a fundamental understanding of genomics and bioinformatics, as well as practical experience of working with specific tools and downstream analyses (Hemstrom et al., 2024). Ecologists and others working in invasion biology can thus benefit greatly from collaborating with experts in bioinformatics, rather than simply using tools readily available without necessarily possessing the requisite background knowledge.

Once the sampling design is complete (see section 2.4), the next step of the bioinformatic pipeline is to select reference samples to compare with your sequenced samples. In some cases, as in population studies, all individuals will be compared with each other, and you may not need reference data. However, in other study designs, for example, if you are comparing across species, you may need to gather already published data, such as reference genomes of known species that are housed in repositories. These repositories are managed by a number of different organizations in different countries, which hold genetic data for the use of others. Many of these organizations share data through the International Nucleotide Sequence Data Collaboration (INSDC). Partners include the National Center for Biotechnology Information (NCBI), the European Molecular Biology Laboratory (EMBL) and the DNA Data Bank of Japan (DDBJ), among others (Karsch-Mizrachi et al., 2025). Along with depositing (and retrieving) genomic data from these repositories, it is crucial to also deposit voucher specimens so that morphological or other identifying characteristics linked to genomic data are available (Buckner *et al*., 2021).

Sequencing entire genomes used to be restricted to model organisms, such as *Arabidopsis* spp., but increasingly, genome sequencing or population genomics of non-model organisms is becoming commonplace because new technologies have made the process more efficient and less expensive (e.g. Stahlke *et al*., 2023). Despite the reductions in genome sequencing costs, each genome sequencing or population genomics project has three main methodological questions to address: (i) how much of the genome needs to be sequenced for each sample; (ii) how well (or at what coverage) the genome needs to be sequenced; and (iii) how many samples need to be sequenced. The answer to these methodological questions depends on the scientific questions, study design and study organism. In any case, there have been huge advances toward sequencing of whole genomes and in population genetic studies that will undoubtedly continue to revolutionize invasion biology research (McCartney *et al*., 2019; Pujolar *et al*., 2022).

2.3.2 Whole-genome sequencing

High-throughput sequencing allows sequencing of whole genomes, providing vast amounts of information for studying biological invasions (McCartney *et al*., 2019; North *et al*., 2021). Whole-genome sequencing projects are used to characterize an organism's invasive potential at multiple scales, from examining patterns in many genes that may identify changes in populations through adaptive evolution, for example, to studies at the individual gene level pinpointing candidate genes involved in invasiveness (McCartney *et al*., 2019). For population biology, whole-genome sequencing provides large amounts of loci and single-nucleotide polymorphisms (SNPs) to compare across species or populations within a species (de Coster *et al*., 2021). These large data sets provide and improve signal, accuracy and power to address many topics in invasion biology, including characterizing invasion history and adaptive evolution (Taylor *et al*., 2021). However, ensuring that the data are accurate helps strengthen the validity of the results.

Error rates among the technologies vary dramatically. Short-read technologies sequence DNA that has been fragmented into short pieces of typically 50–300 base pairs, and the most commonly used methods, including Illumina technologies, have accuracy rates of, on average, 99.5 % (Goodwin *et al*., 2016). Long-read technologies including Pacific Biosciences (PacBio) and Oxford Nanopore technology (ONT) offer vast amounts of data that are accurate and contiguous, which has helped to simplify and enhance genome reconstruction and assembly (Espinosa *et al*., 2024). Reads from PacBio range from 20 to 100 kb, while reads from ONT range from 10 to 100 kb in size. Both approaches have accuracy rates in the middle of the reads of over 99% (Stoler and Nekrutenko, 2021; Espinosa *et al*., 2024). To manage errors in the output data, regions are sequenced multiple times. The number of times a particular nucleotide in the genome is read during sequencing is called the read depth or sequencing depth (Fig. 2.1). Genome coverage is the proportion of the entire genome that is sequenced at a certain depth (but sometimes the term genome coverage is used interchangeably with read depth) (Lindner *et al*., 2013; Sims *et al*., 2014; Desvillechabrol *et al*., 2018).

Read depth is a key metric in genome sequencing that indicates how thoroughly a given region of the genome has been analyzed. Increasing the read depth improves the confidence of the accuracy of each called nucleotide, which is crucial as otherwise errors can be indistinguishable from sequence variation (Sims *et al*., 2014). For example, if a genome has 15× coverage, then on average across the genome, 15 copies of each nucleotide have been sequenced. If a nucleotide at a locus has one of those copies as thymine (T) while the rest of the copies have an adenine (A), it can be inferred with confidence that the actual nucleotide in the organism sequenced is an A, and that the T represents a sequencing error. Thus, read depth affects accuracy of genomic analysis and ensures that when nucleotide variants are detected they represent true differences among individuals rather than sequencing errors, and thus greater sequencing depth generally provides more robust results (Ajay *et al*., 2011). Increasing read depth improves the reliability of the data but also raises the cost of sequencing, so there is often a balance between the desired quality of data and

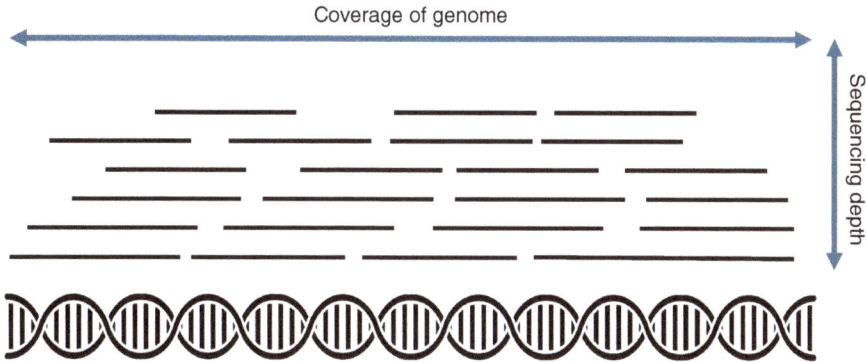

Fig. 2.1. Schematic illustrating genome coverage (the amount of the genome sequenced) and sequencing depth (the number of times a particular nucleotide in the genome is read). Often researchers talk about full coverage at a certain depth. Modified from https://3billion.io/blog/sequencing-depth-vs-coverage (accessed 29 July 2025).

budget constraints, which can impact the number of samples that can be sequenced for a project.

Generally, high sequencing depth of >50–60× coverage is needed for assembling reference genomes, which provide a benchmark sequence of an organism (Fuentes-Pardo and Ruzzante, 2017). With depth of over 50×, and assemblies that integrate short- and long-read technologies, reference genomes can accurately represent the organism and have chromosome-to-chromosome-level resolution. While reference genomes are common for model organisms, increasing numbers of non-model organisms now have reference genomes (27 for invasive species at the time of a review by McCartney *et al.*, 2019). These references enhance further research because new studies can use these benchmarks to identify variation among individuals of the same species. At the other end of sequencing depth, a population genetic study with many individuals might aim for coverage of 2–15× (Lou *et al.*, 2021). These are called "resequencing" studies, as multiple individuals of the same species are sequenced and often compared to a reference genome (when available).

2.3.3 Reduced-representation sequencing for non-model organisms

Reduced-representation sequencing, which targets a small and random portion of the genome across many individuals, can capture thousands of polymorphic markers for discovery of genetic variation among populations and genotyping (Lou *et al.*, 2021). Reduced-representation sequencing is commonly used for two main reasons. First, little to no information is needed about the organism studied, so it is easy to apply to non-model species (Christiansen *et al.*, 2021). Second, because specific areas of the genome are targeted, high-read depth of each locus can be achieved at a lower cost compared with whole-genome sequencing (Davey and Blaxter, 2010; Andrews *et al.*, 2016). There are various methods that fall under reduced-representation sequencing.

One of the most common reduced-representation sequencing methods is restriction-site-associated DNA sequencing (RAD-seq) (Davey and Blaxter, 2010). RAD-seq uses restriction enzymes to cut the genome into fragments, and then adapters and a barcode for each sample are ligated on to the fragments for sequencing (Andrews *et al.*, 2016). Samples are pooled into a single sample that can be sequenced using Illumina and then bioinformatically separated back to individuals with the unique barcodes. With this method, only a small percentage of the genome is sequenced, but it is sequenced at specific sites, with the use of the restriction enzyme, across many individuals that can be compared. The data yields roughly 500–2000 loci although this can vary widely depending on the genome, the type of restriction enzymes used and the size selection step implemented during library preparation (Bresadola *et al.*, 2020). Typically, however, hundreds of thousands of SNPs can be used to assess the relationship between sequenced individuals (Andrews *et al.*, 2016).

2.3.4 Pool sequencing approaches

Pool sequencing, commonly called "pool-seq," is an approach that incorporates whole-genome sequences across pools of 20–100 or more individuals (Schlötterer *et al.*, 2014), for example individuals from a single collection site. In contrast to RAD-seq, pool-seq does not provide individual genotypes, but rather reveals differences between pools of individuals (Dorant *et al.*, 2019). The benefits of pool-seq include enabling the sequencing of many individuals within each of the pools, needing less DNA from each individual within a pool, reducing the cost of sequencing substantially and reducing the amount of time needed for each sequencing project (Anand *et al.*, 2016). The ability to estimate allele frequencies is good with pools of 50 genomes or more. Sequencing pools with too few individuals is not recommended as it can lead to suboptimal results (Schlötterer *et al.*, 2014). However, experimental evolution studies where multiple replicate populations evolve in a context that lends insight into the evolutionary processes that occur during biological invasion have found biologically important patterns using 20 diploid individuals per replicate experimental population, and thus pools of 40 genomes (Weiss-Lehman *et al.*, 2019; Tittes *et al.*, 2024). During preparation for sequencing, it is important to have similar amounts of DNA from each individual in the pool, as different amounts can result in an imbalance that could bias estimates of allele frequency and diversity. To save time and expenses, extracting DNA from pools of samples is possible if all the samples are treated similarly prior to extraction (e.g. sampled live, frozen at −80°C, and then pooled). Overall, pool-seq is a cost-effective method for capturing variation that occurs across populations, whether this variation might be explained by geographical differences, phenotypic variation or ecological differences.

2.3.5 Low-coverage whole-genome sequencing approaches

Low-coverage whole-genome sequencing is another cost-effective technology that can accurately allow population-level screening while keeping some individual genotype information. This approach is best applied to haploid organisms with small genomes. The method is similar to RAD-seq in that library preparation is done separately for each sample, but instead of having a set of small fragments that are sequenced across the population, each individual is sequenced across its entire genome but at a depth as low as $1\times$ (Chat *et al.*, 2022). Typically, at these low levels of sequencing, missing data will occur at the individual level and therefore individual genotypes are not always reliably inferred (Pujolar *et al.*, 2022). Therefore, these data cannot be used to find individual SNPs within an individual but can be used for inferring population-level patterns of variation, including frequencies of more common alleles or linkage disequilibrium, population structure and ancestry relationships (Lou *et al.*, 2021). Furthermore, these data can also be combined with analyses that allow us to infer the true genotype. This can include probabilistic analysis frameworks that integrate genotype uncertainty across the population and make an inference at the individual level (Lou *et al.*, 2021).

2.4 The Crucial Importance of Sampling

Inferences made from genomic data depend strongly on sampling strategy and whether the obtained samples are appropriate for the questions being asked. There is no single best answer to how many and which samples are needed for a study, as biological (e.g. dispersal, clonality, seasonal availability and length of time in a geographical area), financial and logistical constraints vary across systems. The design of a sampling strategy can also benefit from incorporating power analyses and simulation-based approaches, which can help determine the number of samples required to detect patterns such as population structure, gene flow or selection with confidence (Hoban and Schlarbaum, 2014; Qu *et al.*, 2020). Questions to consider in developing a sampling design include: Which individuals should be sampled and from where? How many geographic locations should be sampled? How far apart should they be? Do samples need to be collected through time? If so, at what frequency? Does the design require the presence of an outgroup (a population of a closely related species)? Answers to these sampling design questions will depend on the scientific question and study system.

In an ideal scenario, individuals are sampled randomly, allowing the capture of genetic diversity that represents the larger population. In practice, however, random sampling is rarely possible—particularly in the context of invading organisms—due to both ecological and logistical constraints. Even more so than for native organisms, the full extent of an invader's introduced range is often unknown, and populations may be patchy, transient or restricted to isolated "hot spots," making it difficult to access and sample the population evenly. Additionally, sampling is typically limited to certain times of year, further reducing temporal randomness. These spatial and temporal limitations mean that the genetic variation captured in a study will reflect not the entire population but the specific areas and times at which sampling occurred. As a result, sampling design becomes a critical determinant of the genetic patterns observed and will directly affect inferences from the generated data.

2.4.1 Choosing individuals to sample

Practical approaches to randomize sampling in space can involve first defining a random sampling unit (e.g. random points in the field, forest, lake, region) and then collecting samples from those units (e.g. Hübner *et al.*, 2022). When an area is large, transects are often used for sampling. In this case, a researcher follows a predefined transect (straight line, Z- or W-pattern) and collects samples at fixed distances along the transect. These approaches work well for species with random geographical distribution. If a species is not randomly distributed in the studied ecosystem, more complicated sampling strategies are required to capture a representative sample. For example, stratified sampling is often used in such cases. Stratified sampling is a probability sampling technique that involves dividing the population into subgroups (strata) and then taking samples from each subgroup. The goal of stratified sampling is to ensure that different subgroups within a population are adequately represented in a sample. For example, habitat type (e.g. coastal, forest, grassland) can be used as strata in studies across large geographical regions. In studies with a complex invasion history, the time of invasion can be used as a stratum, for example early invasion sites or recent invasion sites. With this approach, an equal number of samples of an invading organism must be collected within each stratum. Then, at the data analysis step, a researcher can compare populations across different strata. Overall, stratification is a powerful technique that can help to account for potential confounding factors, leading to more accurate and generalized conclusions.

Haphazard sampling, in turn, is a technique where samples are collected arbitrarily and without any specific plan or pattern. Haphazard sampling might involve collecting samples from easily accessible locations, such as plants growing along a roadside or in a public park, without regard to geographical distribution, habitat type or invasion timeline. Haphazard sampling will limit the inferences made from the collected data as it lacks the rigor and representativeness required for more detailed analyses. Nevertheless, it can still be informative, especially in preliminary or exploratory studies where quick and easy data collection is needed. For example, in studies assessing invasion fronts, conducting a survey to assess the geographical extent of a newly reported invasive species and sampling along the way can provide useful preliminary data. At times when only haphazard sampling is possible, its limitations must be taken into account when generating hypotheses and making inferences from the data (e.g. McGoey and Stinchcombe, 2021; VanWallendael *et al.*, 2021).

2.4.2 Spatial considerations in sampling and use of outgroups

To understand the source of invasions, it is necessary to sample several populations from different regions, including both native and non-native ecosystems. This will allow testing of different scenarios of a species' invasion history and will help determine the presence and direction of gene flow among the populations (Quarrell *et al.*,

2018). The presence and directionality of gene flow can unravel recent and/or ancient movement pathways of an invading organism, or its demographic history. Some methods aimed at reconstruction of demographic history of the species require the presence of an outgroup in the data set. An outgroup serves as a reference point that is distantly related to the populations of the studied organism. By comparing the genetic variation of the invading organism with the outgroup, researchers can infer the direction of evolutionary changes, helping to identify the ancestral state of genetic traits. This allows a more accurate reconstruction of the evolutionary history and origins of the invading organism, helping to distinguish between native and introduced populations and to trace the invasion pathways. In cases where the demographic history of the invader is of interest, sampling a closely related species population should be included in the study (Bieker *et al.*, 2022).

2.4.3 Sampling over time

Some questions focus on how populations change over time and thus require repeated sampling. This approach is particularly valuable for investigating evolutionary processes, such as natural selection, genetic drift and migration, which can lead to significant shifts in population structure and genetic composition. By capturing data across different time points, researchers can track how these forces shape the genetic landscape of populations, providing insights into both short-term fluctuations and long-term evolutionary trends.

In addition to answering specific temporal questions, it is important to recognize that populations inherently change over time. This is especially true for organisms with mixed reproductive modes (e.g. many fungal and plant species). Diversity is generally highest right after sexual reproduction occurs, and if mating is followed by a series of asexual cycles, the population genetic diversity will decrease over time. For such species, being sure to sample different locations at comparable times in the life cycle and, if feasible, sampling repeatedly over time, will aid in making robust comparisons among samples. Integrating both spatial and temporal data in

population genomics studies of invasive species can provide a more detailed understanding of their evolutionary history, offering higher resolution insights into the dynamics of biological invasions (Jaspers *et al.*, 2021).

2.4.4 Accounting for underlying population structure

Another important factor to consider during sampling is the existing population structure, as a failure to account for it will lead to misinterpretations of generated data. For example, if individuals from multiple distinct genetic populations are pooled together and analyzed as if they were a single population, the resulting sample may appear to have reduced genetic diversity and increased homozygosity, giving the impression of inbreeding. This occurs even if sexual reproduction takes place within each of the original populations. Scenarios like these may lead to erroneous inferences when evaluating the risks of successful establishment of the invading species in a new location, its rate of spread and the ecological dynamics in a new environment. In studies focused on identifying genomic regions under selection to understand adaptive evolution of invading species, failure to account for the underlying population structure in data analysis may lead to false signals of selection. This will happen due to allele frequency differences between the subpopulations that are actually due to genetic drift rather than selection. Overlooked underlying population structure may also lead to underestimates of the number of invasion events that happened in the past and the number of source locations (Lohmueller, 2014; Hohenlohe *et al.*, 2021). As a result, a species' invasion history can be evaluated incorrectly. To avoid this, pilot studies are usually recommended (Hahn, 2018; Schmidt *et al.*, 2024). They will help us understand the rate of genetic diversity and underlying population structure of species and help in the design of an appropriate sampling strategy. When pilot studies are not possible, population geneticists can help to design a sampling strategy that could account for these potential biases and define what inferences can be expected from the data.

2.4.5 Sampling constraints and marker selection

Sampling can also define the type of genomic markers to be used. For invasive species, there can be a limited number of samples available, especially if it is a recently emerged population or a population in a location that is difficult to access. In such cases, it may be advisable to generate a larger amount of genomic data per sample, such as using RAD-seq or whole-genome sequencing (see section 2.3 on the brief overview of genomic markers in this chapter). These approaches are particularly valuable because many population genomics tools tend to be more robust and yield more reliable insights when the amount of data per sample is high, even if the overall sample size is limited. Thus, the choice of markers and sequencing strategy should be carefully considered in light of the practical realities of sample availability (Qu et al., 2020; Soghigian et al., 2020).

2.4.6 Statistical power and sampling

Sampling strategies should not only reflect the biological realities of the system but also be informed by statistical power considerations. The number of individuals and populations sampled affects the ability to detect patterns such as population structure, gene flow or selection. Empirical and simulation studies suggest that while as few as eight to 12 individuals per population may suffice to detect strong population structure when using thousands of SNPs (Willing et al., 2012; Nazareno et al., 2017), more individuals (20–30) are often needed for estimating migration rates or effective population size with confidence (Hoban et al., 2016). The number of populations sampled also plays a critical role: for example, detecting loci under selection is often more powerful when sampling more populations with fewer individuals each, rather than fewer populations with deeper sampling (Lotterhos and Whitlock, 2015). Tools such as msprime (Baumdicker et al., 2022), Fastsimcoal2 (Excoffier et al., 2013), or R packages like adegenet (Jombart, 2008) and strataG (Archer et al., 2017) can be used to simulate data under different sampling designs and assess the expected power of downstream analyses. Incorporating such tools early in study planning, even with preliminary data or pilot studies, can help to estimate statistical power of different sampling designs.

2.5 Uniting Environmental and Phenotypic Data with Genomics

2.5.1 Genotype–environment associations

Genetic signatures of adaptation to different environments may be detected by evaluating correlations between allele frequencies, for example using SNP data, and environmental variables from the location of origin. For genotype–environment associations to be detected, five conditions must hold, as outlined by Lasky et al. (2023). First, the species in question must be found across varying environmental conditions. Second, individuals must be locally adapted to the environmental conditions at that site. Third, to detect associations with distinct aspects of the environment, the conditions cannot be perfectly correlated across space. For example, if warm winters are always found with sandy soils and hard freezes only occur in areas with clay soils, it will not be possible to disentangle the genetic underpinning of adaptation to soil type from adaptation to temperature. If these environmental correlations are not absolute, however, then carefully considered choices of sample sites to disentangle them can address this. Fourth, local adaptation cannot be perfectly correlated with population structure. For example, populations may be adapted to hard winter freezes, and all those freezes occur in more northern locations. However, if northern locations have an invasive origin that differs from southern locations, then adaptation is correlated with genetic population structure, so it will not be possible to distinguish genomic associations with the environment from that population structure. Again, careful consideration of sampling locations may be able to help disentangle genotype–environment associations from population structure. Fifth, the chance of detecting a correlation will be higher if the genomic basis of local adaptation is relatively simple, such that a few genes have major effects, rather than being polygenic with many genes having small effects (Lasky et al., 2023).

Detecting genotype–environment associations thus requires data from multiple locations that vary by environment. As with all correlative approaches, robust sampling is key. There are two different types of sampling units: individuals within a site, to estimate allele frequencies, and across-sample sites to have a broad range of environments. A minimum of ten individuals within sites will provide estimates of allele frequencies down to the 0.05 level for diploids, but more samples and finer proportions will provide even more precise estimates. Larger sample sizes are needed for pool-seq approaches as discussed above, and for detecting minor (rare) alleles. For minor alleles, the probability of capture depends strongly on allele frequency and sample size (see Fig. 2 in Lasky *et al.*, 2023, for a graphic). At the level of sampling location, a minimum of 20 locations that vary by an environmental factor important to the species in question should be able to reveal strong correlations, but 50 sites or more will provide more power to detect genotype–environment correlations (Selmoni *et al.*, 2020). As noted, sampling locations should be selected carefully with the species' biology in mind, and across multiple environmental axes to disentangle different climatic variables. With appropriate sampling, distinguishing the relative importance of genetic ancestry and differences in selection across different environments can be done robustly (Pita-Aquino *et al.*, 2023).

Statistically, correlations between allele frequencies and environmental variables can be evaluated one SNP and one variable at a time using univariate approaches, or simultaneously across multiple SNPs and environmental variables with multivariate approaches (see Sherpa *et al.*, 2019, for examples of each).

2.5.2 Genotype–phenotype associations

Genomic variation underlying phenotypic traits can be evaluated with genome-wide association studies (GWAS). This is an incredibly powerful tool that has rarely been used in the context of biological invasions (but see Olazcuaga *et al.*, 2020; Turner *et al.*, 2021). Here, the individual is the typical unit of replication for phenotyping, with individuals with different phenotypes needed to detect potential genomic associations.

The phenotypes of interest should be chosen carefully. Continuous phenotypes (e.g. flowering time) can be used, but it is easier to detect associations with phenotypes that are discrete (e.g. flowering or not flowering under a particular light regime).

An alternative to phenotyping individuals is phenotyping at the level of the population. For example, populations can be characterized as either invasive or not (e.g. from the invasive range or from the native range) to search for genotypes associated with invasion, an approach that works particularly well with sequence data from pools of individuals (Olazcuaga *et al.*, 2020).

GWAS has only rarely been applied to biological invasions, perhaps because phenotypes among invasive populations may be relatively homogeneous, precisely because of the invasion. However, much research shows that phenotypes of invasive populations can differ from native ones (e.g. Blumenthal and Hufbauer, 2007; Parker *et al.*, 2013), and GWAS could be used to understand deeply the genomics of evolution during invasion through such comparisons (e.g. Olazcuaga *et al.*, 2020; Turner *et al.*, 2021). This thus represents an area of enormous promise for insights and fruitful collaborations.

2.6   Conclusions

Here, we have provided a background in genomic approaches for ecologists and other biological experts with the aim of facilitating productive collaborations exploring biological invasions. We covered the topics that genomics can provide insights into, such as identifying routes of invasion, and discerning neutral and adaptive evolution during invasion, as well as identifying which environmental selection pressures drive adaptive evolution. We then turned to techniques to address those topics, and in particular how to design robust sampling to address questions of interest. Through understanding the strengths and weaknesses of different sequencing approaches, ecologists and other biological experts can have more productive collaborations with their geneticist colleagues.

References

Aguirre-Liguori, J.A., Ramírez-Barahona, S. and Gaut, B.S. (2021) The evolutionary genomics of species' responses to climate change. *Nature Ecology and Evolution* 5, 1350–1360.DOI: 10.1038/s41559-021-01526-9

Ahrens, C.W., Jordan, R., Bragg, J., Harrison, P.A., Hopley, T. *et al.* (2021) Regarding the F-word: the effects of data filtering on inferred genotype-environment associations. *Molecular Ecology Resources* 21, 1460–1474. DOI: 10.1111/1755-0998.13351

Ajay, S.S., Parker, S.C.J., Abaan, H.O., Fajardo, K.V.F. and Margulies, E.H. (2011) Accurate and comprehensive sequencing of personal genomes. *Genome Research* 21, 1498–1505. DOI: 10.1101/gr.123638.111

Al-Shuhaib, M.B.S. and Hashim, H.O. (2023) Mastering DNA chromatogram analysis in Sanger sequencing for reliable clinical analysis. *Journal of Genetic Engineering and Biotechnology* 21: 115. DOI: 10.1186/s43141-023-00587-6

Ambardar, S., Gupta, R., Trakroo, D., Lal, R. and Vakhlu, J. (2016) High throughput sequencing: an overview of sequencing chemistry. *Indian Journal of Microbiology* 56, 394–404. DOI: 10.1007/s12088-016-0606-4

Anand, S., Mangano, E., Barizzone, N., Bordoni, R., Sorosina, M. *et al.* (2016) Next generation sequencing of pooled samples: guideline for variants' filtering. *Scientific Reports* 6: 33735. DOI: 10.1038/srep33735

Andrews, K.R., Good, J.M., Miller, M.R., Luikart, G. and Hohenlohe, P.A. (2016) Harnessing the power of RADseq for ecological and evolutionary genomics. *Nature Reviews Genetics* 17, 81–92. DOI: 10.1038/nrg.2015.28

Archer, F.I., Adams, P.E. and Schneiders, B.B. (2017) stratag: an R package for manipulating, summarizing and analysing population genetic data. *Molecular Ecology Resources* 17, 5–11. DOI: 10.1111/1755-0998.12559

Battlay, P., Craig, S., Putra, A.R., Monro, K., de Silva, N.P. *et al.* (2025) Rapid parallel adaptation in distinct invasions of *Ambrosia artemisiifolia* is driven by large-effect structural variants. *Molecular Biology and Evolution* 42: msae270. DOI: 10.1093/molbev/msae270

Baumdicker, F., Bisschop, G., Goldstein, D., Gower, G., Ragsdale, A.P. *et al.* (2022) Efficient ancestry and mutation simulation with msprime 1.0. *Genetics* 220: iyab229. DOI: 10.1093/genetics/iyab229

Bieker, V.C., Battlay, P., Petersen, B., Sun, X., Wilson, J. *et al.* (2022) Uncovering the genomic basis of an extraordinary plant invasion. *Science Advances* 8: eabo5115. DOI: 10.1126/sciadv.abo5115

Blumenfeld, A.J., Eyer, P.-A., Husseneder, C., Mo, J., Johnson, L.N.L. *et al.* (2021) Bridgehead effect and multiple introductions shape the global invasion history of a termite. *Communications Biology* 4: 196. DOI: 10.1038/s42003-021-01725-x

Blumenthal, D.M. and Hufbauer, R.A. (2007) Increased plant size in exotic populations: a common-garden test with 14 invasive species. *Ecology* 88, 2758–2765. DOI: 10.1890/06-2115.1

Bock, D.G., Baeckens, S., Kolbe, J.J. and Losos, J.B. (2024) When adaptation is slowed down: genomic analysis of evolutionary stasis in thermal tolerance during biological invasion in a novel climate. *Molecular Ecology* 33: e17075. DOI: 10.1111/mec.17075

Boscolo Agostini, R., Vizzari, M.T., Benazzo, A. and Ghirotto, S. (2025) Disentangling the worldwide invasion process of *Halyomorpha halys* through approximate Bayesian computation. *Heredity* 134, 64–74. DOI: 10.1038/s41437-024-00735-9

Bresadola, L., Link, V., Buerkle, C.A., Lexer, C. and Wegmann, D. (2020) Estimating and accounting for genotyping errors in RAD-seq experiments. *Molecular Ecology Resources* 20, 856–870. DOI: 10.1111/1755-0998.13153

Buckner, J.C., Sanders, R.C., Faircloth, B.C. and Chakrabarty, P. (2021) The critical importance of vouchers in genomics. *eLife* 10: e68264. DOI: 10.7554/eLife.68264

Chat, V., Ferguson, R., Morales, L. and Kirchhoff, T. (2022) Ultra low-coverage whole-genome sequencing as an alternative to genotyping arrays in genome-wide association studies. *Frontiers in Genetics* 12: 790445. DOI: 10.3389/fgene.2021.790445

Christiansen, H., Heindler, F.M., Hellemans, B., Jossart, Q., Pasotti, F. *et al.* (2021) Facilitating population genomics of non-model organisms through optimized experimental design for reduced representation sequencing. *BMC Genomics* 22: 625. DOI: 10.1186/s12864-021-07917-3

Davey, J.W. and Blaxter, M.L. (2010) RADSeq: next-generation population genetics. *Briefings in Functional Genomics* 9, 416–423. DOI: 10.1093/bfgp/elq031 de Coster, W., Weissensteiner, M.H. and Sedlazeck, F.J. (2021) Towards population-scale long-read sequencing. *Nature Reviews Genetics* 22, 572–587. DOI: 10.1038/s41576-021-00367-3

Desvillechabrol, D., Bouchier, C., Kennedy, S. and Cokelaer, T. (2018) Sequana coverage: detection and characterization of genomic variations using running median and mixture models. *GigaScience* 7: giy110. DOI: 10.1093/gigascience/giy110

Dorant, Y., Benestan, L., Rougemont, Q., Normandeau, E., Boyle, B. *et al.* (2019) Comparing Pool-seq, Rapture, and GBS genotyping for inferring weak population structure: the American lobster (*Homarus americanus*) as a case study. *Ecology and Evolution* 9, 6606–6623. DOI: 10.1002/ece3.5240

Ekblom, R. and Wolf, J.B.W. (2014) A field guide to whole-genome sequencing, assembly and annotation. *Evolutionary Applications* 7, 1026–1042. DOI: 10.1111/eva.12178

Espinosa, E., Bautista, R., Larrosa, R. and Plata, O. (2024) Advancements in long-read genome sequencing technologies and algorithms. *Genomics* 116: 110842. DOI: 0.1016/j.ygeno.2024.110842

Estoup, A. and Guillemaud, T. (2010) Reconstructing routes of invasion using genetic data: why, how and so what? *Molecular Ecology* 19, 4113–4130. DOI: 10.1111/j.1365-294X.2010.04773.x

Excoffier, L., Dupanloup, I., Huerta-Sánchez, E., Sousa, V.C. and Foll, M. (2013) Robust demographic inference from genomic and SNP data. *PLOS Genetics* 9: e1003905. DOI: 10.1371/journal.pgen.1003905

Fontaine, M.C., Labbé, F., Dussert, Y., Delière, L., Richart-Cervera, S. *et al.* (2021) Europe as a bridgehead in the worldwide invasion history of grapevine downy mildew, *Plasmopara viticola*. *Current Biology* 31, 2155-2166.e4. DOI: 10.1016/j.cub.2021.03.009

Fraimout, A., Debat, V., Fellous, S., Hufbauer, R.A., Foucaud, J. *et al.* (2017) Deciphering the routes of invasion of *Drosophila suzukii* by means of ABC random forest. *Molecular Biology and Evolution* 34, 980–996. DOI: 10.1093/molbev/msx050

Fuentes-Pardo, A.P. and Ruzzante, D.E. (2017) Whole-genome sequencing approaches for conservation biology: advantages, limitations and practical recommendations. *Molecular Ecology* 26, 5369–5406. DOI: 10.1111/mec.14264

Goodwin, S., McPherson, J.D. and McCombie, W.R. (2016) Coming of age: ten years of next-generation sequencing technologies. *Nature Reviews Genetics* 17, 333–351. DOI: 10.1038/nrg.2016.49

Hahn, M.W. (2018) *Molecular Population Genetics*. Oxford University Press, New York.

Head, S.R., Komori, H.K., LaMere, S.A., Whisenant, T., van Nieuwerburgh, F. *et al.* (2014) Library construction for next-generation sequencing: overviews and challenges. *BioTechniques* 56, 61–77. DOI: 10.2144/000114133

Heather, J.M. and Chain, B. (2015) The sequence of sequencers: the history of sequencing DNA. *Genomics* 107, 1–8. DOI: 10.1016/j.ygeno.2015.11.003.

Hemstrom, W., Grummer, J.A., Luikart, G. and Christie, M.R. (2024) Next-generation data filtering in the genomics era. *Nature Reviews Genetics* 25, 750–767. DOI: 10.1038/s41576-024-00738-6.

Hoban, S. and Schlarbaum, S. (2014) Optimal sampling of seeds from plant populations for *ex-situ* conservation of genetic biodiversity, considering realistic population structure. *Biological Conservation* 177, 90–99. DOI: 10.1016/j.biocon.2014.06.014

Hoban, S., Kelley, J.L., Lotterhos, K.E., Antolin, M.F., Bradburd, G. *et al.* (2016) Finding the genomic basis of local adaptation: pitfalls, practical solutions, and future directions. *American Naturalist* 188, 379–397. DOI: 10.1086/688018

Hohenlohe, P.A., Funk, W.C. and Rajora, O.P. (2021) Population genomics for wildlife conservation and management. *Molecular Ecology* 30, 62–82. DOI: 10.1111/mec.15720

Hübner, S., Sisou, D., Mandel, T., Todesco, M., Maor Matzrafi and Eizenberg, H. (2022) Wild sunflower goes viral: citizen science and comparative genomics allow tracking the origin and establishment of invasive sunflower in the Levant. *Molecular Ecology* 3, 2061–2072. DOI: 10.1111/mec.1638

Hutchison, C.A. 3rd (2007) DNA sequencing: bench to bedside and beyond. *Nucleic Acids Research* 35, 6227–6237. DOI: 10.1093/nar/gkm688.

Jaspers, C., Ehrlich, M., Pujolar, J.M., Künzel, S., Bayer, T. *et al.* (2021) Invasion genomics uncover contrasting scenarios of genetic diversity in a widespread marine invader. *Proceedings of the National Academy of Sciences USA* 118: e2116211118. DOI:

Johri, P., Aquadro, C.F., Beaumont, M., Charlesworth, B., Excoffier, L. *et al.* (2022) Recommendations for improving statistical inference in population genomics. *PLOS Biology* 20: e3001669. DOI: 10.1371/journal.pbio.3001669

Jombart, T. (2008) adegenet: a R package for the multivariate analysis of genetic markers. *Bioinformatics* 24, 1403–1405. DOI: 10.1093/bioinformatics/btn129

Karsch-Mizrachi, I., Arita, M., Burdett, T., Cochrane, G., Nakamura, Y. and Pruitt, K.D. (2025) The international nucleotide sequence database collaboration (INSDC): enhancing global participation. *Nucleic Acids Research* 53, D62–D66. DOI: 10.1093/nar/gkae1058

Kołodziejczyk, J., Fijarczyk, A., Porth, I., Robakowski, P., Vella, N. *et al.* (2025) Genomic investigations of successful invasions: the picture emerging from recent studies. *Biological reviews of the Cambridge Philosophical Society* 100, 1396–1418. DOI: 10.1111/brv.70005

Lasky, J.R., Josephs, E.B. and Morris, G.P. (2023) Genotype–environment associations to reveal the molecular basis of environmental adaptation. *Plant Cell* 35, 125–138. DOI: 10.1093/plcell/koac26

Lawson Handley, L.-J., Estoup, A., Evans, D.M., Thomas, C.E., Lombaert, E. *et al.* (2011) Ecological genetics of invasive alien species. *BioControl* 56, 409–428. DOI: 10.1007/s10526-011-9386-2

Lindner, M.S., Kollock, M., Zickmann, F. and Renard, B.Y. (2013) Analyzing genome coverage profiles with applications to quality control in metagenomics. *Bioinformatics* 29, 1260–1267. DOI: 10.1093/bioinformatics/btt147

Lohmueller, K.E. (2014) The impact of population demography and selection on the genetic architecture of complex traits. *Plos Genetics* 10: e1004379. DOI: 10.1371/journal.pgen.1004379

Lotterhos, K.E. and Whitlock, M.C. (2015) The relative power of genome scans to detect local adaptation depends on sampling design and statistical method. *Molecular Ecology* 24, 1031–1046. DOI: 10.1111/mec.13100

Lou, R.N., Jacobs, A., Wilder, A.P. and Therkildsen, N.O. (2021) A beginner's guide to low-coverage whole genome sequencing for population genomics. *Molecular Ecology* 30, 5966–5993. DOI: 10.1111/mec.16077

McCartney, M.A., Mallez, S. and Gohl, D.M. (2019) Genome projects in invasion biology. *Conservation Genetics* 20, 1201–1222. DOI: 10.1007/s10592-019-01224-x

McGaughran, A., Dhami, M.K., Parvizi, E., Vaughan, A.L., Gleeson, D.M. *et al.* (2024) Genomic tools in biological invasions: current state and future frontiers. *Genome Biology and Evolution* 16: evad230. DOI: 10.1093/gbe/evad230

McGoey, B.V. and Stinchcombe, J.R. (2021) Introduced populations of ragweed show as much evolutionary potential as native populations. *Evolutionary Applications* 14, 1436–1449. DOI: 10.1111/eva.13211

Metzker, M.L. (2010) Sequencing technologies—the next generation. *Nature Reviews Genetics* 11, 31–46. DOI: 10.1038/nrg2626

Nazareno, A.G., Bemmels, J.B., Dick, C.W. and Lohmann, L.G. (2017) Minimum sample sizes for population genomics: an empirical study from an Amazonian plant species. *Molecular Ecology Resources* 17, 1136–1147. DOI: doi: 10.1111/1755-0998.12654

North, H.L., McGaughran, A. and Jiggins, C.D. (2021) Insights into invasive species from whole-genome resequencing. *Molecular Ecology* 30, 6289–6308. DOI: 10.1111/mec.15999

Olazcuaga, L., Loiseau, A., Parrinello, H., Paris, M., Fraimout, A *et al.* (2020) A whole-genome scan for association with invasion success in the fruit fly *Drosophila suzukii* using contrasts of allele frequencies corrected for population structure. *Molecular Biology and Evolution* 37, 2369–2385. DOI: 10.1093/molbev/msaa098

Parker, J.D., Torchin, M.E., Hufbauer, R.A., Lemoine, N.P., Alba, C. *et al.* (2013) Do invasive species perform better in their new ranges? *Ecology* 94, 985–994. DOI: 10.1890/12-1810.1

Parvizi, E., Vaughan, A.L., Dhami, M.K. and McGaughran, A. (2024) Genomic signals of local adaptation across climatically heterogenous habitats in an invasive tropical fruit fly (*Bactrocera tryoni*). *Heredity* 132, 18–29. DOI: 10.1038/s41437-023-00657-y

Pearman, W.S., Urban, L. and Alexander, A. (2022) Commonly used Hardy–Weinberg equilibrium filtering schemes impact population structure inferences using RADseq data. *Molecular Ecology Resources* 22, 2599–2613. DOI: 10.1111/1755-0998.13646

Pita-Aquino, J.N., Bock, D.G., Baeckens, S., Losos, J.B. and Kolbe, J.J. (2023) Stronger evidence for genetic ancestry than environmental conditions in shaping the evolution of a complex signalling trait during biological invasion. *Molecular Ecology* 32, 5558–5574. DOI: 10.1111/mec.17123

Poland, T.M. and Rassati, D. (2019) Improved biosecurity surveillance of non-native forest insects: a review of current methods. *Journal of Pest Science* 92, 37–49. DOI: 10.1007/s10340-018-1004-y

Pujolar, J.M., Limborg, M.T., Ehrlich, M. and Jaspers, C. (2022) High throughput SNP chip as cost effective new monitoring tool for assessing invasion dynamics in the comb jelly *Mnemiopsis leidyi*. *Frontiers in Marine Science* 9: 1019001. DOI: 10.3389/fmars.2022.1019001

Qu, W., Liang, N., Wu, Z., Zhao, Y. and Chu, D. (2020) Minimum sample sizes for invasion genomics: empirical investigation in an invasive whitefly. *Ecology and Evolution* 10, 38–49. DOI: 10.1002/ece3.5677

Quarrell, S.R., Arabi, J., Suwalski, A., Veuille, M., Wirth, T. and Allen, G.R. (2018) The invasion biology of the invasive earwig, *Forficula auricularia* in Australasian ecosystems. *Biological Invasions* 20, 1553–1565. DOI: 10.1007/s10530-017-1646-3

Sanger, F., Nicklen, S. and Coulson, A.R. (1977) DNA sequencing with chain-terminating inhibitors. *Proceedings of the National Academy of Sciences USA* 74, 5463–5467. DOI: 10.1073/pnas.74.12.5463

Šarhanová, P., Pfanzelt, S., Brandt, R., Himmelbach, A. and Blattner, F.R. (2018) SSR-seq: genotyping of microsatellites using next-generation sequencing reveals higher level of polymorphism as compared to traditional fragment size scoring. *Ecology and Evolution* 8, 10817–10833. DOI: 10.1002/ece3.4533

Schlötterer, C., Tobler, R., Kofler, R. and Nolte, V. (2014) Sequencing pools of individuals—mining genome-wide polymorphism data without big funding. *Nature Reviews Genetics* 15, 749–763. DOI: 10.1038/nrg3803

Schmidt, T.L., Thia, J.A. and Hoffmann, A.A. (2024) How can genomics help or hinder wildlife conservation? *Annual Review of Animal Biosciences* 12, 45–68. DOI: 10.1146/annurev-animal-021022-051810

Schulz, A.N., Mech, A.M., Ayres, M.P., Gandhi, K.J.K., Havill, N.P. *et al.* (2021) Predicting non-native insect impact: focusing on the trees to see the forest. *Biological Invasions* 23, 3921–3936. DOI: 10.1007/s10530-021-02621-5

Selmoni, O., Vajana, E., Guillaume, A., Rochat, E. and Joost, S. (2020) Sampling strategy optimization to increase statistical power in landscape genomics: a simulation-based approach. *Molecular Ecology Resources* 20, 154–169. DOI: 10.1111/1755-0998.13095

Shendure, J. and Ji, H. (2008) Next-generation DNA sequencing. *Nature Biotechnology* 26, 1135–1145. DOI: doi: 10.1038/nbt1486

Sherpa, S., Blum, M.G.B. and Després, L. (2019) Cold adaptation in the Asian tiger mosquito's native range precedes its invasion success in temperate regions. *Evolution* 73, 1793–1808. DOI: 10.1111/evo.13801

Sims, D., Sudbery, I., Ilott, N.E., Heger, A. and Ponting, C.P. (2014) Sequencing depth and coverage: key considerations in genomic analyses. *Nature Reviews Genetics* 15, 121–132. DOI: 10.1038/nrg3642

Slatko, B.E., Gardner, A.F. and Ausubel, F.M. (2018) Overview of next-generation sequencing technologies. *Current Protocols in Molecular Biology* 122: e59. DOI: 10.1002/cpmb.59.

Soghigian, J., Gloria-Soria, A., Robert, V., Le Goff, G., Failloux, A. and Powell, J.R. (2020) Genetic evidence for the origin of *Aedes aegypti*, the yellow fever mosquito, in the southwestern Indian Ocean. *Molecular Ecology* 29, 3593–3606. DOI: 10.1111/mec.15590.

Stahlke, A.R., Chang, J., Tembrock, L.R., Sim, S.B., Chudalayandi, S. *et al.* (2023) A chromosome-scale genome assembly of a *Helicoverpa zea* strain resistant to *Bacillus thuringiensis* Cry1Ac insecticidal protein. *Genome Biology and Evolution* 15: evac131. DOI: 10.1093/gbe/evac131

Stoler, N. and Nekrutenko, A. (2021) Sequencing error profiles of Illumina sequencing instruments. *NAR Genomics and Bioinformatics* 3: lqab019. DOI: 10.1093/nargab/lqab019

Taylor, R.S., Jensen, E.L., Coltman, D.W., Foote, A.D. and Lamichhaney, S. (2021) Seeing the whole picture: what molecular ecology is gaining from whole genomes. *Molecular Ecology* 30, 5917–5922. DOI: 10.1111/mec.16282

Tittes, S., Weiss-Lehman, C., Kane, N.C., Hufbauer, R.A., Emery, N.C. and Melbourne, B.A. (2024) Evolution is more repeatable in the introduction than range expansion phase of colonization. *Evolution Letter* 8, 351–360. DOI: 10.1093/evlett/qrad063

Turner, K.G., Ostevik, K.L., Grassa, C.J. and Rieseberg, L.H. (2021) Genomic analyses of phenotypic differences between native and invasive populations of diffuse knapweed (*Centaurea diffusa*). *Frontiers in Ecology and Evolution* 8: 577635. DOI: 10.3389/fevo.2020.577635

van Kleunen, M., Weber, E. and Fischer, M. (2010) A meta-analysis of trait differences between invasive and non-invasive plant species. *Ecology Letters* 13, 235–245. DOI: 10.1111/j.1461-0248.2009.01418.x

VanWallendael, A., Alvarez, M. and Franks, S.J. (2021) Patterns of population genomic diversity in the invasive Japanese knotweed species complex. *American Journal of Botany* 108, 857–868. DOI: 10.1002/ajb2.1653

Wee, Y., Bhyan, S.B., Liu, Y., Lu, J., Li, X. and Zhao, M. (2019) The bioinformatics tools for the genome assembly and analysis based on third-generation sequencing. *Briefings in Functional Genomics* 18, 1–12. DOI: 10.1093/bfgp/ely037

Weiss-Lehman, C., Tittes, S., Kane, N.C., Hufbauer, R.A. and Melbourne, B.A. (2019) Stochastic processes drive rapid genomic divergence during experimental range expansions. *Proceedings of the Royal Society B: Biological Sciences* 286: 20190231. DOI: 10.1098/rspb.2019.0231

Willing, E.-M., Dreyer, C. and van Oosterhout, C.J. (2012) Estimates of genetic differentiation measured by F_{ST} do not necessarily require large sample sizes when using many SNP markers. *PLOS ONE* 7: e42649. DOI: 10.1371/journal.pone.0042649

Zamudio, K.R., Bell, R.C. and Mason, N.A. (2016) Phenotypes in phylogeography: species' traits, environmental variation, and vertebrate diversification. *Proceedings of the National Academy of Sciences USA* 113, 8041–8048. DOI: 10.1073/pnas.1602237113

3 Methods for Evaluating Invasion Potential Based on Genomic Data

Amy L. Vaughan[1,2], Elahe Parvizi[3], Manpreet K. Dhami[1]* and Angela McGaughran[3]*

[1]*Biocontrol and Molecular Ecology, Manaaki Whenua Landcare Research Group, Bioeconomy Science Institute, Lincoln, Aotearoa, New Zealand;* [2]*Kura Ngahere School of Forestry, University of Canterbury, Christchurch, New Zealand;* [3]*Te Aka Mātuatua/School of Science, University of Waikato, Hamilton, New Zealand*

Abstract

Our current understanding of the genomic mechanisms that underscore the success – or failure – of biological invasions is limited. As genomic tools and bioinformatic software rapidly advance, we have an unparalleled opportunity to apply new and emerging methodologies to progress our understanding of what makes a successful invader. Current research demonstrates strong linkages between genomic variation and successful biological invasions. This chapter explores genomic methods for evaluating invasion potential. It first focuses on the current scope of population-based and comparative genomic analyses and their ability to result in genome-informed management tools before delving into deep learning tools that are beginning to leverage genome data for prediction and pattern detection to answer fundamental questions pertaining to invasions. Throughout, this chapter discusses the promise of new technologies, weighted by their current limitations, to promote constructive experimental design for data interoperability as a key consideration for future research.

3.1 Introduction

Biological invasions are commonly characterized by the establishment and spread of non-native species across a wide range of potentially novel environmental conditions. Thus, rapid post-invasion responses to new challenges can directly contribute to invasion success (Battlay *et al.*, 2023). Although phenotypic, behavioural and ecological responses are well-characterized in many taxonomic systems (e.g. Strayer *et al.*, 2006; Pyper *et al.*, 2024), our understanding of the role of genomic factors in biological invasion is less advanced (McGaughran *et al.*, 2024). Yet such understanding can bolster management outcomes, with an emergent research area focusing on the use of genomic data to investigate invasion potential. As high-throughput genome sequencing is becoming increasingly affordable, and associated analytical approaches continue to be developed, such goals are becoming more realizable. Collectively, these tools are providing new opportunities to investigate long-standing questions in invasion biology through a genomic lens (Pélissié *et al.*, 2022), such as whether natural selection targets new mutations or standing variation, how the distribution and characteristics (e.g. regulatory, coding, structural) of adaptive loci vary across the genome, and whether genetic responses in new environments evolve independently or in parallel across different populations and species.

*Corresponding author: dhamim@landcareresearch.co.nz or angela.mcgaughran@waikato.ac.nz

© CAB International 2025. *Invasion Genomics* (Eds D. Bock and M. Rius)
DOI: 10.1079/9781800626263.0003

This chapter discusses the promise and limitations of genomic tools and methods for evaluating invasion potential, with a special focus on those that hold the promise of predicting invasion success, that is, the ability of invasive species to establish and spread in novel environments. It begins with the use of population genomic methods to: (i) identify genomic signatures of post-invasion adaptation; (ii) quantify gene flow in the invasive range; and (iii) understand the range limits of invasive species. The next section considers the use of comparative genomic methods in invasion biology. It then explores the relevance of population and comparative genomics for translating new knowledge towards genome-informed management applications. Finally, this chapter evaluates the application of new pattern-searching methods trained on genomic data, focusing on the use of large language models and machine learning for predicting invasion potential (Fig. 3.1).

3.2 Population Genomics

3.2.1 Genomic signatures of post-invasion adaptation

Methods to identify genomic signatures of post-invasion adaptation can leverage comparisons between invasive- and native-range populations to identify shifts in the site frequency spectrum (e.g. Tajima's D), population summary statistics (e.g. genome-wide fixation index (F_{ST}), nucleotide diversity (π)) and/or patterns of linkage disequilibrium (reviewed by Weigand and Leese, 2018). Specific genes and alleles responsible for local adaptation can also be detected using genome-scanning techniques (including genotype–environment association analyses and differentiation outlier methods; reviewed by Hoban *et al.*, 2016) on divergent invasive/native populations. The choice of these methods depends primarily on the spatial and temporal scale of

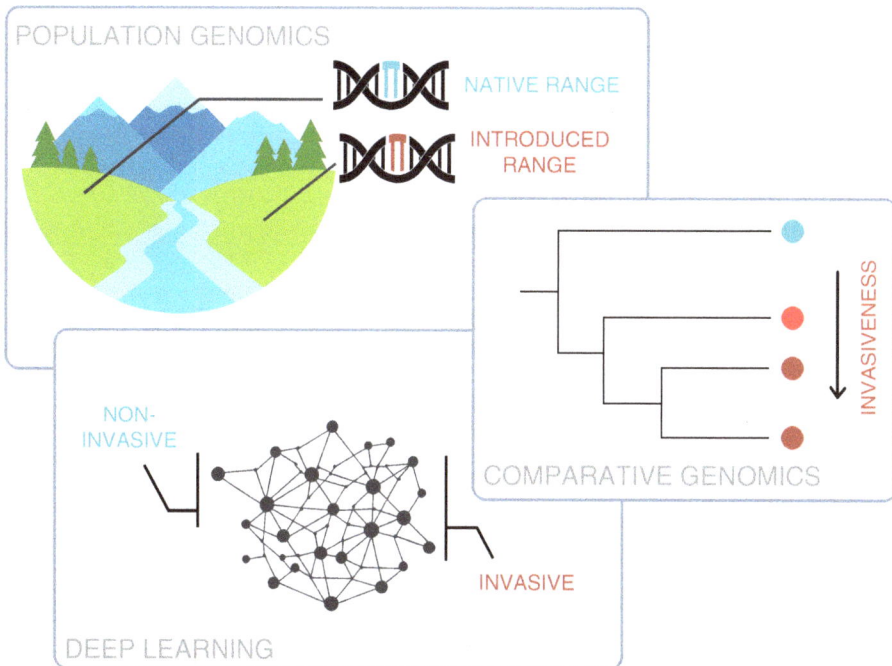

Fig. 3.1. Conceptual figure illustrating the three main approaches discussed in this chapter in a biological invasions context: population genomics – comparing populations of the same species from both the introduced and native ranges; comparative genomics – comparing species across an invasiveness gradient; and deep learning – using large language models and machine learning to predict invasion potential.

local adaptation, the type of high-throughput sequencing data and the availability of genomic resources, such as contiguous annotated reference genomes, phased data (or haplotype information) and recombination maps (Hoban *et al.*, 2016; Weigand and Leese, 2018; North *et al.*, 2021).

Applying and interpreting these methods requires caution, as non-equilibrium demographic conditions can lead to substantial variation in loci across different environments even if none of them is under selection (François *et al.*, 2016). This is especially important for genome-scanning methods on invasive species because biological invasions are characterized by non-equilibrium demographic conditions. For example, demographic bottlenecks (which occur when invasions initiate with only a few colonizers; Estoup *et al.*, 2016) and allele surfing due to strong genetic drift at the leading edge of an expanding population (especially in species with limited dispersal abilities; Paulose and Hallatschek, 2020) can each result in correlated random fluctuations in variant frequencies that can lead to false associations with invasive status (Olazcuaga *et al.*, 2020). One genome-scanning method that addresses such confounding effects involves the use of a relatedness matrix among samples to account for covariance in allele frequencies among populations (implemented in the software BayPass; Coop *et al.*, 2010; Gautier, 2015). The recent development of the BayPass model's contrast statistic, C_2, enables the comparison of allele frequencies corrected for shared demographic events between native and invasive populations (Olazcuaga *et al.*, 2020). This method can be used to identify adaptive signals common or specific to various invasion routes, providing a robust genome-scanning approach for association with invasion success (e.g. Yang *et al.*, 2022; Parvizi *et al.*, 2023).

Genome scans can also provide insights into the repeatability of post-invasion adaptation (e.g. van Boheemen and Hodgins, 2020; Pélissié *et al.*, 2022), by identifying how many, and which, loci show shifts in allele frequency across multiple pairs of populations, and whether parallel adaptive patterns arise from *de novo* mutations or selection on standing genetic variation (Lee and Coop, 2017; Pélissié *et al.*, 2022). Lists of adaptive candidate loci for different pairwise population comparisons can be assessed to determine whether any overlap, indicative of parallel adaptation, surpasses certain null expectations (e.g. null-*W* method; Yeaman *et al.*, 2016; van Boheemen and Hodgins, 2020). Alternatively, a recent method, PicMin, uses statistical theory to quantify the extent of non-randomness in measures of adaptation for individual genes across multiple populations (Booker *et al.*, 2023). This approach does not rely on arbitrary thresholds for classifying loci as adaptive/non-adaptive; it takes advantage of continuous distributions of summary statistics (e.g. obtained from sliding-window analysis), enhancing statistical power to detect genomic regions involved in repeated adaptation (Booker *et al.*, 2023). This ability extends to loci with weak adaptive signals, enabling the method's use in recent invasion scenarios, where populations may experience weak selection and individual genome scans suffer from insufficient power to detect parallel adaptation.

Genome scans and other methods for detecting genomic signatures of post-invasion adaptation can be limited by the choice of high-throughput sequencing method. For example, tests of selection that use haplotype blocks to identify adaptive loci require whole-genome resequencing (WGR) data that is preferably long read and has been mapped against a contiguous reference genome to obtain reliable haplotype frequencies across chromosomes or long genomic scaffolds (North *et al.*, 2021). Meanwhile, genome scans based on data obtained from reduced-representation sequencing are likely to miss many loci involved in local adaptation, except where linkage disequilibrium is exceptionally high (e.g. in chromosomal inversions; Twyford and Friedman, 2015; Hoban *et al.*, 2016). As well as choice of sequencing method, studies targeting the genomic basis of post-invasion adaptation require a comprehensive sampling design that includes different native and invasive populations and encompasses various occupied geographical ranges to enable identification of the extent of relative allele frequency change associated with the invasion process.

3.2.2 Quantifying gene flow in invasive species ranges

Invasive species often disperse over long distances, and invasive populations can experience repeated introductions, which can directly affect patterns of genetic diversity across introduced ranges to potentially enhance adaptation and invasion success (Smith *et al.*, 2020; Parvizi *et al.*, 2022). Understanding the extent and directionality of gene flow across the invasive range can thus offer insights into the potential of different populations to rapidly adapt to new environmental constraints. Gene flow is also important in the context of repeated local adaptation because the associated spread of adaptive alleles can cause convergent adaptation (Lee and Coop, 2017).

Genomic analyses of gene flow in invasive species can be limited by assumptions that populations are in migration–drift equilibrium. For example, estimates of connectivity based on the genetic distances between pairs of populations are often not applicable to populations along an invasion front (Fitzpatrick *et al.*, 2012; Pepin *et al.*, 2022), and methods that use measures of genetic differentiation to estimate directional gene flow among populations (e.g. Sundqvist *et al.*, 2016) may not be reliable when applied to invasive populations that are not at equilibrium. Future studies could combine simulations with empirical data to gain a more comprehensive understanding of the effectiveness and reliability of such methods in an invasive species context.

Some demographic processes, such as founder and bridgehead events, may weaken genetic differentiation between invasive populations, making it more challenging to estimate their connectivity (Tsutsui *et al.*, 2000; Parvizi *et al.*, 2022). If such processes are known to feature in an invasive species' evolutionary history, migration may be difficult to estimate, and it is best to avoid assignment-based analytical methods (e.g. BayesAss; Meirmans, 2014). Such methods make assumptions about the underlying population model and/or assume that migration rates are low (Wilson and Rannala, 2003; Meirmans, 2014), which can decrease the reliability of assigning individuals to specific populations and lead to inaccurate inference of migration rates (Meirmans, 2014). In such cases, exploratory analyses of population structure and admixture can provide a more reliable, if descriptive, understanding of population connectivity. Alternatively, deep learning approaches to infer per-generation dispersal distances from genomic data (Smith *et al.*, 2023), or techniques that combine genomic data with spatial information to measure relative (versus absolute) migration rates, can be

helpful (Petkova *et al.*, 2016; Marcus *et al.*, 2021). However, the reliability of these approaches in the context of invasive species is not yet clear.

3.2.3 Understanding range limits of invasive species

In many ways, invasion biology is about understanding why some species are widespread, while others are geographically restricted. Elucidating the mechanisms that facilitate range expansion is therefore a key focus, and central to this is the use of model-based approaches that infer the potential for a species to spread spatially (Hastings *et al.*, 2005).

Species distribution models (SDMs; see Chapter 11, this volume) are a commonly employed method for predicting range limits across time and space using environmental data. Genomic data can be used to complement classic SDMs, and can increase their accuracy (Sotka *et al.*, 2018). For example, Hudson *et al.* (2021) united SDMs with population genomic data from the range-restricted invasive ascidian *Pyura praeputialis* to identify suitable habitat adjacent to its current introduced range and found that SDMs were only accurate when genomic data were incorporated. Genome-informed SDMs can also relate high-resolution population structure to spatial landscape features (Razgour *et al.*, 2014; Marcer *et al.*, 2016) and, importantly, can be used to help predict future range expansions (Ikeda *et al.*, 2017; Sotka *et al.*, 2018). In *P. praeputialis*, for example, the genome-informed SDMs indicated that small changes in ocean currents or shipping routes may have a major impact on current distribution patterns along the South American coastline (Hudson *et al.*, 2021). Despite their promise, genome-informed SDMs are not overly common and their output models are often not validated (e.g. using hindcasting, assuming historical genomic data can be obtained) (see Chapter 11, this volume, for more information).

3.3 Comparative Genomics

In contrast to population genomics, comparative genomic methods identify similarities and differences in interspecific evolution and variation by examining diversification among genes and species and identifying novel genomic variants across taxa.

Typically, comparative genomic analyses are used to predict gene function or variation within clades, as well as to identify regions of variation between closely related species. Methods such as OrthoFinder (Emms and Kelly, 2019) or OrthoMCL (Li *et al.*, 2003) can be used to infer orthologous (conserved) and unique gene families among taxa, while annotation of orthologues and analyses of evolutionary rates (e.g. CAFE 5; Mendes *et al.*, 2020) can provide information about the function of genes and their associated temporal dynamics, respectively. Comparative analysis can also identify structural variants, such as inversion polymorphisms and transposable elements (TEs). For example, common structural variant callers (using paired and split short-read WGR) include Lumpy (Layer *et al.*, 2014), DELLY (Rausch *et al.*, 2012) and Manta (Chen *et al.*, 2016), and these can be used individually or in tandem to generate a consensus list (Stuart *et al.*, 2023). For TEs, pipelines such as the extensive *de novo* TE annotator (EDTA) – created from compiling robust TE caller programmes – can be used to generate high-quality non-redundant TE libraries (Ou *et al.*, 2019).

In an invasive species context, comparisons between invasive and native (or 'more invasive' and 'less invasive') species can identify the genomic architecture associated with invasive success. In particular, structural variants have been identified as playing a key role in adaptation during biological invasion (although they can also be disadvantageous in new environments; see Roesti *et al.*, 2022). For example, in the European green crab, *Carcinus maenas*, one inferred chromosomal inversion has been repeatedly associated with cold tolerance and may be critical to its spread across climatic gradients within the introduced range (Stapley *et al.*, 2015; Tepolt *et al.*, 2022). TEs may also facilitate rapid adaptation after exposure to stress events (Stapley *et al.*, 2015). Meanwhile, comparative analysis of more versus less invasive species can help to determine whether invasion is facilitated by standing genetic variation or rapid co-option of existing variants (or genes) towards new functions. For example, recent research on species of Leptinotarsa (leaf beetles) has demonstrated the role of standing genetic variation in pesticide resistance for the invasive Colorado potato beetle, *Leptinotarsa decemlineata* (Cohen *et al.*, 2021). Finally, comparative genomic and/or transcriptomic methods are beginning to identify putative signatures that correlate with genomic traits that are necessary for success in biological invaders. This includes genes associated with invasiveness in Leptinotarsa (Cohen *et al.*, 2021), and detoxification in Lepidoptera (Thrimawithana *et al.*, 2022) and brown marmorated stink bug, *Halyomorpha halys* (Bansal and Michel, 2018). Relevant to detoxification, the expansion of cytochrome p450 gene families is a common trend across lineages for signals of invasiveness (e.g. Pearce *et al.*, 2017; Bansal and Michel, 2018; Wenger *et al.*, 2020; Thrimawithana *et al.*, 2022) that also supports the role of co-opted gene function versus *de novo* mutation in facilitating invasion.

While new findings are providing intriguing insights into interspecific trends of genome-driven invasion potential, cost and other limitations in whole-genome sequencing restrict the evaluation of cross-taxa comparative genomics, instead favouring cheaper alternatives, such as reduced-representation sequencing (e.g. Friel *et al.*, 2021) and WGR (Scheben *et al.*, 2017; North *et al.*, 2021). However, high-quality reference genomes are an important tool in the exploration of fundamental questions in invasion genomics (and evolutionary biology more generally), and reference biases can have complicated downstream effects on data analysis. For example, rare variants not represented in a reference genome may be lost in low-coverage sequencing – a particular challenge for historical sequence data, where low sequence coverage can perpetuate reference bias (Günther and Nettelblad, 2019). In their under-representation of invasive species, model organisms also exacerbate reference bias in genomic data repositories, resulting in a lack of functional annotations for invasive species that limits our understanding of their evolutionary dynamics.

Under-representation of genomic data for invasive species is a problem more generally. In fact, a search of publication records in Scopus using the term 'invasive species*comparative genomics' identified only 56 studies for eukaryotic organisms (as at 29th September 2025; Fig. 3.2). Among these, insects were the most well-studied group of invasive species (23/56

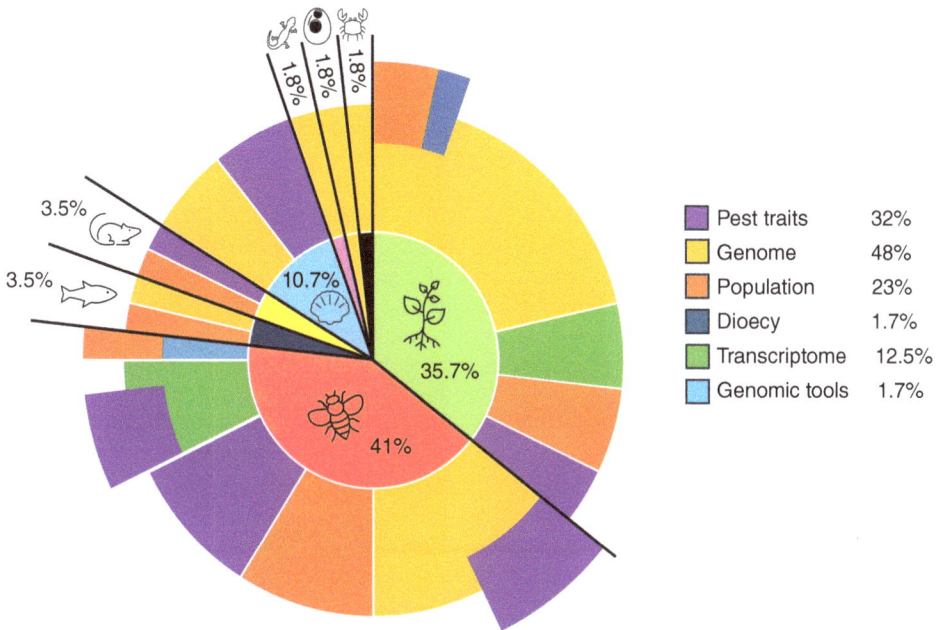

Fig. 3.2. Comparative genomic studies are under-represented for invasive species. The inner ring denotes the taxon group associated with study results (i.e. insect, plant, mammal or mollusc). The outer rings show the study aim and secondary aim (if any) in the context of invasion, as per the colour key. Pest traits: beneficial traits associated with niche adaptation in invasive species; genome: studies for the purpose of recording a species' genomic composition; population: comparative genomic study between invasive species and native conspecifics or other invasive populations; dioecy: genomic profiling of sex genes in related dioecious plants; transcriptome: studies generating transcriptome data; genomic tools: studies with the aim to develop genomic-based tools). Percentages denote the total number of studies with classifiers in either the main or secondary study aims.

studies), reflecting likely taxonomic biases in our understanding of comparative invasion genomics. Interestingly, 17 (30%) of these studies had aims associated with identifying 'pest' traits of invasive species or populations, but the experimental design was not typically centred on comparative genomics between invasive versus non-invasive dynamics (only 8/56 studies; Fig. 3.2).

Restricted study designs (e.g. that focus on only a subset of species or traits) can limit the potential reuse of genomic data, which is an ongoing objective of comparative genomic approaches aimed at identifying common patterns across invasive species. Associated with this, recent research has shown that data reuse is often restricted by incompleteness of the spatiotemporal metadata that accompanies the deposition of genomic data into public repositories (Toczydlowski

et al., 2021; Vaughan et al., 2023). In an invasion context, comprehensive metadata associated with genomes that delineate native versus expanded range and pre- versus post-invasion events are necessary for further characterization of invasive species and their adaptive success (e.g. Fox et al., 2013).

3.4   Applications Towards Genome-informed Invasive Species Management

The mechanisms that underpin intraspecific differentiation in invasion potential can be exploited for invasive species management. In particular, adaptive strategies of invasive populations can be characterized by identifying population-specific evolutionary mechanisms that confer invasion success (e.g. in the invasive versus native range), and using this information to develop targeted control approaches (see also Chapter 12, this volume). For example, Pélissié *et al.* (2022) identified different insecticide resistance mechanisms across agricultural regions using genome-wide analysis of multiple Colorado potato beetle populations. Although these mechanisms involved similar genetic pathways, they differed in specific genes – findings that can be used to inform the future development of population-specific gene knockdown via RNA interference for effective pest management. Identifying population-specific adaptive genes can also prove valuable for the development of genetic markers for efficient field surveillance and monitoring, such as detection of the distribution and frequency of resistance alleles across time and space (Thia *et al.*, 2021).

Understanding population-level demographic processes, such as gene flow and dispersal rates, can be useful for implementing strategies to restrict reintroductions from high-risk source populations or for limiting the spread of highly adaptive alleles, such as those involved in pesticide resistance and other traits associated with stress tolerance (Thia *et al.*, 2021, 2023). Explicit incorporation of spatial genomic clusters derived from population genomic data into SDMs can also inform management strategies by enhancing the predictability of further expansion of species in introduced ranges (e.g. Hudson *et al.*, 2021). SDMs that integrate adaptive genomic variation across geographical regions can further enable the estimation of genomic-niche indices for a more comprehensive assessment of invasion risks. For example, in the invasive ascidian *Molgula manhattensis*, Chen *et al.* (2023) used the genomic-niche index to compare the invasion potential of genetically structured populations across different geographical regions and climate-change scenarios, forecasting which genetic cluster had higher invasion risk for future management approaches.

As comparative genomics focuses on the interspecific variation of genes and species diversity, these analyses can potentially unlock patterns of genomic architecture for applications of downstream genome-informed management programmes. Harrop *et al.* (2020), with the publication of three Hymenoptera genomes,

acknowledged the importance of these resources for the identification of species-specific targets for gene drives, biocontrol and/or monitoring. Efforts to determine genomic differences between native- and invasive-range populations (e.g. with the common wasp, *Vespula vulgaris*; Lester *et al.*, 2020) allow an even narrower target range to specifically eliminate off-target native populations from eradication or control programmes. In addition, phylogenomic approaches have been used to identify species-specific markers for polymerase chain reaction (PCR)-based rapid identification tools, as demonstrated in the fruit fly *Drosophila suzukii* (Murphy *et al.*, 2015). Overall, current research demonstrates the applicability of population and comparative genomics to invasive species management, illustrating how genomic understanding can lead to better control strategies.

3.5 Emergent Approaches: Large Pattern Deep Learning Search Methods

With the rapidly expanding volume of 'big data', the success of deep learning tools, such as large language models (LLMs) and machine learning (ML), has also accelerated. This chapter finishes with a discussion of these emergent approaches in genomics, with a view towards how they might push forward the identification of invasion potential.

3.5.1 LLMs for genomics

Recent developments in natural language processing through the training of LLMs on vast volumes of data have allowed the prediction of text output to a high degree of accuracy. In a genetic context, LLMs are trained on biological 'tokens' corresponding to a text-based matrix, such as DNA bases or gene annotations. For example, genome-scale language models (genSLMs) were used in the severe acute respiratory syndrome coronavirus 2 (SARS-CoV-2) pandemic, trained on upwards of 110 million gene sequences, to rapidly and efficiently identify genomic variants of concern (Zvyagin *et al.*, 2023).

The processes employed by these models include both transformer and non-transformer-based LLMs (Vaswani *et al.*, 2017), both of which have been used in recent studies for genomics-based platforms. Transformer-based-LLM models (e.g. Geneformer, Theodoris *et al.*, 2023; Enformer, Avsec *et al.*, 2021) use self-attention mechanisms that weight the importance of sequential information to allow the model to consider the entire data context simultaneously, resulting in efficient modelling of long-range relationships in data. Transformer-based models have been applied to short DNA sequences in pre-trained models, such as DNABERT (Ji *et al.*, 2021), to predict promoters and splice sites in DNA sequences. They have also been used to improve gene expression prediction in novel deep learning models – methods that can integrate long-range interactions within the genome for effective predictions. Conversely, non-transformer-based LLMs, such as long short-term memory models (Luo, 2017), have historically shown high efficacy in capturing sequential dependencies in data. To date, non-transformer-based LLMs have been used in a genomic pre-trained network to predict functional effects on gene variants, with high levels of success in the model organism *Arabidopsis thaliana* (Benegas *et al.*, 2023).

Current use of both kinds of LLM models extends primarily from function to annotation and prediction of functional gene candidates and traits. However, their current training and use on a variety of genomic and transcriptomic databases opens the door to a plethora of biological applications. For example, genSLMs could be used to determine variants related to invasion potential using weighted traits on the genome. However, by definition, LLMs require *a priori* access to and training on a high volume of data. Thus, limitations in the availability of genomic and transcriptomic resources associated with invasive species currently restricts further advancements within this field.

3.5.2 ML for predicting invasion potential

Deep learning technologies are currently being used to detect invasive species, monitor eradication efforts via the analysis of satellite imagery

and geospatial long short-term memory network models (Xiao *et al.*, 2018; Lake *et al.*, 2022; Elias, 2023) and predict the geographical origin of samples based on their genetic profile (e.g. Locator; Battey *et al.*, 2020).

An emergent ML application is the prediction of invasion likelihood based on genomic signatures, whereby various sources of omics data (e.g. transcriptomes, genomes, outlier SNPs, microbiomes) in a traits-based matrix could be analysed in a hierarchical framework (using training and test data) to determine the degree of invasiveness of a novel invasive species. For example, random forest is a widely applied ML method in biological sciences (e.g. Huang and Boutros, 2016) that can partition data based on high cardinality (Breiman, 2001; Han *et al.*, 2021; Quist *et al.*, 2021). To function accurately (i.e. without randomized partitioning) in the context described here, large amounts of genomic data would be needed, with sufficient 'more invasive' and 'less invasive' (or native and non-invasive) representatives to find genuine patterns within the data (e.g. Huang *et al.*, 2021). However, again, limitations in the availability of genomic and metadata resources for invasive species (Matheson and McGaughran, 2022; Vaughan *et al.*, 2023) currently constrain the development of accurate, precision tools for predicting inherent invasion risk. Such issues are exacerbated by debates within the scientific community around a consensus definition of invasive status and associated terminology (Colautti and MacIsaac, 2004).

3.6 Conclusions

Genomic studies have begun to reveal spatial and temporal invasion dynamics, including the impact of selection on the successful establishment of pest and invasive species and the genomic architecture of traits that underpin invasiveness. While some of the field's analytical tools (e.g. in population genomics) require tailoring for invasive species to better take into account invasion-associated demographic events, emergent methods in deep learning hold great promise for our future ability to predict invasion potential. However, the field broadly suffers from a poor availability of high-resolution and broad-scale genomic data and associated metadata. Moving forward, invasion genomics will benefit from increased data availability and greater integration of traditional genomic approaches with novel computational technologies to evaluate invasion potential.

Acknowledgements

This work was partially funded by Genomics Aotearoa, a New Zealand Ministry of Business, Innovation and Employment-funded research platform (Grant GA 2102 awarded to A.M. and M.K.D.).

References

Avsec, Ž., Agarwal, V., Visentin, D., Ledsam, J.R., Grabska-Barwinska, A. *et al.* (2021) Effective gene expression prediction from sequence by integrating long-range interactions. *Nature Methods* 18, 1196–1203. DOI: 10.1038/s41592-021-01252-x

Bansal, R. and Michel, A. (2018) Expansion of cytochrome P450 and cathepsin genes in the generalist herbivore brown marmorated stink bug. *BMC Genomics* 19: 60. DOI: 10.1186/s12864-017-4281-6

Battey, C.J., Ralph, P.L. and Kern, A.D. (2020) Predicting geographic location from genetic variation with deep neural networks. *eLife* 9: e54507. DOI: 10.7554/eLife.54507

Battlay, P., Wilson, J., Bieker, V.C., Lee, C., Prapas, D. *et al.* (2023) Large haploblocks underlie rapid adaptation in the invasive weed *Ambrosia artemisiifolia*. *Nature Communications* 14: 1717. DOI: 10.1038/s41467-023-37303-4

Benegas, G., Batra, S.S. and Song, Y.S. (2023) DNA language models are powerful predictors of genome-wide variant effects. *Proceedings of the National Academy of Sciences USA* 120: e2311219120. DOI: 10.1073/pnas.2311219120

Booker, T.R., Yeaman, S. and Whitlock, M.C. (2023) Using genome scans to identify genes used repeatedly for adaptation. *Evolution* 77, 801–811. DOI: 10.1093/evolut/qpac063

Breiman, L. (2001) Random forests. *Machine Learning* 45, 5–32. DOI: 10.1023/A:1010933404324

Chen, X., Schulz-Trieglaff, O., Shaw, R., Barnes, B., Schlesinger, F., Källberg, M. *et al.* (2016) Manta: rapid detection of structural variants and indels for germline and cancer sequencing applications. *Bioinformatics* 32, 1220–1222. DOI: 10.1093/bioinformatics/btv710

Chen, Y., Gao, Y., Huang, X., Li, S., Zhang, Z. and Zhan, A. (2023) Incorporating adaptive genomic variation into predictive models for invasion risk assessment. *Environmental Science and Ecotechnology* 18: 100299. DOI: 10.1016/j.ese.2023.100299

Cohen, Z.P., Brevik, K., Chen, Y.H., Hawthorne, D.J., Weibel, B.D. and Schoville, S.D. (2021) Elevated rates of positive selection drive the evolution of pestiferousness in the Colorado potato beetle (*Leptinotarsa decemlineata*, Say). *Molecular Ecology* 30, 237–254. DOI: 10.1111/mec.15703

Colautti, R.I. and MacIsaac, H.J. (2004) A neutral terminology to define 'invasive' species. *Diversity and Distributions* 10, 135–141. DOI: 10.1111/j.1366-9516.2004.00061.x

Coop, G., Witonsky, D., di Rienzo, A. and Pritchard, J.K. (2010) Using environmental correlations to identify loci underlying local adaptation. *Genetics* 185, 1411–1423. DOI: 10.1534/genetics.110.114819

Elias, N. (2023) Deep learning methodology for early detection and outbreak prediction of invasive species growth. In: *IEEE/CVF Winter Conference on Applications of Computer Vision (WACV)*, Waikoloa, Hawaii. IEEE, New York, pp. 6324–6332,

Emms, D.M. and Kelly, S. (2019) OrthoFinder: phylogenetic orthology inference for comparative genomics. *Genome Biology* 20: 238. DOI: 10.1186/s13059-019-1832-y

Estoup, A., Ravigné, V., Hufbauer, R., Vitalis, R., Gautier, M. and Facon, B. (2016) Is there a genetic paradox of biological invasion? *Annual Review of Ecology, Evolution, and Systematics* 47, 51–72. DOI: 10/1146/annurev-ecolsys-121415-032116

Fitzpatrick, B.M., Fordyce, J.A., Niemiller, M.L. and Reynolds, R.G. (2012) What can DNA tell us about biological invasions? *Biological Invasions* 14, 245–253. DOI: 10.1007/s10530-011-0064-1

Fox, S.E., Preece, J., Kimbrel, J.A., Marchini, G.L., Sage, A. *et al.* (2013) Sequencing and *de novo* transcriptome assembly of *Brachypodium sylvaticum* (Poaceae). *Applications in Plant Sciences* 1: 1200011. DOI: 10.3732/apps.1200011

François, O., Martins, H., Caye, K. and Schoville, S.D. (2016) Controlling false discoveries in genome scans for selection. *Molecular Ecology* 25, 454–469. DOI: 10.1111/mec.13513

Friel, J., Bombarely, A., Fornell, C.D., Luque, F. and Fernández-Ocaña, A.M. (2021) Comparative analysis of genotyping by sequencing and whole-genome sequencing methods in diversity studies of *Olea europaea* L. *Plants* 10: 2514. DOI: 10.3390/plants10112514

Gautier, M. (2015) Genome-wide scan for adaptive divergence and association with population-specific covariates. *Genetics* 201, 1555–1579. DOI: 10.1534/genetics.115.181453

Günther, T. and Nettelblad, C. (2019) The presence and impact of reference bias on population genomic studies of prehistoric human populations. *PLoS Genetics* 15: e1008302. DOI: 10.1371/journal.pgen.1008302

Han, S., Williamson, B.D. and Fong, Y. (2021) Improving random forest predictions in small datasets from two-phase sampling designs. *BMC Medical Informatics and Decision Making* 21: 322. DOI: 10.1186/s12911-021-01688-3

Harrop, T.W.R., Guhlin, J., McLaughlin, G.M., Permina, E., Stockwell, P. *et al.* (2020) High-quality assemblies for three invasive social wasps from the *Vespula* genus *G3: Genes, Genomes, Genetics* 10, 3479–3488. DOI: 10.1534/g3.120.401579

Hastings, A., Cuddington, K., Davies, K.F., Dugaw, C.J., Elmendorf, S. *et al.* (2005) The spatial spread of invasions: new developments in theory and evidence. *Ecology Letters* 8, 91–101. DOI: 10.1111/j.1461-0248.2004.00687.x

Hoban, S., Kelley, J.L., Lotterhos, K.E., Antolin, M.F., Bradburd, G. *et al.* (2016) Finding the genomic basis of local adaptation: pitfalls, practical solutions, and future directions. *American Naturalist* 188, 379–397. DOI: 10.1086/688018

Huang, B.F.F. and Boutros, P.C. (2016) The parameter sensitivity of random forests. *BMC Bioinformatics* 17: 331. DOI: 10.1186/s12859-016-1228-x

Huang, C., Lang, K., Qian, W.-Q., Wang, S.-P., Cao, X.-M. *et al.* (2021) InvasionDB: a genome and gene database of invasive alien species. *Journal of Integrative Agriculture* 20, 191–200. DOI: 10.1016/S2095-3119(20)63231-2.

Hudson, J., Castilla, J.C. and Teske, P.R. (2021) Genomics-informed models reveal extensive stretches of coastline under threat by an ecologically dominant invasive species. *Proceedings of the National Academy of Sciences USA* 188: e2022169118. DOI: 10.1073/pnas.2022169118

Ikeda, D.H., Max, T.L., Allan, G.J., Lau, M.K., Shuster, S.M. and Whitham, T.G. (2017) Genetically informed ecological niche models improve climate change predictions. *Global Change Biology* 23, 164–176. DOI: 10.1111/gcb.13470

Ji, Y., Zhou, Z., Liu, H. and Davuluri, R.V. (2021) DNABERT: pre-trained Bidirectional Encoder Representations from Transformers model for DNA-language in genome. *Bioinformatics* 37, 2112–2120. DOI: 10.1093/bioinformatics/btab083

Lake, T.A., Briscoe Runquist, R.D. and Moeller, D.A. (2022) Deep learning detects invasive plant species across complex landscapes using Worldview-2 and Planetscope satellite imagery. *Remote Sensing in Ecology and Conservation* 8, 875–889. DOI: 10.1002/rse2.288

Layer, R.M., Chiang, C., Quinlan, A.R. and Hall, I.M. (2014) LUMPY: a probabilistic framework for structural variant discovery. *Genome Biology* 15: R84. DOI: 10.1186/gb-2014-15-6-r84

Lee, K.M. and Coop, G. (2017) Distinguishing among modes of convergent adaptation using population genomic data. *Genetics* 207, 1591–1619. DOI: 10.1534/genetics.117.300417

Lester, P.J., Bulgarella, M., Baty, J.W., Dearden, P.K., Guhlin, J. and Kean, J.M. (2020) The potential for a CRISPR gene drive to eradicate or suppress globally invasive social wasps. *Scientific Reports* 10: 12398. DOI: 10.1038/s41598-020-69259-6

Li, L., Stoeckert, C.J. Jr and Roos, D.S. (2003) OrthoMCL: identification of ortholog groups for eukaryotic genomes. *Genome Research* 13, 2178–2189. DOI: 10.1101/gr.1224503

Luo, Y. (2017) Recurrent neural networks for classifying relations in clinical notes. *Journal of Biomedical Informatics* 72, 85–95. DOI: 10.1016/j.jbi.2017.07.006

Marcer, A., Méndez-Vigo, B., Alonso-Blanco, C. and Picó, F.X. (2016) Tackling intraspecific genetic structure in distribution models better reflects species geographical range. *Ecology and Evolution* 6, 2084–2097. DOI: 10.1002/ece3.2010

Marcus, J., Ha, W., Barber, R.F. and Novembre, J. (2021) Fast and flexible estimation of effective migration surfaces. *eLife* 10: e61927. DOI: 10.7554/eLife.61927

Matheson, P. and McGaughran, A. (2022) Genomic data is missing for many highly invasive species, restricting our preparedness for escalating incursion rates. *Scientific Reports* 12: 13987. DOI: 10.1038/s41598-022-17937-y

McGaughran, A., Dhami, M.K., Parvizi, E., Vaughan, A.L., Gleeson, D.M. *et al.* (2024) Genomic tools in biological invasions: current state and future frontiers. *Genome Biology and Evolution* 16: evad230. DOI: 10.1093/gbe/evad230

Meirmans, P.G. (2014) Nonconvergence in Bayesian estimation of migration rates. *Molecular Ecology Resources* 14, 726–733. DOI: 10.1111/1755-0998.12216

Mendes, F.K., Vanderpool, D., Fulton, B. and Hahn, M.W. (2020) CAFE 5 models variation in evolutionary rates among gene families. *Bioinformatics* 36, 5516–5518. DOI: 10.1093/bioinformatics/btaa1022

Murphy, K.A., Unruh, T.R., Zhou, L.M., Zalom, F.G., Shearer, P.W. *et al.* (2015) Using comparative genomics to develop a molecular diagnostic for the identification of an emerging pest *Drosophila suzukii*. *Bulletin of Entomological Research* 105, 364–372. DOI: 10.1017/s0007485315000218

North, H.L., McGaughran, A. and Jiggins, C.D. (2021) Insights into invasive species from whole-genome resequencing. *Molecular Ecology* 30, 6289–6308. DOI: 10.1111/mec.15999

Olazcuaga, L., Loiseau, A., Parrinello, H., Paris, M., Fraimout, A. *et al.* (2020) A whole-genome scan for association with invasion success in the fruit fly *Drosophila suzukii* using contrasts of allele frequencies corrected for population structure. *Molecular Biology and Evolution* 37, 2369–2385. DOI: 10.1093/molbev/msaa098

Ou, S., Su, W., Liao, Y., Chougule, K., Agda, J.R.A. *et al.* (2019) Benchmarking transposable element annotation methods for creation of a streamlined, comprehensive pipeline. *Genome Biology* 20: 275. DOI: 10.1186/s13059-019-1905-y

Parvizi, E., Dhami, M.K., Yan, J. and McGaughran, A. (2022) Population genomic insights into invasion success in a polyphagous agricultural pest, *Halyomorpha halys*. *Molecular Ecology* 32, 138–151. DOI: 10.1111/mec.16740

Parvizi, E., Vaughan, A.L., Dhami, M.K. and McGaughran, A. (2023) Genomic signals of local adaptation across climatically heterogenous habitats in an invasive tropical fruit fly (*Bactrocera tryoni*). *Heredity* 132, 18–29. DOI: 10.1038/s41437-023-00657-y

Paulose, J. and Hallatschek, O. (2020) The impact of long-range dispersal on gene surfing. *Proceedings of the National Academy of Sciences USA* 117, 7584–7593. DOI: 10.1073/pnas.1919485117

Pearce, S.L., Clarke, D.F., East, P.D., Elfekih, S., Gordon, K.H.J. *et al.* (2017) Genomic innovations, transcriptional plasticity and gene loss underlying the evolution and divergence of two highly polyphagous and invasive *Helicoverpa* pest species. *BMC Biology* 15: 63. DOI: 10.1186/s12915-017-0402-6

Pélissié, B., Chen, Y.H., Cohen, Z.P., Crossley, M.S., Hawthorne, D.J. *et al.* (2022) Genome resequencing reveals rapid, repeated evolution in the Colorado potato beetle. *Molecular Biology and Evolution* 39: msac016. DOI: 10.1093/molbev/msac016

Pepin, K.M., Davis, A.J., Epanchin-Niell, R.S., Gormley, A.M., Moore, J.L. *et al.* (2022) Optimizing management of invasions in an uncertain world using dynamic spatial models. *Ecological Applications* 32: e2628. DOI: 10.1002/eap.2628

Petkova, D., Novembre, J. and Stephens, M. (2016) Visualizing spatial population structure with estimated effective migration surfaces. *Nature Genetics* 48, 94–100. DOI: 10.1038/ng.3464

Pyper, N.R., Painting, C.J. and McGaughran, A. (2024) Home and away: the role of intraspecific behavioural variation in biological invasion. *New Zealand Journal of Zoology* 51, 151–174. DOI: 10.1080/03014223.2024.2336035

Quist, J., Taylor, L., Staaf, J. and Grigoriadis, A. (2021) Random forest modelling of high-dimensional mixed-type data for breast cancer classification. *Cancers* 13: 991. DOI: 10.3390/cancers13050991

Rausch, T., Zichner, T., Schlattl, A., Stütz, A.M., Benes, V. and Korbel, J.O. (2012) DELLY: structural variant discovery by integrated paired-end and split-read analysis. *Bioinformatics* 28, i333–i339. DOI: 10.1093/bioinformatics/bts378

Razgour, O., Rebelo, H., Puechmaille, S.J., Juste, J., Ibáñez, C. *et al.* (2014) Scale-dependent effects of landscape variables on gene flow and population structure in bats. *Diversity and Distributions* 20, 1173–1185. DOI: 10.1111/ddi.12200

Roesti, M., Gilbert, K.J. and Samuk, K. (2022) Chromosomal inversions can limit adaptation to new environments. *Molecular Ecology* 31, 4435–4439. DOI: 10.1111/mec.16609

Scheben, A., Batley, J. and Edwards, D. (2017) Genotyping-by-sequencing approaches to characterize crop genomes: choosing the right tool for the right application. *Plant Biotechnology Journal* 15, 149–161. DOI: 10.1111/pbi.12645

Smith, A.L., Hodkinson, T.R., Villellas, J., Catford, J.A., Csergő, A.M. *et al.* (2020) Global gene flow releases invasive plants from environmental constraints on genetic diversity. *Proceedings of the National Academy of Sciences USA* 117, 4218–4227. DOI: 10.1073/pnas.1915848117

Smith, C.C.R., Tittes, S., Ralph, P.L. and Kern, A.D. (2023) Dispersal inference from population genetic variation using a convolutional neural network. *Genetics* 224: iyad068. DOI: 10.1093/genetics/iyad068

Sotka, E.E., Baumgardner, A.W., Bippus, P.M., Destombe, D., Duermit, E.A. *et al.* (2018) Combining niche shift and population genetic analyses predicts rapid phenotypic evolution during invasion. *Evolutionary Applications* 11, 781–793. DOI: 10.1111/eva/12592

Stapley, J., Santure, A.W. and Dennis, S.R. (2015) Transposable elements as agents of rapid adaptation may explain the genetic paradox of invasive species. *Molecular Ecology* 24, 2241–2252. DOI: 10.1111/mec.13089

Strayer, D.L., Eviner, V.T., Jeschke, J.M. and Pace, M.L. (2006) Understanding the long-term effects of species invasions. *Trends in Ecology and Evolution* 21, 645–651. DOI: 10.1016/j.tree.2006.07.007

Stuart, K.C., Hofmeister, N.R., Zichello, J.M. and Rollins, L.A. (2023) Global invasion history and native decline of the common starling: insights through genetics. *Biological Invasions* 25, 1291–1316. DOI: 10.1007/s10530-022-02982-5

Sundqvist, L., Keenan, K., Zackrisson, M., Prodöhl, P. and Kleinhans, D. (2016) Directional genetic differentiation and relative migration. *Ecology and Evolution* 6, 3461–3475. DOI: 10.1002/ece3.2096

Tepolt, C.K., Grosholz, E.D., de Rivera, C.E. and Ruiz, G.M. (2022) Balanced polymorphism fuels rapid selection in an invasive crab despite high gene flow and low genetic diversity. *Molecular Ecology* 31, 55–69. DOI: 10.1111/mec.16143

Theodoris, C.V., Xiao, L., Chopra, A., Chaffin, M.D., Al Sayed, Z.R. *et al.* (2023) Transfer learning enables predictions in network biology. *Nature* 618, 616–624. DOI: 10.1038/s41586-023-06139-9

Thia, J.A., Hoffmann, A.A. and Umina, P.A. (2021) Empowering Australian insecticide resistance research with genetic information: the road ahead. *Austral Entomology* 60, 147–162. DOI: 10.1111/aen.12512

Thia, J.A., Korhonen, P.K., Young, N.D., Gasser, R.B., Umina, P.A. *et al.* (2023) The redlegged earth mite draft genome provides new insights into pesticide resistance evolution and demography in its invasive Australian range. *Journal of Evolutionary Biology* 36, 381–398. DOI: 10.1111/jeb.14144

Thrimawithana, A.H., Wu, C., Christeller, J.T., Simpson, R.M., Hilario, E. *et al.* (2022) The genomics and population genomics of the light brown apple moth, *Epiphyas postvittana*, an invasive tortricid pest of horticulture. *Insects* 13: 264. DOI: 10.3390/insects13030264.

Toczydlowski, R.H., Liggins, L., Gaither, M.R., Anderson, T.J., Barton, R.L. *et al.* (2021) Poor data stewardship will hinder global genetic diversity surveillance. *Proceedings of the National Academy of Sciences USA* 118: e2107934118. DOI: 10.1073/pnas.2107934118

Tsutsui, N.D., Suarez, A.V., Holway, D.A. and Case, T.J. (2000) Reduced genetic variation and the success of an invasive species. *Proceedings of the National Academy of Sciences USA* 97, 5948–5953. DOI: 10.1073/pnas.10011039

Twyford, A.D. and Friedman, J. (2015) Adaptive divergence in the monkey flower *Mimulus guttatus* is maintained by a chromosomal inversion. *Evolution* 69, 1476–1486. DOI: 10.1111/evo.12663

van Boheemen, L.A. and Hodgins, K.A. (2020) Rapid repeatable phenotypic and genomic adaptation following multiple introductions. *Molecular Ecology* 29, 4102–4117. DOI: 10.1111/mec.15429

Vaswani, A., Shazeer, N., Parmar, N., Uszkoreit, J., Jones, L. *et al.* (2017) Attention is all you need. In: Guyon, I., von Lexburg, U., Bengio, S., Wallach, H., Fergus, R. *et al.* (eds) *31st Conference on Neural Information Processing Systems (NIPS 2017), Long Beach, CA, USA*. Available at: https://papers.nips.cc/paper/2017 (accessed 22 July 2025).

Vaughan, A., Parvizi, E., Matheson, P., McGaughran, A. and Dhami, M. (2023) Current stewardship practices in invasion biology limit the value and secondary use of genomic data. *Molecular Ecology Resources* 25: e13858. DOI: 10.1111/1755-0998.13858

Weigand, H. and Leese, F. (2018) Detecting signatures of positive selection in non-model species using genomic data. *Zoological Journal of the Linnean Society* 184, 528–583. DOI: 10.1093/zoolinnean/zly007

Wenger, J.A., Cassone, B.J., Legeai, F., Johnston, J.S., Bansal, R. *et al.* (2020) Whole genome sequence of the soybean aphid, *Aphis glycines*. *Insect Biochemistry and Molecular Biology* 123: 102917. DOI: 10.1016/j.ibmb.2017.01.005

Wilson, G.A. and Rannala, B. (2003) Bayesian inference of recent migration rates using multilocus genotypes. *Genetics* 163, 1177–1191. DOI: 10.1093/genetics/163.3.1177

Xiao, Y., Greiner, R. and Lewis, M.A. (2018) Evaluation of machine learning methods for predicting eradication of aquatic invasive species. *Biological Invasions* 20, 2485–2503. DOI: 10.1007/s10530-018-1715-2

Yang, F., Crossley, M.S., Schrader, L., Dubovskiy, I.M., Wei, S.J. and Zhang, R. (2022) Polygenic adaptation contributes to the invasive success of the Colorado potato beetle. *Molecular Ecology* 31, 5568–5580. DOI: 10.1111/mec.16666

Yeaman, S., Hodgins, K.A., Lotterhos, K.E., Suren, H., Nadeau, S. *et al.* (2016) Convergent local adaptation to climate in distantly related conifers. *Science* 353, 1431–1433. DOI: 10.1126/science.aaf78

Zvyagin, M., Brace, A., Hippe, K., Deng, Y., Zhang, B. *et al.* (2023) GenSLMs: genome-scale language models reveal SARS-CoV-2 evolutionary dynamics. *International Journal of High Performance Computing Applications* 37, 683–705. DOI: 10.1177/10943420231201154

4 Structural Variants and Transposable Elements as Facilitators of Rapid Evolutionary Change in Invasive Populations

Katarina C. Stuart[1,2]*, Anna W. Santure[1] and Lee A. Rollins[2]
[1]*School of Biological Sciences, University of Auckland, Auckland, New Zealand;*
[2]*Evolution & Ecology Research Centre, University of New South Wales, Sydney, NSW, Australia*

Abstract

Invasion genomics quantifies genetic diversity within populations to understand structure, demography, dispersal and adaptation during invasion. Routinely, large numbers of single-nucleotide polymorphisms (SNPs) are used as a proxy for genome-wide diversity. However, non-SNP variants, including structural variants (SVs) and transposable elements (TEs), represent a previously understudied area of invasion genomics that may play an important role in invasive species' success. This chapter explores SVs and TEs, their impact on rapid evolution and their role in biological invasions. Emphasis is placed on the deleterious nature of SVs and TEs, their interaction with invasion processes such as demography and selection regime shifts, as well as their unique contributions – SVs to recombination landscapes and TEs to novel variation and gene regulation. Finally, we offer practical recommendations for working with these complicated genomic features and suggest key future directions for understanding how they contribute to the success of biological invasions.

4.1 Introduction

Invasion genomics encompasses a diverse range of analytical approaches, including analysis of population structure, demographics, dispersal and adaptation. Ultimately, the study of these processes relies on quantification of the genetic diversity, or variation, that is present within focal individuals or populations. Relevant genetic information may range from a single base pair to megabases of variable DNA content, or even the duplication of entire chromosomes or genomes (Scherer *et al.*, 2007). Historically, genetic variation has been identified and quantified by focusing on one or a few sites in the genome, for example using gene sequences or panels of microsatellite markers (Avise, 2004). The transition to newer sequencing technologies enabled the detection of large numbers of single-nucleotide polymorphisms (SNPs), which can capture genome-wide diversity. More recently, the increased yield and decreased cost of genome sequencing, including from long-read technologies, has facilitated the genome-wide study of

*Corresponding author: Katc.stuart@gmail.com

© CAB International 2025. *Invasion Genomics* (Eds D. Bock and M. Rius)
DOI: 10.1079/9781800626263.0004

more complex genomic variants within populations, including non-SNP genetic variants such as structural variants (SVs; alterations in the DNA structure due to sequence deletions, duplications, inversions, insertions or translocations, ranging from tens of bases to megabases in length) and repeat elements, such as transposable elements (TEs; mobile genetic elements also known as transposons) (Scherer *et al.*, 2007). While patterns in one type of genetic variant are often assumed to reflect others, this may not always be the case (e.g. Lerat *et al.*, 2019).

Non-SNP variants are difficult to identify, owing to their sequence complexity and co-location with (or identity as) repeat elements (Collins *et al.*, 2020; Munasinghe *et al.*, 2023). However, with the increasing affordability of both short- and long-read data, there is an exciting opportunity to characterize non-SNP variants more broadly. As our ability to quantify and study more complex genetic variants expands, so does our appreciation of the important role they play in evolution within the natural world, and their particular importance to invasion biology (e.g. Dennenmoser *et al.*, 2019; Mérel *et al.*, 2021; Raingeval *et al.*, 2023). Research in this space will undoubtedly reveal important discoveries, particularly regarding how standing and novel genetic variation contribute to adaptation following introduction to new environments. Development of techniques for sequencing and downstream bioinformatic analysis is rapid and ongoing (Liu *et al.*, 2024), and as such, in this chapter, we will not focus on specific technologies. Rather, we will explore the relevant background information needed to understand SV and TE dynamics, and explore existing examples of their role in, and make predictions about, their impact on biological invasions. We will also provide a general framework for working with these elusive but intriguing genomic features, and suggest key future directions within invasion genomics.

4.2 An Overview of Types of Non-SNP Genetic Variation in Populations

The term 'non-SNP variation' describes intraspecific variation that is larger than 1 bp, and may also encompass chromosome or whole-genome-level changes (Fig. 4.1a) (Mérot *et al.*,

2020). Broadly, these fall into five categories: deletions, insertions, duplications, inversions and translocations, based on their sequence identity relative to a reference sequence. In this chapter, we will be focusing primarily on two main classes of non-SNP genetic variation: SVs (Fig. 4.1a) and TEs (Fig. 4.1b). We have chosen to highlight these variant classes because they (i) represent two of the more diverse classes of non-SNP variants; (ii) are near ubiquitous across all taxa groups; and (iii) have potential diverse and large functional impacts (Alkan *et al.*, 2011; Chalopin *et al.*, 2015). Hence, SVs and TEs have broad applicability to understanding invasion processes. These two variant classes are partial subsets of one another: only some TEs will vary across individuals of a population and thus be identified/classified as SVs and, conversely, not all SVs within a population are due to movement of TEs (Fig. 4.1c). Because SVs and TEs arise via different molecular mechanisms and have different mutation rates (Biémont and Vieira, 2006; Belyeu *et al.*, 2021a), segregating these diverse signals within an amalgamated variant data set may elucidate more nuanced patterns of diversity or evolution within genomes (Munasinghe *et al.*, 2023).

The large nature of many SVs means that they can introduce massive changes and rearrangements within the genome and, unlike SNPs, can affect multiple genes (Scott *et al.*, 2021). Broadly, the creation of SVs occurs through recombination-, repair- or replication-associated processes. The meiotic process of recombination may introduce variation, either through errors during recombination with a homologous region, or by accidentally recombining with other regions of the genome (Carvalho and Lupski, 2016). During repair of DNA damage, SVs may arise through the improper rejoining of a double-stranded DNA break (Currall *et al.*, 2013). Finally, processes like replication slippage and DNA template switching may result in errors during replication (Carvalho and Lupski, 2016). The improper execution of these mechanisms, when occurring in germline cells, will result in heritable genetic variation. Most of our knowledge of the mechanisms of SV creation has been within the context of human diseases, which has demonstrated the huge diversity and impact that this group of non-SNP variants can have on an organism's phenotype (Collins *et al.*, 2020).

(a) Non-SNP variants

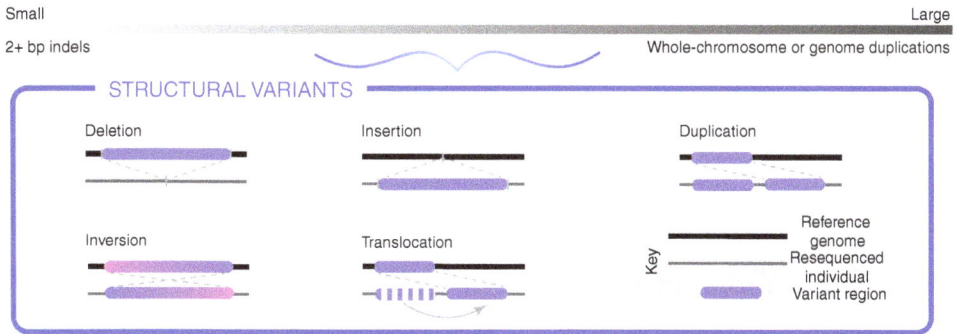

(b) Repeat variants

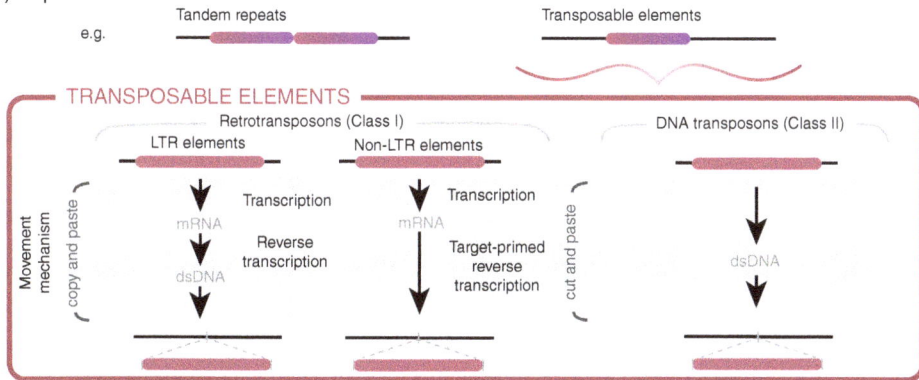

(c) SV and TE overlap classifications

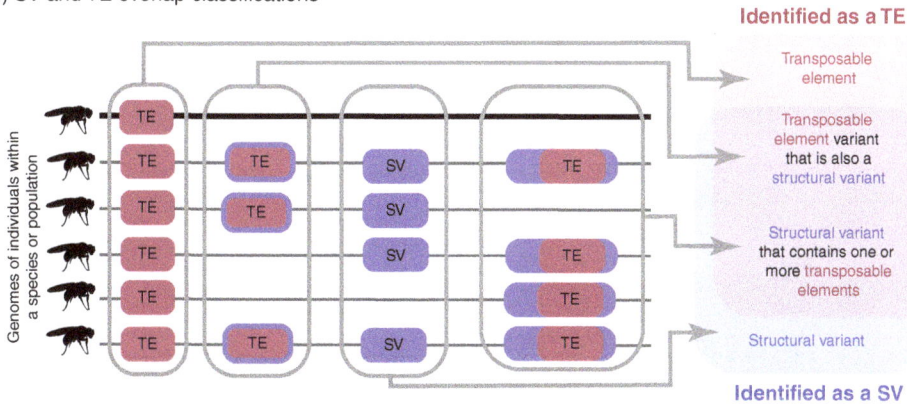

Fig. 4.1. Diagram of the major subgroups of non-single-nucleotide polymorphism (non-SNP) variants and how their identities may overlap. (a) Non-SNP variants, including a scale depicting the size range of structural variants (SVs) as well as diagrams of the different types. (b) Repeat variants, with a focus on transposable elements (TEs) and their different subclassifications. LTR, long terminal repeat. (c) Diagram of the potential ways in which SV and TE classifications may overlap. TEs are identified by their sequence identity and their mechanism of movement, while SVs are defined by their sequence relative to the reference, and arise through recombination-, repair- and replication-based molecular mechanisms, and thus sometimes SV and TE classification may overlap. The black line represents the reference genome, and the grey lines are the resequenced individuals.

TEs are mobile genetic elements within their host's genome. Genomic TE content varies widely across taxonomic groups (from <10% to >50%; Chalopin *et al.*, 2015). TEs have complex interactions with genome evolution and function (Bourque *et al.*, 2018), and may appear as both tandem and interspersed repeats within the genome. TEs are classified by their mechanism of movement, with retrotransposons (Class I) using 'copy and paste' mechanisms via RNA intermediates, while DNA transposons (Class II) use 'cut and paste' mechanisms via DNA intermediates (Finnegan, 1989). Within these TE classes, subclasses are defined based on their method of back-integration into the genome, and these are subdivided further into superfamilies and families based on phylogenetic sequence identity and overlap (Fig. 4.1b) (Bourque *et al.*, 2018). Due to the predictable and phylogenetic sequence characteristics of TEs, mutational substitutions within their sequences may be used to age TEs relative to one another, and thus the relative time points of TE family expansions and contractions can be identified (Bourque *et al.*, 2018).

4.3 Structural Variants and Invasion Processes

The importance of studying SVs within a population genomics context is increasingly being recognized (Wellenreuther *et al.*, 2019; Mérot *et al.*, 2020). Examining SVs in conjunction with SNPs enables a more thorough assessment of inter-individual genetic variation, potentially unveiling new facets of, for example, adaptive divergence within a population (e.g. Catanach *et al.*, 2019; Stuart *et al.*, 2023). While in some cases SNP variants will reflect signals of adaptation, there are documented instances where novel patterns emerge only when analysing non-SNP variants, by, for example, revealing previously hidden genotype–environment associations, such as those seen in the range-expanding Asian honeybee, *Apis cerana* (Li *et al.*, 2024). To understand the unique perspective that SVs can provide us as we study invasive populations, we must first consider the differential properties and impact that SVs and SNPs may have, particularly in relation to invasion-relevant processes such as population demographic changes and rapid adaptation. First and foremost, SVs by definition are larger. The larger a variant is, the higher the likelihood that it will have a functional impact on the organism (Collins *et al.*, 2020), and larger variants are more likely to accumulate smaller SVs and SNPs inside them, which will be jointly inherited by offspring (Mahmoud *et al.*, 2019). SVs can alter genomic sequence content in several different ways; they may create new or remove existing genetic variation (e.g. deletions, duplications and insertions) or create new rearrangements of existing genomic sequence information (e.g. translocations) (Carvalho and Lupski, 2016). Additionally, SVs arise at a lower mutation rate (Belyeu *et al.*, 2021a), and consequently are much less common than SNPs, although their size means that cumulative base-pair coverage across the genome is often many times that of SNPs (e.g. Catanach *et al.*, 2019). Finally, the presence of an SV in heterozygous form can inhibit recombination (Morgan *et al.*, 2017) or lead to unbalanced meiotic products if recombination is successful (Mérot *et al.*, 2020). These unique properties of SVs have important implications for population genetic processes within the context of invasions and their role in invasion success (Table 4.1).

4.3.1 Fitness impacts of SVs within invasive populations

The distribution of fitness effects of SNPs has been well characterized, with a majority of novel variants having neutral or slightly deleterious impacts on fitness (Bank *et al.*, 2014). In contrast to these single base-pair changes, the size of SVs means that their impacts are likely to be more extreme than SNPs, with many fewer neutral variants, and thus are more likely to be deleterious (Fig. 4.2a) (Barton and Zeng, 2018; Hämälä *et al.*, 2021; Scott *et al.*, 2021). For example, SVs characterized in *Drosophila melanogaster* were found to frequently contribute to complex traits but were generally rare, indicative of deleterious effects (Chakraborty *et al.*, 2019). Conversely, a novel SV may confer a larger adaptive advantage than might be expected from a single SNP change, or a formerly deleterious SV

Table 4.1. Examples of specific invasion systems where a range of different non-SNP variants have facilitated adaptive phenotypic changes within introduced ranges that have very likely contributed to the success of the biological invasion.

Species	Variant type	Details	Reference(s)
Fruit fly (*Drosophila melanogaster*)	TEs: P elements, I elements	Introgression of TEs from closely related species/strains are thought to have driven the spread of this species	Bucheton *et al.* (1992); Engels (1992)
Annual ragweed (*Ambrosia artemisiifolia*)	SV: haploblocks caused by inversions	Large haploblocks have driven parallel adaptation in this invasive weed across independent ranges	Battlay *et al.* (2023)
Brown planthopper (*Nilaparvata lugens*)	SV: gene duplication	Duplication and neofunctionalization of a gene directly led to insecticide resistance	Zimmer *et al.* (2018)
Roughfruit amaranth (*Amaranthus tuberculatus*)	SV: gene duplication	Glyphosate-resistant resistance has evolved independently numerous times through numerous duplication events	Kreiner *et al.* (2019)
European green crab (*Carcinus maenas*)	SV: inversion	A balanced inversion polymorphism being maintained in a highly variable environment is likely to have contributed to their success and range expansion	Tepolt *et al.* (2022)

SNP, single-nucleotide polymorphism; SV, structural variant; TE, transposable element.

may become adaptive under a new selection regime (Gaut *et al.*, 2018). Both of these cases may elicit rapid evolutionary change within introduced ranges, either through genetic novelty, selection regime shift or a combination of both. Furthermore, different types of SVs (e.g. inversions in contrast to duplications) have different mutation rates and will impact the genome differently (Barton and Zeng, 2018), so the proportion of each SV type is important to consider when quantifying SV dynamics within an invasive population.

4.3.2 SVs and invasion population demography

Small founding numbers of individuals and reduced genetic diversity typify many invasive populations. As with SNPs, a vast majority of SVs within introduced ranges will be present within the pool of standing genetic variants in the native range, such as is seen in the invasive Colorado potato beetle, *Leptinotarsa decemlineata* (Cohen *et al.*, 2023). Typically, only a small number of variants arise through novel mutations

within introduced ranges, such as in glyphosate-resistant populations of roughfruit amaranth, *Amaranthus tuberculatus* (Kreiner *et al.*, 2019). A small effective population size during the invasion lag phase allows stochastic processes like drift to drive further rare allele loss. However, during this period of smaller effective population size, there is reduced selection efficacy, and mildly deleterious mutations not lost by drift may rise to higher frequencies (Edmonds *et al.*, 2004). The opposing forces of rare allele loss and reduced selection efficacy are balanced in outcrossing populations (Simons *et al.*, 2014), resulting in individuals carrying similar numbers of deleterious variants pre- and post-bottleneck (Gaut *et al.*, 2018), but with the deleterious alleles rising to higher proportions in the population following a bottleneck. Thus, previously rare SV alleles may contribute more to trait variation within bottlenecked populations (Lohmueller, 2014), including invasive populations.

Increased SV allele frequencies, and a resulting increase in heterozygous individuals, may lead to reduced population-wide recombination. This in turn could lead to increased mutational load (accumulated deleterious mutations within

INVASION RELEVANT EVOLUTIONARY QUESTION	GENOMIC SIGNAL OR CONCEPT	POTENTIAL TEST INVASIVE SYSTEM

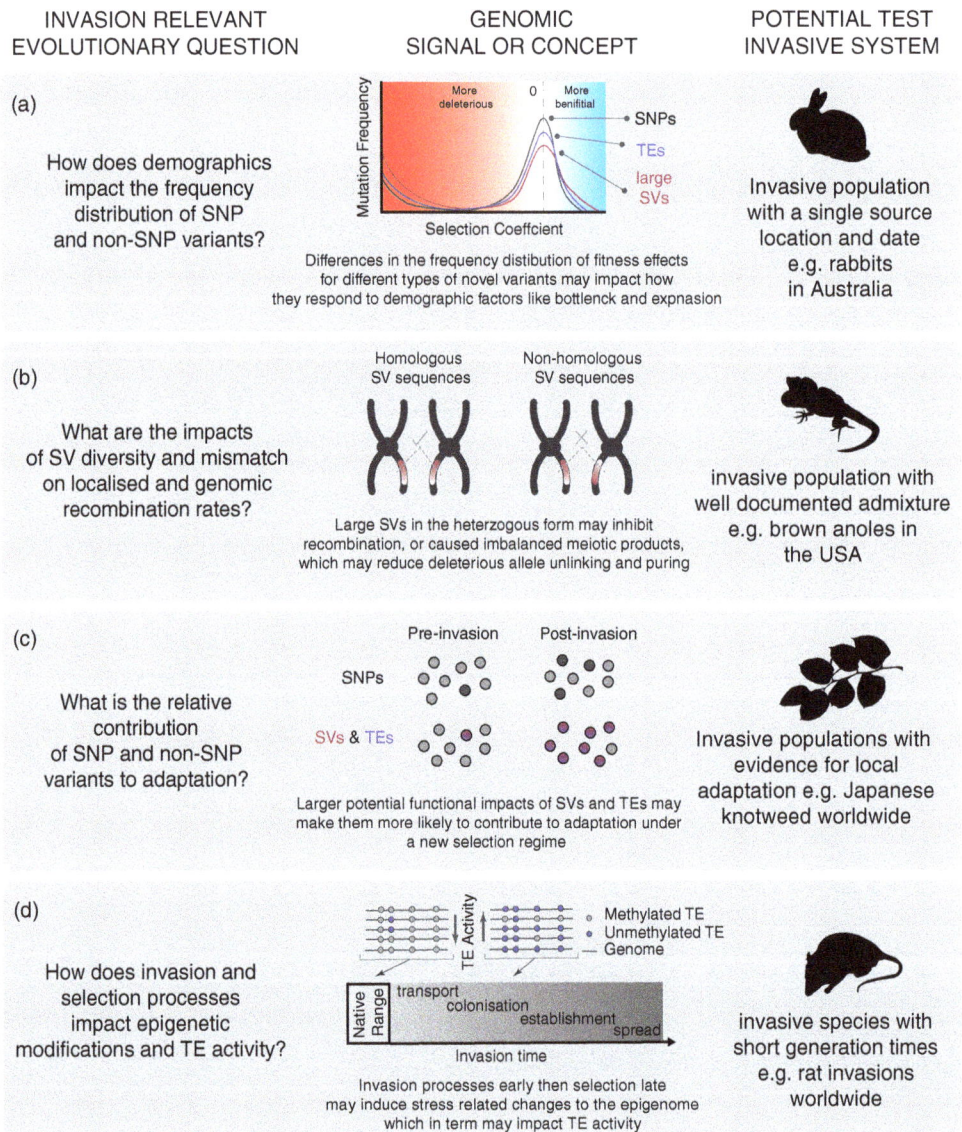

(a)

How does demographics impact the frequency distribution of SNP and non-SNP variants?

Differences in the frequency distibution of fitness effects for different types of novel variants may impact how they respond to demographic factors like bottlenck and expnasion

Invasive population with a single source location and date e.g. rabbits in Australia

(b)

What are the impacts of SV diversity and mismatch on localised and genomic recombination rates?

Large SVs in the heterzogous form may inhibit recombination, or caused imbalanced meiotic products, which may reduce deleterious allele unlinking and puring

invasive population with well documented admixture e.g. brown anoles in the USA

(c)

What is the relative contribution of SNP and non-SNP variants to adaptation?

Larger potential functional impacts of SVs and TEs may make them more likely to contribute to adaptation under a new selection regime

Invasive populations with evidence for local adaptation e.g. Japanese knotweed worldwide

(d)

How does invasion and selection processes impact epigenetic modifications and TE activity?

Invasion processes early then selection late may induce stress related changes to the epigenome which in term may impact TE activity

invasive species with short generation times e.g. rat invasions worldwide

Fig. 4.2. Suggested key future directions for the comparative study of single-nucleotide polymorphism (SNP) and non-SNP variants relevant to invasive systems. (a) Differences in the theoretical distribution of fitness effects for novel SNPs, structural variants (SVs) and transposable elements (TEs), with the frequency of neutral variants (selection coefficient of 0) scaling with the impact of each variant group. Investigations into the implications of this prediction could be completed in invasive populations with simple introduction histories that have not experienced admixture or multiple introduction locations (e.g. Alves *et al.*, 2022). (b) A conceptual diagram of how non-homologous SV sequences, in this case caused by a large sequence inversion appearing in the heterozygous form, may inhibit recombination. Investigating the role of SV heterozygosity on the recombination landscape could be investigated in invasive populations with high levels of admixture (e.g. Kolbe *et al.*, 2004). (c) Identifying the proportion of adaptive change occurring within an invasive range that is directly attributable to TEs, SVs and SNPs alone would be best undertaken within an invasive population where adaptive phenotypes have been well characterized (e.g. Yuan *et al.*, 2024). (d) Multi-generational investigations into the interacting role of epigenetic changes and TE activity in enabling invasion success may be best undertaken using invasive species with short generation times such as rats (e.g. Abdelkrim *et al.*, 2005).

an individual or a population) across generations because a lack of recombination means new deleterious variants arising within SVs cannot be purged. Within introduced ranges, it is not unusual for invasive populations to be characterized by strong admixture due to multiple introductions of distinct genetic lineages (e.g. Kolbe *et al.*, 2004). Individuals possessing heterozygous SVs within admixed invasive populations may provide opportunities to investigate how recombination suppression contributes to mutational load and adaptive potential (Fig. 4.2b). Furthermore, within highly inbred bottlenecked populations, it is also possible that higher SV allelic frequencies may result in a homozygous state of a previously rare variant, presenting an opportunity for once recombination-suppressed regions of the genome to undergo meiotic recombination, increasing the efficacy of selection on genetic variants that have arisen within rare larger SVs. In investigating the behaviour of SVs within small populations, invasion and conservation science (Kardos *et al.*, 2021) have similar interests: to determine how interactions between SVs and demography impact a population's adaptive potential. Discoveries in both fields will undoubtedly complement one another.

Ultimately, invasive populations are characterized by expansion, in terms of both absolute geographical range and population size. As with bottlenecks, decreasing the effective population size at expanding range edges reduces the efficacy of selection. This issue may be further compounded in invasive populations through increasing expansion load, which is the increased mutational burden present in geographically expanding populations due to genetic drift (Peischl *et al.*, 2015). While the rate of adaptation within demographically expanding populations is higher when recombination is present, a lack of recombination may increase the likelihood of high-fitness mutations moving from the core to the range edge (Peischl *et al.*, 2015). Thus, the SV load of a population may impact both the SV- and SNP-based adaptive processes via the influence that SVs have on recombination. Finally, large effective population sizes increase the likelihood of parallel selection (selection of the same genomic region or variant in independent populations) of large-effect alleles (MacPherson and Nuismer, 2017). This has implications for the role SVs may play in evolution

for invasive species where fairly large founder populations have been introduced into independent locations such that much of the source genetic variation is retained. Across the bottlenecks and expansions that characterize invasions, we may consider an important parallel in studies on SV genetic variants and demographics in crop species, because the stages of domestication (Gaut *et al.*, 2018) are similar in an invasion (Kolar and Lodge, 2001).

4.3.3 SVs as facilitators of rapid adaptation and selection

SVs are known to occur non-randomly throughout the genome (Mérot *et al.*, 2020), and may harbour other SVs and SNPs that can contribute to their impact (Cohen *et al.*, 2023). Genetic variation present as SVs may affect gene function or regulation by duplicating genomic content, deleting or inserting novel sequence, or rearranging existing variation (Mortazavi *et al.*, 2022). Examples of each occur in a wide range of invasive systems, emphasizing the unique ways in which SVs may contribute to adaptation under new adaptive regimes.

Gene duplications in particular are known sources of novel genetic variation, because they create redundancy and enable the evolution of secondary functions (Magadum *et al.*, 2013). Duplications are a major mechanism through which insecticide resistance arises in many species of invasive pests such as the brown planthopper, *Nilaparvata lugens* (Zimmer *et al.*, 2018), and have been associated with invasive potential by facilitating tolerance to increased environmental variability across a range of invasive invertebrate species (Makino and Kawata, 2019). Furthermore, whole-genome duplication events such as those seen in the invasive Atlantic salmon, *Salmo salar* L., provide redundancy and buffer the impact of mutations to key gene regions, enabling the accumulation of SVs (Bertolotti *et al.*, 2020) and contributing to their impact.

Genetic diversity can result from the inclusion of a completely novel genomic sequence via insertion, or can be a result of the new sequence generated at the SV breakends of insertions, inversions, translocations, deletions and duplications (Berdan *et al.*, 2023). While many insertions can be attributed to the action of TEs (discussed more specifically below), non-TE insertions and other types of SV have been found to play a role in generating novel genetic diversity by, for example, deleting or inserting exons; this may result in alternative isoforms of a gene (e.g. insecticide resistance in *L. decemlineata*; Cohen *et al.*, 2023).

Genomic rearrangements in the form of translocations or inversions may unite clusters of genetic variants that now must be jointly inherited due to low rates or a lack of recombination – creating 'genomic islands of divergence' (Yeaman, 2013). These are sometimes referred to as haploblocks (or supergenes), and they can facilitate rapid adaptation by ensuring the joint inheritance of multiple beneficial alleles contained within a larger variant, explaining adaptive patterns in, for example, the invasive annual ragweed, *Ambrosia artemisiifolia* (Battlay *et al.*, 2023). Large inversions can disrupt concerted evolution (i.e. gene families evolving non-independently of one another, such as heat-shock protein response genes in *D. melanogaster*; Puig Giribets *et al.*, 2019) and better facilitate rapid adaptive change.

The impact of an SV differs across both size and SV type; for example, inversions that simply rearrange genetic material locally are more common (Berdan *et al.*, 2023), probably because they are generally less deleterious than other SVs of similar size (Scott *et al.*, 2021). However, smaller-scale rearrangements are also important because they may, for example, rearrange the *cis*-regulatory elements of genes (Mortazavi *et al.*, 2022).

While many of the above examples involve directional selection, both alleles of an SV may also be maintained within a population under balancing selection, an evolutionary process that maintains genetic diversity within populations by favouring the persistence of multiple variants of a gene over time. An example of this is the inversion polymorphism varying with a latitudinal environmental cline in the invasive European green crab, *Carcinus maenas*, where both forms of the variant are maintained under balancing selection (Tepolt *et al.*, 2022). Polymorphism for large-effect SVs can therefore be maintained across a selection gradient (e.g. an environmental cline), with migrational load (the maladaptation in individuals moving out of

their genotype-matched environment) acting against fixation of either SV allele (Wellenreuther and Bernatchez, 2018). Other inversion polymorphism examples are *Drosophila subobscura* (Balanyá *et al.*, 2006) and *D. melanogaster* (Kapun *et al.*, 2016), with balancing selection over spatial (dispersal) or temporal (varying climate) scales maintaining the polymorphisms, respectively. Balancing selection within native ranges may maintain SNP genetic diversity that facilitates parallel rapid adaptation within invasive ranges, as has been demonstrated in invasive populations of the *Eurytemora affinis* copepod complex (Stern and Lee, 2020). This phenomenon could also apply to SVs. The probability of parallel genetic evolution from standing genetic variation scales proportionally with an allele's phenotypic impact (MacPherson and Nuismer, 2017), meaning that SVs are likely to play a larger role in parallel evolution than SNPs. Ultimately, SVs represent a huge component of hidden genetic diversity that needs to be characterized to better understand the evolutionary dynamics of invasive populations (Fig. 4.2c).

4.4 Transposable Elements and Invasion

When studying the population dynamics of SVs, it is important to remember that many of them may be specifically of TE origin (Fig. 4.1) (Mérot *et al.*, 2020). Many of the features that differentiate SVs from SNPs within an invasion genomics context can also be extended to TEs. However, in order to understand the role TEs play in populations, it may be necessary to consider SVs and TEs as separate variant types due to several key differences between them (Bourgeois and Boissinot, 2019). First, TEs are more likely to contribute to novel variation than SNPs and SVs, because they have a much higher mutation rate than SNPs and SVs (Biémont and Vieira, 2006). TEs occupy a smaller size range than SVs, generally spanning 100–10,000 bp (Wells and Feschotte, 2020), meaning that we may expect their impacts to be more extreme than those of SNPs but less than those of larger SVs (Fig. 4.2a). TEs, because of their nature as self-replicating units, contain specific motifs that could cause transcription and so are capable of reactivating old or creating new genes (Bourque *et al.*, 2018). The genome's mechanism for silencing the activity (replication) of TEs is altering DNA accessibility with epigenetic markers, and thus there is a direct interaction between the TE content of a genome (the mobilome) and the epigenome (the chemical and protein additions to the DNA sequence, usually reversible, that can influence gene expression and regulation without changes to the underlying DNA sequence). Because of this, TEs have different population dynamics to SNPs and SVs as they may be under selection from both natural selection and host genome factors employed in the regulation of TEs (Guio and González, 2019). Within the context of invasive populations, TEs should be considered for their unique contributions as sources of novel genetic diversity, as well as their role as regulators within their hosts (Fig. 4.2).

Finally, we note here that in TE research, the term 'invasion' is often used to describe the process of TE sequence movement within their host genome and should not be confused with biological invasions of populations. Within this section, we will use the term activity instead to refer to this process of TE multiplication or excision within host genomes.

4.4.1   TEs as sources of novelty

TEs may act as important sources of both standing genetic variation within native ranges and novel genetic variation within invasive ranges. There has long been interest in the role TEs play in invasion success because stress (presumably experienced by invasive populations during invasion processes; Fig. 4.2d) is a known trigger of TE activity and has been posited to potentially play a role in increasing adaptive potential throughout the invasion process (Stapley *et al.*, 2015). Early comparative genomics on *Drosophila* spp. proposed that the colonization process may promote an increase in TEs within the genome, through disrupting or alleviating the constraining conditions on TE activity in the genome during range expansion (Vieira *et al.*, 1999). Studies in *Arabidopsis thaliana* (Jiang *et al.*, 2023) and *Drosophila suzukii* (Mérel *et al.*, 2021) found that range expansion was associated with a drop in effective population size and thus selection strength, and caused an increase in TE load.

From this, we may expect to see larger TE loads on the range edge of expanding populations, where effective population sizes are lower. While this has not directly been tested, there is evidence for changes to the epigenome via increased methylation of CpG sites in the DNA of the house sparrow, *Passer domesticus*, at range edges (Hanson *et al.*, 2022), and greater changes to DNA methylation in stress-exposed cane toad (*Rhinella marina*) offspring of range edge parents, compared with those of range core parents (Sarma *et al.*, 2021). While increased TE load is more likely to have negative impacts on fitness, for the most part it has the potential to increase the genetic variants present in the population and induce regulatory changes in their host (discussed below).

The novel genetic variation generated from TE activity is less likely than SVs to have an impact at the organism level (Scott *et al.*, 2021), which is partially a result of preferences in introduction sites for many TE families to be in non-coding regions where the novel insertion will not immediately be purged (Wells and Feschotte, 2020). Nevertheless, across the tree of life, TEs have been found to provide the raw materials for evolution (Pourrajab and Hekmatimoghaddam, 2021), indicating that they play an important role in population adaptation. Furthermore, for populations that encounter closely related species, introgression (the transfer of genetic material between species or populations through repeated hybridization and backcrossing) may lead to the inclusion of novel interspecific SNP and SV content at specific sites within a genome. In comparison, introgression of a TE may then lead to rapid duplication of that TE in a genome that has not evolved to silence it, and novel TE activity may in turn introduce genomic changes outside the specifically introgressed regions (Bucheton *et al.*, 1992; Engels, 1992). TE insertions can also passively encourage more SVs to arise within a genome through recombination of non-homologous regions (Klein and O'Neill, 2018). These novel variants can contribute massively to the functional diversity of a species and may be responsible for large portions of genomic divergence, such as that seen in the invasive ant *Cardiocondyla obscurior* (Schrader *et al.*, 2014). Investigating selection in TEs across different invasive populations is a promising avenue for enquiry (Fig. 4.2c), and because TE families occur in phylogenies and may be aged, this presents interesting additional avenues for investigations into their responses to selection and their role in evolution (Villanueva-Cañas *et al.*, 2017).

4.4.2 TEs as regulators

The role of epigenetics in invasion is discussed elsewhere (Chapter 7, this volume), so here we will briefly cover the unique role TEs play in epigenetic regulation. TEs may be agents of rapid evolutionary change via alterations in the epigenome that change gene expression, resulting in increased host plasticity. Interactions between TEs and the epigenome may form an important part of functional variation that remains hidden in the native range but gets released within the invasive ranges to quickly alter the phenotype of an organism (Marin *et al.*, 2020). While stress can trigger the generation of novel variants, what must happen first is that the TE is unmasked, a process that has been identified as a mechanism of rapid and repeated adaptation in, for example, *A. thaliana* (Raingeval *et al.*, 2023). Altered TE expression levels under stress have been demonstrated to occur under experimental conditions in the invasive cane toad *R. marina* (Ludwig *et al.*, 2021). TEs may also impact their host not just through direct transcriptional effects but also because the epigenetic silencing or unmasking of TEs may alter the expression levels in nearby genes (Choi and Lee, 2020). There is some evidence that the imperfect genomic mechanisms responsible for resetting the epigenome in mammalian germ cells that contribute to epigenetic heritability may have a TE bias, in which many TE-associated markers 'escape' their transgenerational reprogramming, and may be responsible for unaccounted trait heritability (van Calker and Serchov, 2021). Thus, TEs have the potential to play an important role in the genetic, epigenetic and plastic responses to novel selection regimes within invasive populations.

4.5 Studying Non-SNP Variants Within an Invasion Genomics Context

There are many promising avenues of enquiry into the role that non-SNP variants play in the success and dynamics of invasive populations and species.

4.5.1 Study considerations

Invasive populations present important opportunities to understand the dynamics of non-SNP variants, and conversely, studying non-SNP variants will increase our understanding of invasion processes. As our ability to identify complex, genome-wide variants increases, so too does the need to carefully consider the nuance in the types of non-SNP variants we are choosing to study. Non-SNP variant subclassifications have been artificially produced within the context of specific data or study types, which has resulted in overlap and redundancy (Liehr, 2021). Researchers seeking to understand the impact of non-SNP variants will have to navigate literature in which similar phenomena may be presented very differently and the use of nuanced, overlapping and occasionally redundant terms that have changed over time or across fields as technologies and fields develop.

It is pragmatic to consider the biological questions being asked. Researchers will need to carefully consider which classes of non-SNP variants can be found using their sequence data and chosen identification method, and whether further discrimination of variant classes may be necessary within the biological scope of their study. Depending on the focal study question, it may be most important to consider the different variant classes, to focus on their size or to consider both aspects (e.g. Munasinghe *et al.*, 2023). For example, large deletions are likely to be more deleterious than other variants of similar size because they can remove non-redundant vital genomic content (Petrov, 2002). Additionally, it is important to bear in mind that the identity of a variant may be entirely defined by the reference genome (e.g. an insertion could also be an absence of a deletion) and so choice of reference genome, or even the use of a pangenome (i.e. the entire set of genetic variants present in a species, including the core genome (genetic regions shared by all individuals) and the accessory genome (genetic regions present in some but not all individuals)), may be very important. Within the context of invasive populations, this means that without knowledge of the ancestral allele state, the identity of the alternative allele should be inferred carefully, because genetic divergence between native and invasive ranges cannot all be attributed to changes in the latter (Stuart *et al.*, 2022).

4.5.2 Data considerations

Although some SVs may be identified using data types like reduced-representation sequencing or RNA sequencing, whole-genome resequencing is required for comprehensive genome-wide *de novo* complex variant discovery (Alkan *et al.*, 2011). SV detection from whole-genome sequencing data involves looking for paired signals of discontinuity in read-pairs, split reads and read depth of sequence reads aligned to a reference genome. Short-read sequencing allows the identification of smaller non-SNP variants up to tens of kilobases, while long-read sequencing is required for resolving large or more complex genomic variation (Mahmoud *et al.*, 2019); however, large and very complex variants remain challenging to identify (Vendrell-Mir *et al.*, 2019; Wold *et al.*, 2023). Assembly-based discovery methods may be used, but these rely on comparisons between individual reference genomes (a higher cost investment) (Alkan *et al.*, 2011) and would be best used to interrogate broad interspecific hypotheses about the role of non-SNP variants in biological invasions. Manual curation to validate identified variants may be required (Belyeu *et al.*, 2021b).

TE identification can involve the use of generic databases of TE family sequences, although the use of curated species-specific TE libraries is recommended (Goubert *et al.*, 2022). Unlike SVs, whose sequences are non-specific and defined only in relation to a reference genome, TEs are identified based on their sequence characteristics. While it may appear that TE identification should be more straightforward than SV identification, TEs may occur many times within a genome (e.g. the most common TE family in humans is Alu, which appears over 1 million times within the human genome; Lander *et al.*, 2001). Although a genome will contain many TEs that are present in all individuals within that taxon lineage, TE movements can lead to TE sequences being lost with respect to the reference genome, or gained as a novel TE insertion, and hence are variable across individuals (Fig. 4.1c). Considerable sequencing depth is then needed at the flanking regions around a TE in order for it to be confidently placed within the genome of a resequenced individual (Vendrell-Mir *et al.*, 2019).

Variation in quality and length of raw sequence data will have a greater impact on SV and TE discovery compared with SNPs, meaning that accounting for raw sequence data variation even within a single study is needed (Navarro-Dominguez *et al.*, 2022), and strong batch effects (Stuart *et al.*, 2023) may necessitate extra consideration when combining separate data sets. It is worth noting that the best practices for SV and TE discovery are evolving rapidly, and consulting recent benchmarking literature for your specific sequencing technology is recommended.

4.5.3 The next frontiers in non-SNP invasion genomics

The role of non-SNP variants in invasion genomics is complex. We need more studies that directly compare SNP and non-SNP patterns to address specific questions within singular species or populations, which will help to resolve broad patterns that characterize invasions (Fig. 4.2). In addition to these studies that aim to profile evolutionarily relevant patterns within populations, some extra lines of questioning we should consider for drawing broad conclusions include the following:

1. Is there confirmation bias (e.g. are we finding patterns just because we are seeking them)? We need to compare invasive and non-invasive populations.
2. Are existing findings related to population size? We need to compare invasive and heavily managed populations that have been reduced in size to determine this.
3. How much do these patterns differ over taxonomic groups, and with genome and repeat content? We need to compare patterns across species, taking into account phylogenetic non-independence of genome structure and content.

Because non-SNP variants are less likely to result in neutral variation, they are likely to play a disproportionate role in facilitating invasions throughout each invasion stage. Understanding the unique perspective they provide, and making sure to discriminate between different types of variants, will undoubtedly reveal many interesting discoveries in the field of invasion genomics.

References

Abdelkrim, J., Pascal, M., Calmet, C. and Samadi, S. (2005) Importance of assessing population genetic structure before eradication of invasive species: examples from insular Norway rat populations. *Conservation Biology* 19, 1509–1518. DOI: 10.1111/j.1523-1739.2005.00206.x

Alkan, C., Coe, B.P. and Eichler, E.E. (2011) Genome structural variation discovery and genotyping. *Nature Reviews Genetics* 12, 363–376. DOI: 10.1038/nrg2958

Alves, J.M., Carneiro, M., Day, J.P., Welch, J.J., Duckworth, J.A. *et al.* (2022) A single introduction of wild rabbits triggered the biological invasion of Australia. *Proceedings of the National Academy of Sciences USA* 119: e2122734119. DOI: 10.1073/pnas.2122734119

Avise, J.C. (2004) *Molecular Markers, Natural History, and Evolution*, 2nd edn. Sinauer Associates, Sunderland, Massachusetts.

Balanyá, J., Oller, J.M., Huey, R.B., Gilchrist, G.W. and Serra, L. (2006) Global genetic change tracks global climate warming in *Drosophila subobscura*. *Science* 313, 1773–1775. DOI: 10.1126/science.1131002

Bank, C., Ewing, G.B., Ferrer-Admettla, A., Foll, M. and Jensen, J.D. (2014) Thinking too positive? Revisiting current methods of population genetic selection inference. *Trends in Genetics* 30, 540–546. DOI: 10.1016/j.tig.2014.09.010

Barton, H.J. and Zeng, K. (2018) New Methods for Inferring the Distribution of Fitness Effects for INDELs and SNPs. *Molecular Biology and Evolution* 35, 1536–1546. DOI: 10.1093/molbev/msy054

Battlay, P., Wilson, J., Bieker, V.C., Lee, C., Prapas, D., Petersen, B., Craig, S., van Boheemen, L., Scalone, R., de Silva, NP. *et al.* (2023) Large haploblocks underlie rapid adaptation in the invasive weed *Ambrosia artemisiifolia*. *Nature Communications* 14: 1717. DOI: 10.1038/s41467-023-37303-4

Belyeu, J.R., Brand, H., Wang, H., Zhao, X., Pedersen, B.S. *et al.* (2021a) *De novo* structural mutation rates and gamete-of-origin biases revealed through genome sequencing of 2,396 families. *American Journal of Human Genetics* 108, 597–607. DOI: 10.1016/j.ajhg.2021.02.012

Belyeu, J.R., Chowdhury, M., Brown, J., Pedersen, B.S., Cormier, M.J. *et al.* (2021b) Samplot: a platform for structural variant visual validation and automated filtering. *Genome Biology* 22: 161. DOI: 10.1186/s13059-021-02380-5

Berdan, E.L., Barton, N.H., Butlin, R., Charlesworth, B., Faria, R. *et al.* (2023) How chromosomal inversions reorient the evolutionary process. *Journal of Evolutionary Biology* 36, 1761–1782. DOI: 10.1111/jeb.14242

Bertolotti, A.C., Layer, R.M., Gundappa, M.K., Gallagher, M.D., Pehlivanoglu, E. *et al.* (2020) The structural variation landscape in 492 Atlantic salmon genomes. *Nature Communications* 11: 5176. DOI: 10.1038/s41467-020-18972-x

Biémont, C. and Vieira, C. (2006) Junk DNA as an evolutionary force. *Nature* 443, 521–524. DOI: 10.1038/443521a

Bourgeois, Y. and Boissinot, S. (2019) On the population dynamics of junk: a review on the population genomics of transposable elements. *Genes* 10: 419. DOI: 10.3390/genes10060419

Bourque, G., Burns, K.H., Gehring, M., Gorbunova, V., Seluanov, A. *et al.* (2018) Ten things you should know about transposable elements. *Genome Biology* 19: 199. DOI: 10.1186/s13059-018-1577-z

Bucheton, A., Vaury, C., Chaboissier, M.-C., Abad, P., Pélisson, A. and Simonelig, M. (1992) I elements and the *Drosophila* genome. *Genetica* 86, 175–190. DOI: 10.1007/BF00133719

Carvalho, C.M.B. and Lupski, J.R. (2016) Mechanisms underlying structural variant formation in genomic disorders. *Nature Reviews Genetics* 17, 224–238. DOI: 10.1038/nrg.2015.25

Catanach, A., Crowhurst, R., Deng, C., David, C., Bernatchez, L. and Wellenreuther, M. (2019) The genomic pool of standing structural variation outnumbers single nucleotide polymorphism by threefold in the marine teleost *Chrysophrys auratus*. *Molecular Ecology* 28, 1210–1223. DOI: 10.1111/mec.15051

Chakraborty, M., Emerson, J.J., Macdonald, S.J. and Long, A.D. (2019) Structural variants exhibit widespread allelic heterogeneity and shape variation in complex traits. *Nature Communications* 10: 4872. DOI: 10.1038/s41467-019-12884-1

Chalopin, D., Naville, M., Plard, F., Galiana, D. and Volff, J.-N (2015) Comparative analysis of transposable elements highlights mobilome diversity and evolution in vertebrates. *Genome Biology and Evolution* 7, 567–580. DOI: 10.1093/gbe/evv005

Choi, J.Y. and Lee, Y.C.G. (2020) Double-edged sword: the evolutionary consequences of the epigenetic silencing of transposable elements. *PLOS Genetics* 16: e1008872. DOI: 10.1371/journal.pgen.1008872

Cohen, Z.P., Schoville, S.D. and Hawthorne, D.J. (2023) The role of structural variants in pest adaptation and genome evolution of the Colorado potato beetle, *Leptinotarsa decemlineata* (Say). *Molecular Ecology* 32, 1425–1440. DOI: 10.1111/mec.16838

Collins, R.L., Brand, H., Karczewski, K.J., Zhao, X., Alföldi, J. *et al.* (2020) A structural variation reference for medical and population genetics. *Nature* 581, 444–451. DOI: 10.1038/s41586-020-2287-8

Currall, B.B., Chiang, C., Talkowski, M.E. and Morton, C.C. (2013) Mechanisms for structural variation in the human genome. *Current Genetic Medicine Reports* 1, 81–90. DOI: 10.1007/s40142-013-0012-8

Dennenmoser, S., Sedlazeck, F.J., Schatz, M.C., Altmüller, J., Zytnicki, M. and Nolte, A.W. (2019) Genome-wide patterns of transposon proliferation in an evolutionary young hybrid fish. *Molecular Ecology* 28, 1491–1505. DOI: 10.1111/mec.14969

Edmonds, C.A., Lillie, A.S. and Cavalli-Sforza, L.L. (2004) Mutations arising in the wave front of an expanding population. *Proceedings of the National Academy of Sciences USA* 101, 975–979. DOI: 10.1073/pnas.0308064100

Engels, W.R. (1992) The origin of P elements in *Drosophila melanogaster*. *BioEssays* 14, 681–686. DOI: 10.1002/bies.950141007

Finnegan, D.J. (1989) Eukaryotic transposable elements and genome evolution. *Trend in Genetics* 5, 103–107. DOI: 10.1016/0168-9525(89)90039-5

Gaut, B.S., Seymour, D.K., Liu, Q. and Zhou, Y. (2018) Demography and its effects on genomic variation in crop domestication. *Nature Plants* 4, 512–520. DOI: 10.1038/s41477-018-0210-1

Goubert, C., Craig, R.J., Bilat, A.F., Peona, V., Vogan, A.A. and Protasio, A.V. (2022) A beginner's guide to manual curation of transposable elements. *Mobile DNA* 13: 7. DOI: 10.1186/s13100-021-00259-7

Guio, L. and González, J. (2019) New insights on the evolution of genome content: population dynamics of transposable elements in flies and humans. In: Anisimova, M. (ed.) *Evolutionary Genomics: Statistical and Computational Methods. Methods in Molecular Biology, Vol.* 1910. Springer, New York, pp. 505–530. DOI: 10.1007/978-1-4939-9074-0_16

Hämälä, T., Wafula, E.K., Guiltinan, M.J., Ralph, P.E., dePamphilis, C.W. and Tiffin, P. (2021) Genomic structural variants constrain and facilitate adaptation in natural populations of *Theobroma cacao*, the chocolate tree. *Proceedings of the National Academy of Sciences USA* 118: e2102914118. DOI: 10.1073/pnas.2102914118

Hanson, H.E., Wang, C., Schrey, A.W., Liebl, A.L., Ravinet, M. *et al.* (2022) Epigenetic potential and DNA methylation in an ongoing house sparrow (*Passer domesticus*) range expansion. *American Naturalist* 200, 662–674. DOI: 10.1086/720950

Jiang, J., Xu, Y.-C., Zhang, Z.-Q., Chen, J.-F., Niu, X.-M. *et al.* (2023) Forces driving transposable element load variation during *Arabidopsis* range expansion. *Plant Cell* 36, 840-862. DOI: 10.1093/plcell/koad296

Kapun, M., Fabian, D.K., Goudet, J. and Flatt, T. (2016) Genomic evidence for adaptive inversion clines in *Drosophila melanogaster*. *Molecular Biology and Evolution* 33, 1317–1336. DOI: 10.1093/molbev/msw016

Kardos, M., Armstrong, E.E., Fitzpatrick, S.W., Hauser, S., Hedrick, P.W. *et al.* (2021) The crucial role of genome-wide genetic variation in conservation. *Proceedings of the National Academy of Sciences USA* 118: e2104642118. DOI: 10.1073/pnas.2104642118

Klein, S.J. and O'Neill, R.J. (2018) Transposable elements: genome innovation, chromosome diversity, and centromere conflict. *Chromosome Research* 26, 5–23. DOI: 10.1007/s10577-017-9569-5

Kolar, C.S. and Lodge, D.M. (2001) Progress in invasion biology: predicting invaders. *Trends in Ecology and Evolution* 16, 199–204. DOI: 10.1016/S0169-5347(01)02101-2

Kolbe, J.J., Glor, R.E., Schettino, L.R., Lara, A.C., Larson, A. and Losos, J.B. (2004) Genetic variation increases during biological invasion by a Cuban lizard. *Nature* 431, 177–181. DOI: 10.1038/nature02807

Kreiner, J.M., Giacomini, D.A., Bemm, F., Waithaka, B., Regalado, J. *et al.* (2019) Multiple modes of convergent adaptation in the spread of glyphosate-resistant *Amaranthus tuberculatus*. *Proceedings of the National Academy of Sciences USA* 116, 21076–21084. DOI: 10.1073/pnas.1900870116

Lander, E.S., Linton, L.M., Birren, B., Nusbaum, C., Zody, M.C. *et al.* (2001) Initial sequencing and analysis of the human genome. *Nature* 409, 860–921. DOI: 10.1038/35057062

Lerat, E., Goubert, C., Guirao-Rico, S., Merenciano, M., Dufour, A.-B. *et al.* (2019) Population-specific dynamics and selection patterns of transposable element insertions in European natural populations. *Molecular Ecology* 28, 1506–1522. DOI: 10.1111/mec.14963

Li, Y., Yao, J., Sang, H., Wang, Q., Su, L. *et al.* (2024) Pan-genome analysis highlights the role of structural variation in the evolution and environmental adaptation of Asian honeybees. *Molecular Ecology Resources* 24: e13905. DOI: 10.1111/1755-0998.13905

Liehr, T. (2021) Repetitive elements in humans. *International Journal of Molecular Sciences* 22: 2072. DOI: 10.3390/ijms22042072

Liu, Y.H., Luo, C., Golding, S.G., Ioffe, J.B. and Zhou, X.M. (2024) Tradeoffs in alignment and assembly-based methods for structural variant detection with long-read sequencing data. *Nature Communications* 15: 2447. DOI: 10.1038/s41467-024-46614-z

Lohmueller, K.E. (2014) The impact of population demography and selection on the genetic architecture of complex traits. *PLOS Genetics* 10: e1004379. DOI: 10.1371/journal.pgen.1004379

Ludwig, A., Schemberger, M.O., Gazolla, C.B., de Moura Gama, J., Duarte, I. *et al.* (2021) Transposable elements expression in *Rhinella marina* (cane toad) specimens submitted to immune and stress challenge. *Genetica* 149, 335–342. DOI: 10.1007/s10709-021-00130-w

MacPherson, A. and Nuismer, S.L. (2017) The probability of parallel genetic evolution from standing genetic variation. *Journal of Evolutionary Biology* 30, 326–337. DOI: 10.1111/jeb.13006

Magadum, S., Banerjee, U., Murugan, P., Gangapur, D. and Ravikesavan, R. (2013) Gene duplication as a major force in evolution. *Journal of Genetics* 92, 155–161. DOI: 10.1007/s12041-013-0212-8

Mahmoud, M., Gobet, N., Cruz-Dávalos, D.I., Mounier, N., Dessimoz, C. and Sedlazeck, F.J. (2019) Structural variant calling: the long and the short of it. *Genome Biology* 20: 246. DOI: 10.1186/s13059-019-1828-7

Makino, T. and Kawata, M. (2019) Invasive invertebrates associated with highly duplicated gene content. *Molecular Ecology* 28, 1652–1663. DOI: 10.1111/mec.15019

Marin, P., Genitoni, J., Barloy, D., Maury, S., Gibert, P. *et al.* (2020) Biological invasion: The influence of the hidden side of the (epi)genome. *Functional Ecology* 34, 385–400. DOI: 10.1111/1365-2435.13317

Mérel, V., Gibert, P., Buch, I., Rodriguez Rada, V., Estoup, A. *et al.* (2021) The worldwide invasion of *Drosophila suzukii* is accompanied by a large increase of transposable element load and a small number of putatively adaptive insertions. *Molecular Biology and Evolution* 38, 4252–4267. DOI: 10.1093/molbev/msab155

Mérot, C., Oomen, R.A., Tigano, A. and Wellenreuther, M. (2020) A roadmap for understanding the evolutionary significance of structural genomic variation. *Trends in Ecology and Evolution* 35, 561–572. DOI: 10.1016/j.tree.2020.03.002

Morgan, A.P., Gatti, D.M., Najarian, M.L., Keane, T.M., Galante, R.J. *et al.* (2017) Structural variation shapes the landscape of recombination in mouse. *Genetics* 206, 603–619. DOI: 10.1534/genetics.116.197988

Mortazavi, M., Ren, Y., Saini, S., Antaki, D., St Pierre, C.L. *et al.* (2022) SNPs, short tandem repeats, and structural variants are responsible for differential gene expression across C57BL/6 and C57BL/10 substrains. *Cell Genomics* 2: 100102. DOI: 10.1016/j.xgen.2022.100102.

Munasinghe, M., Read, A., Stitzer, M.C., Song, B., Menard, C. *et al.* (2023) Combined analysis of transposable elements and structural variation in maize genomes reveals genome contraction outpaces expansion. *PLOS Genetics* 19: e1011086. DOI: 10.1371/journal.pgen.1011086

Navarro-Dominguez, B., Chang, C.-H., Brand, C.L., Muirhead, C.A., Presgraves, D.C. and Larracuente, A.M. (2022) Epistatic selection on a selfish Segregation Distorter supergene – drive, recombination, and genetic load. *eLife* 11: e78981. DOI: 10.7554/eLife.78981

Peischl, S., Kirkpatrick, M. and Excoffier, L. (2015) Expansion load and the evolutionary dynamics of a species range. *American Naturalist* 185, E81–E93. DOI: 10.1086/680220

Petrov, D.A. (2002) Mutational equilibrium model of genome size evolution. *Theoretical Population Biology* 61, 531–544. DOI: 10.1006/tpbi.2002.1605

Pourrajab, F. and Hekmatimoghaddam, S. (2021) Transposable elements, contributors in the evolution of organisms (from an arms race to a source of raw materials). *Heliyon* 7: e06029. DOI: 10.1016/j.heliyon.2021.e06029

Puig Giribets, M., García Guerreiro, M.P., Santos, M., Ayala, F.J., Tarrío, R. and Rodríguez-Trelles, F. (2019) Chromosomal inversions promote genomic islands of concerted evolution of Hsp70 genes in the *Drosophila subobscura* species subgroup. *Molecular Ecology* 28, 1316–1332. DOI: 10.1111/mec.14511

Raingeval, M., Leduque, B., Baduel, P., Edera, A., Roux, F. *et al.* (2023) Retrotransposon-driven environmental regulation of FLC leads to adaptive response to herbicide. *Nature Plants* 10, 1672–1681. DOI: 10.1038/s41477-024-01807-8.

Sarma, R.R., Crossland, M.R., Eyck, H.J.F., DeVore, J.L., Edwards, R.J. *et al.* (2021) Intergenerational effects of manipulating DNA methylation in the early life of an iconic invader. *Philosophical Transactions of the Royal Society B: Biological Sciences* 376: 20200125. DOI: 10.1098/rstb.2020.0125

Scherer, S.W., Lee, C., Birney, E., Altshuler, D.M., Eichler, E.E. *et al.* (2007) Challenges and standards in integrating surveys of structural variation. *Nature Genetics* 39, S7–S15. DOI: 10.1038/ng2093

Schrader, L., Kim, J.W., Ence, D., Zimin, A., Klein, A. *et al.* (2014) Transposable element islands facilitate adaptation to novel environments in an invasive species. *Nature Communications* 5: 5495. DOI: 10.1038/ncomms6495

Scott, A.J., Chiang, C. and Hall, I.M. (2021) Structural variants are a major source of gene expression differences in humans and often affect multiple nearby genes. *Genome Research* 31, 2249–2257. DOI: 10.1101/gr.275488.121

Simons, Y.B., Turchin, M.C., Pritchard, J.K. and Sella, G. (2014) The deleterious mutation load is insensitive to recent population history. *Nature Genetics* 46, 220–224. DOI: 10.1038/ng.2896

Stapley, J., Santure, A.W. and Dennis, S.R. (2015) Transposable elements as agents of rapid adaptation may explain the genetic paradox of invasive species. *Molecular Ecology* 24, 2241–2252. DOI: 10.1111/mec.13089

Stern, D.B. and Lee, C.E. (2020) Evolutionary origins of genomic adaptations in an invasive copepod. *Nature Ecology and Evolution* 4, 1084–1094. DOI: 10.1038/s41559-020-1201-y

Stuart, K.C., Edwards, R.J., Sherwin, W.B. and Rollins, L.A. (2023) Contrasting patterns of single nucleotide polymorphisms and structural variation across multiple invasions. *Molecular Biology and Evolution* 40: msad046. DOI: 10.1093/molbev/msad046

Stuart, K.C., Sherwin, W.B., Austin, J.J., Bateson, M., Eens, M. *et al.* (2022) Historical museum samples enable the examination of divergent and parallel evolution during invasion. *Molecular Ecology* 31, 1836–1852. DOI: 10.1111/mec.16353

Tepolt, C.K., Grosholz, E.D., de Rivera, C.E. and Ruiz, G.M. (2022) Balanced polymorphism fuels rapid selection in an invasive crab despite high gene flow and low genetic diversity. *Molecular Ecology* 31, 55–69. DOI: 10.1111/mec.16143

van Calker, D. and Serchov, T. (2021) The "missing heritability"-problem in psychiatry: is the interaction of genetics, epigenetics and transposable elements a potential solution? *Neuroscience and Biobehavioral Reviews* 126, 23–42. DOI: 10.1016/j.neubiorev.2021.03.019

Vendrell-Mir, P., Barteri, F., Merenciano, M., González, J., Casacuberta, J.M. and Castanera, R. (2019) A benchmark of transposon insertion detection tools using real data. *Mobile DNA* 10: 53. DOI: 10.1186/s13100-019-0197-9

Vieira, C., Lepetit, D., Dumont, S. and Biémont, C. (1999) Wake up of transposable elements following *Drosophila simulans* worldwide colonization. *Molecular Biology and Evolution* 16, 1251–1255. DOI: 10.1093/oxfordjournals.molbev.a026215

Villanueva-Cañas, J.L., Rech, G.E., de Cara, M.A.R. and González, J. (2017) Beyond SNPs: how to detect selection on transposable element insertions. *Methods in Ecology and Evolution* 8, 728–737. DOI: 10.1111/2041-210X.12781

Wellenreuther, M. and Bernatchez, L. (2018) Eco-evolutionary genomics of chromosomal inversions. *Trends in Ecology and Evolution* 33, 427–440. DOI: 10.1016/j.tree.2018.04.002

Wellenreuther, M., Mérot, C., Berdan, E. and Bernatchez, L. (2019) Going beyond SNPs: the role of structural genomic variants in adaptive evolution and species diversification. *Molecular Ecology* 28, 1203–1209. DOI: 10.1111/mec.15066

Wells, J.N. and Feschotte, C. (2020) A field guide to eukaryotic transposable elements. *Annual Review of Genetics* 54, 539–561. DOI: 10.1146/annurev-genet-040620-022145

Wold, J.R., Guhlin, J.G., Dearden, P.K., Santure, A.W. and Steeves, T.E. (2023) The promise and challenges of characterizing genome-wide structural variants: a case study in a critically endangered parrot. *Molecular Ecology Resources* 25: e13783. DOI: 10.1111/1755-0998.13783

Yeaman, S. (2013) Genomic rearrangements and the evolution of clusters of locally adaptive loci. *Proceedings of the National Academy of Sciences USA* 110, E1743–E1751. DOI: 10.1073/pnas.1219381110

Yuan, W., Pigliucci, M. and Richards, C.L. (2024) Rapid phenotypic differentiation in the iconic Japanese knotweed *s.l.* invading novel habitats. *Scientific Reports* 14: 14640. DOI: 10.1038/s41598-024-64109-1

Zimmer, C.T., Garrood, W.T., Singh, K.S., Randall, E., Lueke, B. *et al.* (2018) Neofunctionalization of duplicated P450 genes drives the evolution of insecticide resistance in the brown planthopper. *Current Biology* 28, 268–274.e5. DOI: 10.1016/j.cub.2017.11.060

5 Using Genomics to Understand How Invaders May Adapt: A Marine Perspective

Frédérique Viard[1]*[†] and Carolyn Tepolt[2]*[†]

[1]*ISEM, University of Montpellier, CNRS, IRD, Montpellier, France;* [2]*Department of Biology, Woods Hole Oceanographic Institution, Woods Hole, Massachusetts, USA*

Abstract

Adaptation is essential for non-native species to establish and spread in a new range. This can include pre-adaptation, when a new population is founded by a source that has evolved to thrive under those same conditions, and post-introduction adaptation, when rapid adaptive changes occur after introduction. Research into these processes in introduced species is still relatively rare, especially in marine systems, despite the extraordinary evolutionary diversity of marine life. Genomic tools are improving our ability to understand the nature of the adaptive changes occurring in introduced populations and the conditions under which they arise. Based on theoretical considerations and empirical studies, this chapter shows that selection on standing genetic variation and admixture processes (including hybridization) are pivotal mechanisms by which non-native species can genetically adapt to new environments, illustrated here with marine case studies.

5.1 Introduction

Non-indigenous species (NIS) are species translocated by human activities to a new range where they subsequently become established. On a global scale, there are currently approximately 240,000 marine species described (WoRMS Editorial Board, 2025), 0.5% of which have been reported as NIS in one or several marine ecoregions, a number likely to be underestimated (Bailey *et al.*, 2020). More important than the absolute number is the trend, with marine NIS increasing steadily since the mid-20th century; for example, since 1970 the introduction of NIS into European seas has increased from six to 21 NIS per year (Zenetos *et al.*, 2022). This increase is facilitated by accelerating rates of shipping and maritime trade, which are the primary introduction vectors in the marine realm and are expected to lead to a three- to 20-fold increase in invasion risk by 2050 (Sardain *et al.*, 2019). From an ecological and evolutionary perspective, human-driven species translocations have major consequences. By definition, they break down natural barriers to dispersal, redrawing biogeographical boundaries and disrupting ecosystems (Capinha *et al.*, 2015). The success of so many NIS in abiotic and biotic environments in which they did not evolve also raises questions about the nature of adaptation and the processes by which these species not only survive in a new range but thrive and often spread across extensive environmental gradients (Gribben and Byers, 2020).

---

*Corresponding authors: frederique.viard@umontpellier.fr and ctepolt@whoi.edu
[†]Both authors contributed equally.

© CAB International 2025. *Invasion Genomics* (Eds D. Bock and M. Rius)
DOI: 10.1079/9781800626263.0005

At least initially, the success of NIS relies largely on a match between a species' biological requirements (e.g. physiological, ecological) and the characteristics of its introduced environment (Facon *et al.*, 2006). Aside from environmental changes in the introduced range that may make it more hospitable (Stachowicz *et al.*, 2002; Johnston *et al.*, 2017), there are a number of potential mechanisms by which an introduced species itself may cope with a new environment. First, the NIS may be "pre-adapted" to conditions in the introduced range. For example, species already inhabiting and adapted to urbanized environments such as ports may be poised to thrive in similar habitats in their introduced range (Hufbauer *et al.*, 2012; Briski *et al.*, 2018). Second, NIS may be able to acclimatize to local conditions through phenotypic plasticity (Smith, 2009), including via epigenetic mechanisms (Carneiro and Lyko, 2020), for example the ascidian *Botryllus schlosseri* (Gao *et al.*, 2022), or via microbiome variations (e.g. seagrasses; Aires *et al.*, 2021), which may enhance NIS persistence and establishment in a range of different environments. Finally, NIS may adapt genetically during and after introduction (Tepolt, 2015). This last process is the most challenging to investigate, as detailed below, and has been studied primarily in terrestrial species (e.g. Prentis *et al.*, 2008; Colautti *et al.*, 2015; Hodgins *et al.*, 2025). With the rise of genomic tools and approaches, it has become increasingly tractable to explore its importance in other systems, including the marine realm. Marine species represent far more deep evolutionary diversity than terrestrial species (May, 1994) but historically have been underrepresented in evolutionary research. Here, we present the broad principles of NIS adaptive evolution in the context of the complexity and diversity of the marine environment. In this chapter, we will focus on post-introduction genetic adaptation to ask three main questions: How can genomics shed light on adaptive processes, particularly in NIS? What is the evidence for rapid evolutionary change reported so far? And what are the processes involved?

5.2   The Advantages of Using Genomic Tools to Study Adaptation

Robust evidence for adaptive evolution requires proof of phenotypes or traits that are genetically heritable and that confer fitness advantages. The study carried out by Sotka *et al.* (2018) on the red seaweed *Gracilaria vermiculophylla*, which combined a common-garden experiment, genetic analyses and ecological niche modeling, is one of the rare examples that showed rapid phenotypic evolution likely explained by genetic adaptation during invasion in the marine realm. However, for the vast majority of marine species, these experiments are difficult or impossible to carry out due to the prevalence of complex life cycles, notably bentho–pelagic species that alternate between a free-living larval stage and a benthic adult stage. Experimental approaches are even more difficult when considering that second generations should ideally be used to avoid maternal effects. Lastly, there is an added level of complication with NIS, for which ethical considerations prevent reciprocal transplants. As a consequence, indirect approaches based on genetic methods are much more tractable in these systems.

Genome scanning is the most commonly used indirect approach, identifying loci showing patterns that are unexpected under neutral assumptions (see the seminal paper by Beaumont and Nichols, 1996). Applied to NIS, this method looks for loci showing extreme population genetic differentiation values between introduced populations or between native and introduced populations. Genotype–environment association studies are related methods looking for correlations between allelic frequencies and environmental variables. However, these indirect approaches have their own pitfalls and limitations (Bierne *et al.*, 2013; Lotterhos and Whitlock, 2015; Hoban *et al.*, 2016). The genome-scanning approach can be a double-edged sword in many introduced marine species (see Viard *et al.*, 2016), as they have properties that should facilitate selection, notably a large effective population size. However, the traits of interest (i.e. those involved in invasion success) are most likely to be determined by many genes and/or by different types of variation (e.g. chromosomal architecture, transposable elements) that can be difficult to detect or characterize with genome scans, particularly when based on a limited number of loci or short-read sequencing. For both marine and non-marine introduced species, genetic bottlenecks and repeated introductions may also complicate the identification of signatures of natural selection (Poh *et al.*, 2014). In addition, introduced populations are typically not at demographic equilibrium, violating the assumptions of many genome-scanning approaches and potentially resulting in an excess of false-positive selection candidates (Lotterhos and Whitlock, 2014).

While genetic approaches have their pitfalls, recent technological innovations can help to ameliorate these limitations, notably by vastly increasing genomic coverage (North *et al.*, 2021). The first studies relied on small handfuls of markers, often microsatellite markers (e.g. Riquet *et al.*, 2013). It was truly looking for a needle in a haystack! With reduced-representation sequencing (RRS) approaches, hundreds or thousands of loci can be screened for outliers and to detect subtle evolutionary patterns. As an example, a panel of 312 transcriptome-derived single-nucleotide polymorphisms (SNPs) was unable to detect gene flow between the native sea squirt *Ciona intestinalis* and its introduced congener *Ciona robusta* (Bouchemousse *et al.*, 2016), while later restriction-site-associated DNA sequencing (RAD-seq) involving 13,603 SNPs identified introgression of *C. robusta* genes in a small region of chromosome 5 in the native species (Le Moan *et al.*,2021). In addition to RRS, whole-genome sequencing (WGS) can also provide more direct evidence of genetic changes (North *et al.*, 2021) and identify genes putatively involved in the adaptive process, notably when they involve structural variants that may be key for rapid adaptation (Mérot *et al.*, 2020; Pokrovac and Pezer, 2022). For instance, the adaptive nature of the *C. robusta* introgression cited above was determined thanks to WGS and haplotype-based tests that showed that this genomic region is under recent positive selection and involves a copy-number variant (Fraïsse *et al.*, 2022).

Despite their benefits, RRS and WGS studies are still rare in marine NIS. Surprisingly, a search on Web of Science (29 December 2023) for papers using RRS or WGS in marine invasive species returned only 34 relevant papers, with the following search string: (TS=("Radseq" OR "genome" OR "genome-wide" OR "Rad-Seq" OR "Rad-sequencing" OR "GBS" OR "genotyping-by-sequencing" OR "restriction-site-associated DNA sequencing" OR "SNP" OR "SNPs" OR "single nucleotide polymorphism" OR "single nucleotide polymor-

phism")) AND TS=(Marine) AND TS=("invasive" or "nonindigenous" or "introduced" or "non-indigenous"). Most ($N = 20$) investigated neutral processes, such as introduction pathways or number of introduction events, in various taxa. The remaining 14 studies looked for loci under selection (Table 5.1). Examining these studies reveals three limitations that should inform future research efforts. First, only three studies specifically aimed to identify evidence of post-introduction adaptive processes. Second, half examined the introduced range only, which limits our ability to identify adaptive variation that evolved in the native range, and in turn to distinguish between pre-adaptation (with no adaptive changes after introduction) and rapid adaptation after introduction due to selection on standing genetic variation. Third, almost none of them included temporal replicates, which are crucial for understanding the processes at play, such as the rise of local adaptation versus global positive selection. We may expect that with increasing genomic resources, currently lacking for most marine species (Matheson and McGaughran, 2022), more studies will examine these introduced adaptive processes in marine systems.

5.3 Native Range: How Is (Adaptive) Diversity Generated in Marine Systems?

Standing genetic variation originating in a species' native range provides much of the substrate for rapid adaptation after introduction (Prentis *et al.*, 2008; Kołodziejczyk *et al.*, 2025). Understanding how this standing variation is generated and distributed, therefore, is a critical first step in exploring the processes behind post-introduction adaptation. Marine species are often characterized by high fecundity and large effective population sizes, allowing them to generate and maintain high levels of genetic diversity relative to smaller populations (Kimura and Crow, 1964; Barry *et al.*, 2022). In turn, selection tends to be more effective in larger populations than in small ones, which can be more dominated by genetic drift (Wright, 1931). Dispersal plays a key role in determining how

Table 5.1. Examples of studies looking for adaptive processes in marine non-indigenous species (NIS), using standard genome-scanning methods and/or genotype–environment association (GEA) studies with more than 500 single-nucleotide polymorphisms (SNPs) produced by genotyping-by-sequencing (GBS), restriction-site associated DNA sequencing (RAD-seq), mRNA sequencing (RNA-seq), or whole-genome sequencing (WGS).

NIS (Class)	Type of adaptation	Sampling range	No. of SNPs	No. of outliers	Functions of annotated genes	Variable (GEA)	Reference
Molgula manhattensis (Ascidiacea)	Pre- or post-introduction	Introduced	6,635 (RAD-seq)	109 SNPs	14 genes: related to salinity	Salinity	Chen *et al.* (2021)
Pyura praeputialis (Ascidiacea)	Pre- or post-introduction	Introduced and native	1,205 (GBS)	65 SNPs	NA	Temperature, chlorophyll, salinity	Hudson *et al.* (2021)
Mytilus galloprovincialis (Bivalvia)	Pre- or post-introduction	Introduced and native	56,856 (WGS)	105 SNPs	8 genes (including HSP and GTP binding)	Temperature	Han and Dong (2020)
Mytilus galloprovincialis (Bivalvia)	Post-introduction (hybridization with natives)	Introduced and native	8,902–11,906 (RNA-seq)	NA	3 genes: HSP90 and oxidative stress-related proteins	NA	Popovic *et al.* (2021)
Crassostrea gigas (Bivalvia)	Post-introduction (domestication)	Introduced and native	16,942 (RAD-seq)	36 SNPs (top outliers)	36 genes: calcium signaling, resilience to environmental stressors	NA	Sutherland *et al.* (2020)
Rhithropanopeus harrisii (Malacostraca)	Pre- or post-introduction	Introduced and native	1,013 (RAD-seq)	34 SNPs (temporal sampling)	5 genes (temporal): growth and stress responses, collagen biosynthesis	NA	Forsström *et al.* (2017)
Carcinus maenas (Malacostraca)	Pre- or post-introduction	Introduced	9,137 (RAD-seq)	41 SNPs	NA	Temperature	Jeffery *et al.* (2018)
Carcinus maenas (Malacostraca)	Pre-introduction	Introduced and native	10,809 (RNA-seq)	1,563 SNPs	NA	NA	Tepolt and Palumbi (2015)
Carcinus maenas (Malacostraca)	Pre- and post-introduction	Introduced and native	10,790 (RNA-seq)	104 SNPs (72 in one cluster)	NA	Temperature	Tepolt and Palumbi (2020)
Carcinus maenas (Malacostraca)	Post-introduction	Introduced	9,376 (RNA-seq)	1 cluster (168 SNPs)	Thermal physiology	Temperature	Tepolt *et al.* (2022)
Undaria pinnatifida (Phaeophyceae)	Post-introduction (habitats and farming)	Introduced	10,615 (RAD-seq)	59 SNPs (1 in farms)	NA	NA	Guzinski *et al.* (2018)
Siganus rivulatus (Teleostei)	Pre-introduction	Introduced	27,836 (RAD-seq)	15 SNPs	6 genes, including osmoregulation	NA	Azzurro *et al.* (2022)
Stephanolepis diaspros (Teleostei)	Pre-introduction	Introduced	12,344 (RAD-seq)	22 SNPs	5 genes, including osmoregulation	NA	Azzurro *et al.* (2022)
Pterois volitans (Teleostei)	Pre- or post-introduction	Introduced	12,759 (RAD-seq)	24 SNPs	7 genes, including membrane proteins	Distance	Bors *et al.* (2019)

HSP, heat-shock protein; NA, not applicable.

adaptive diversity is maintained and partitioned across space (Palumbi and Pinsky, 2014). Marine species span an extremely broad range of dispersal distances, in part because the ocean represents a complex three-dimensional physical environment that facilitates often highly effective oceanographically mediated transport of propagules (Álvarez-Noriega *et al.*, 2020). Many benthic species have a two-part life cycle, with the larval early life stages being substantially more vagile than adults. This is especially pronounced for species like barnacles and oysters, which may disperse for weeks to months in the zooplankton before settling down to a sessile adulthood (Palumbi and Pinsky, 2014). Other marine species have a short or non-existent larval phase but are able to disperse highly effectively as adults through mechanisms including drifting, rafting, and hitchhiking (Winston, 2012). Both strategies are well represented in the most prominent introduced marine species.

Dispersal interacts with environmental gradients to shape the type of adaptive diversity maintained in populations. When dispersal is limited relative to environmental change, an individual offspring is very likely to end up in an environment similar to that of its parents (Sanford and Kelly, 2011). In such systems, high population structure is likely and local adaptation in the classical sense is favored, with distinct genetic adaptations in each population well suited to their specific environmental constraints (Kawecki and Ebert, 2004). By contrast, when offspring disperse further than the scale of environmental change, an individual's adult environment is decoupled from its environment of origin. In such cases, populations are typically less structured and the maintenance of balanced polymorphism is favored, resulting in less overall differentiation among sites but more adaptive "flexibility" in the metapopulation (Sanford and Kelly, 2011). This distinction has implications for the type of adaptive diversity that evolves in a system, with lower dispersal and higher genetic structure often associated with polygenic adaptation of many unlinked genes contributing to a phenotype that is optimized for a particular environment (Shi *et al.*, 2023). In the case of open systems, where individuals disperse across a broad environmental mosaic, fewer alleles of large effect may be favored (Tigano and Friesen, 2016). Successful genetic variants in these high-dispersal systems, which may disproportionately represent inversion polymorphisms and other regions of reduced recombination, can have high selection coefficients, which permit effective spatial balancing selection (Tigano and Friesen, 2016; Gilbert and Whitlock, 2017). Empirical evidence in species where the scale of dispersal exceeds the scale of environmental change suggests that inversions and "simple" unlinked SNP variants both play important adaptive roles, with the same inversions largely under parallel selection across environments while specific SNPs vary from region to region (Campbell *et al.*, 2021; Koch *et al.*, 2022).

5.4 Invasion Process: How Does Diversity Change During and After Invasion?

The genetic variation that evolved in a species' native range passes through a series of filters during the invasion process, further influencing its adaptive "options." The specific source population will determine the pool of native range diversity that is being drawn from, while population structure and gene flow in the native range will dictate how much distinct sources differ from each other. Different sources may have a huge influence on the starting pools of adaptive diversity for species with low gene flow and high local adaptation, while they may play only a minor role in species with high gene flow and low population structure (Fig. 5.1). As pointed out in the preceding section, in high-dispersal systems adaptive diversity may be concentrated in a few loci of large effect, frequently regions of reduced recombination such as chromosomal inversions (Tigano and Friesen, 2016). If this specific variation is introduced, it may help invasive populations rapidly adapt to their new environments, even if highly bottlenecked at neutral loci (Tepolt *et al.*, 2022). However, the same reduced recombination that allows multiple co-adapted polymorphisms to be inherited together may also hinder the development of novel variation, limiting or eliminating this adaptive advantage in environments that differ substantially from those in which this variation evolved (Roesti *et al.*, 2022).

After introduction, a series of filters in an invasive population's first few generations

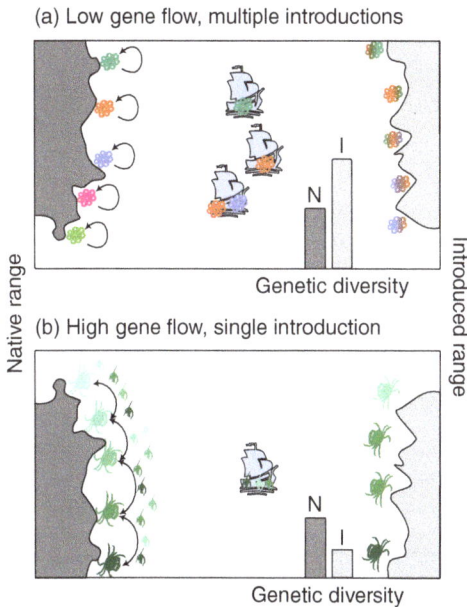

Fig. 5.1. Two modes of adaptation after invasion. (a) In species with low unassisted gene flow and high local adaptation in the native range, such as many introduced tunicates, multiple introductions from genetically diverse source populations may result in admixture in the introduced range, which creates novel allelic combinations on which selection can act. In many cases, introduced populations can have higher overall genetic diversity than native populations. (b) In species with high gene flow and low local adaptation in the native range, such as European green crabs, a single introduction may provide the substrate for adaptation in the introduced range despite loss of genetic diversity. Bars indicate relative mean genetic diversity in a single population in the native (N) and introduced (I) ranges. Silhouette images from PhyloPic contributors B. Duygu Özpolat, Mario Quevedo and Stuart Humphries.

determines how much of the genetic variation in the founding population is retained in the invasive range. Diversity can be lost at the very beginning of the process if transported individuals represent only a subset of native range genetic diversity, it may further be lost during transit if not all transported individuals survive the journey, and finally additional diversity is frequently lost in early establishment, as not all individuals who are transported to a new range will successfully reproduce there (Wilson *et al.*, 2009). In some cases, these filters result in

significant genetic bottlenecks in invasive populations. Classically, decreased genetic diversity and increased inbreeding has deleterious impacts on population persistence (Frankham, 2005). However, a number of marine invasive species have been highly successful despite substantial losses of diversity, leading to the concept of the "genetic paradox of invasions" (Tepolt and Palumbi, 2015; Le Cam *et al.*, 2020; Jaspers *et al.*, 2021).

Finally, some vectors provide repeated opportunities for introduction, often from multiple sources (Wilson *et al.*, 2009). Genetic diversity may be very low in a population derived from a single, small founder population, but repeated introductions may result in diversity in introduced populations that is as high or higher than in any given native range population (Fig. 5.1) (Roman and Darling, 2007, Rius *et al.*, 2015). Multiple introductions from diverse sources may provide introduced populations with a rich substrate of adaptive novelty, and the nature and genetic consequences of this type of admixture are a major area of interest (see below).

5.5 Mixture and Admixture: A Basis for Adaptive Evolution in the Introduced Range

In many successfully established marine NIS, large numbers of individuals and repeated introduction events had been documented (Roman and Darling, 2007; Rius *et al.*, 2015; Viard *et al.*, 2016). This high propagule pressure increases the probability of establishment success by, for instance, decreasing negative Allee effects (i.e. reduced fitness at low density), buffering species against environmental stochasticity and alleviating inbreeding depression. Although there are exceptions (e.g. the invasive seaweed *Sargassum muticum*; Le Cam *et al.*, 2020), higher propagule pressure usually also translates into more genetic diversity on which selection can act (Fig. 5.1a). Multiple introductions from diverse areas of the native range allows an introduced species to draw on a pool of adaptive diversity that evolved in a wider range of environments than a single source, and may provide an advantage in novel environments. In addition, if the native range is highly structured, multiple introductions from diverse sources may result in the local coexistence of diverse genetic lineages, particularly for those taxa that form complexes of cryptic lineages or subspecies (Harper *et al.*, 2022). In sexually reproducing species, recombination can then create novel diversity for selection to draw on in the invasive range, as illustrated by the admixture between the two invasive lineages of the European green crab, *Carcinus maenas*, in their contact zone in the northwest Atlantic (Jeffery *et al.*, 2017).

Besides admixture between genetically differentiated lineages within a given species, hybridization between an introduced species and a native congener can also occur, leading to various outcomes (Fig. 5.2), including the rise of novel lineages adapted to this new habitat (Viard *et al.*, 2020).

Hybridization between distinct accepted species has long been recognized as a major determinant of invasiveness in plants (Schierenbeck and Ellstrand, 2009, and references therein). More recently, it been increasingly documented in marine environments, including in animals (e.g. coral reef fish: Tea *et al.*, 2020),

and anthropogenic hybridization (i.e. "the breakdown of reproductive isolation between two species as a result of human action", as coined by McFarlane and Pemberton, 2019) is likely to be more common than yet documented. Introgression can give rise to new taxa in the case of extreme and stable admixture as documented by SNP analyses of mussel populations established in (and confined to) ports of the North Atlantic whose genomes were shown to be composed of 60–70% of the introduced Mediterranean species *Mytilus galloprovincialis* and 30–40% of the Atlantic species *Mytilus edulis*, following their genome-wide admixture (Simon *et al.*, 2020; F. Touchard *et al.*, unpublished data). Such stable genome-wide admixture, without local fixation of parental alleles, is in agreement with multigenic models involving epistatic effects between many loci (Simon *et al.*, 2018) and the idea that traits involved in adaptation are likely to be polygenic and involve many loci of small effect (see above).

While anthropogenic hybridization can thus favor the adaptation of the introduced

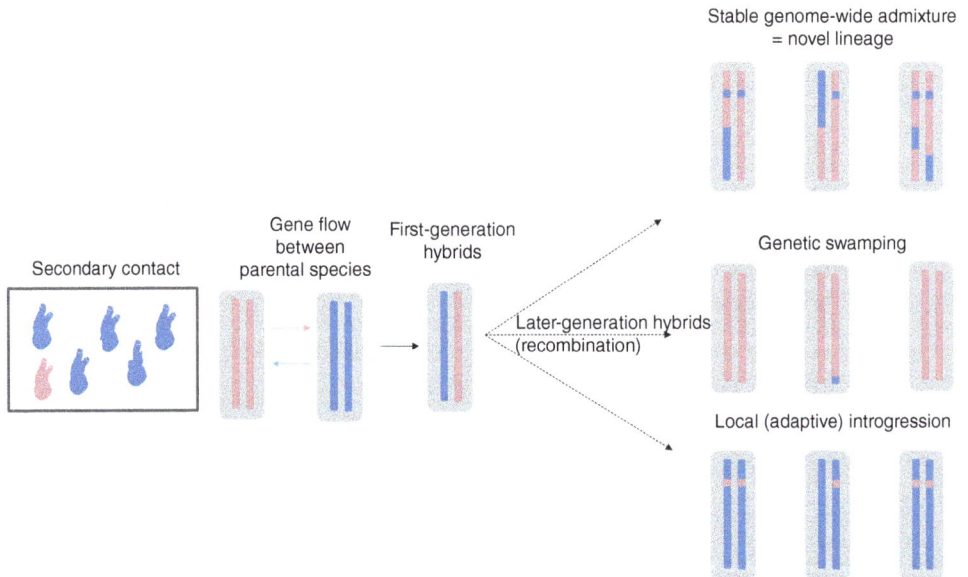

Fig. 5.2. Adaptation following hybridization. Secondary contact between an introduced non-indigenous species (pink) and a local congener (blue) can be followed by gene flow, if the species are not fully reproductively isolated. Recombination then leads to: (i) admixture of both parental genomes (top right), which can give rise to a novel locally adapted lineage; (ii) extinction of the local species following genetic swamping (middle right); or (iii) introgression of adaptive genes, here from the introduced to the native species (bottom right).

species to novel environments, this process can work both ways. In some cases, hybridization of introduced species can provide greater adaptive diversity to the natives with which they introgress, as in the case of the tunicate *C. intestinalis* in which adaptive introgression from its alien congener *C. robusta* has been documented (Touchard *et al.*, 2024, and references therein). Introgression of beneficial alleles from the introduced species may even help native species to outcompete the introduced ones. Introgression can also allow NIS genes to remain in the introduced range even if the NIS itself declines or disappears. For instance, genome-wide studies of the Gulf killifish, *Fundulus grandis*, showed evidence of adaptive introgression from its congener *Fundulus heteroclitus*, which is native to the Atlantic but not observed in the Gulf of Mexico, suggesting that it had once been introduced and then disappeared (Oziolor *et al.*, 2019). Besides altering evolutionary trajectories of native and introduced congeners, gene introgression could thus also be a lasting legacy of invading species.

5.6 The Nature of Adaptation in Marine Invasion

Given all the complexities and possibilities outlined above, what do initial genomics studies on introduced marine species (and systems with similar dynamics) say about the nature of rapid adaptation? First, beneficial *de novo* mutations are unlikely to arise or come to fixation frequently enough to play a substantial role in rapid adaptation, especially as newly founded populations typically start off small and geographically limited and may remain so for some generations (Sakai *et al.*, 2001). While there is some evidence for adaptation from *de novo* mutations in natural systems facing strong selection, including in invasive *Arabidopsis thaliana* populations (Exposito-Alonso *et al.*, 2018), this is unlikely to be a major avenue for many NIS. Selection is thus most likely to act on standing genetic variation already present in introduced populations, which may be enhanced by the admixture that is increasingly commonly observed in marine invasions (see above), or by pre-introduction introgression (Popovic *et al.*, 2021). This standing variation can take many forms: large suites

of minor point mutations; mutations at key metabolic genes; insertions and deletions; tandem repeats; transposable elements; and larger-scale structural variation including chromosomal inversions and gene duplications. The nature of the standing genetic variation available to selection in an introduced population reflects what has evolved in the native range, shaped by population size and dispersal strategy, and filtered through the invasion process.

A large group of successful marine invaders are characterized by high gene flow across native ranges that span broad environmental gradients (Gaither *et al.*, 2013), potentially predisposing them to evolve standing variation of large-effect variants that can provide the substrate for rapid post-introduction adaptation (Fig. 5.1b) (Sanford and Kelly, 2011; Tigano and Friesen, 2016; Gilbert and Whitlock, 2017; Mérot *et al.*, 2020). Chromosomal inversions, an important example of this type of variation, have been implicated in adaptation in an increasing number of marine systems, including crabs, tunicates, fish and mollusks (Barth *et al.*, 2017; Satou *et al.*, 2021; Koch *et al.*, 2022; Tepolt *et al.*, 2022). For example, in the European green crab, *C. maenas*, a putative inversion polymorphism that evolved in the native range is strongly associated with cold tolerance across Europe and North America (Tepolt and Palumbi, 2020). This same region has reformed a latitudinal cline in the rapidly expanding 35-year-old introduced northeast Pacific population and is the only genetic marker identified as under strong selection in that region (Tepolt *et al.*, 2022). However, large-effect variants may be the low-hanging fruit of adaptive diversity, easy to identify especially in the absence of highly complete genome coverage (e.g. Koch *et al.*, 2022). Better genomic resources for marine invaders will be critical in identifying the relative roles of different types of variation in the adaptation process, especially given recent work demonstrating equal importance for small, independent variants in shaping adaptation (Campbell *et al.*, 2021; Koch *et al.*, 2022).

Alternatively, anthropogenic dispersal (e.g. biofouling on ship hulls, aquaculture) can create connectivity shortcuts and facilitate long-distance dispersal events in species with limited unassisted dispersal ability (Hudson *et al.*, 2016; Touchard *et al.*, 2023). These include many important invasive biofouling species, such as algae and tunicates, and can act to mix genetic variation that evolved in a range of populations and, in extreme cases, homogenize genetic signatures across ports well connected by biofouling vectors (Lacoursière-Roussel *et al.*, 2012; Guzinski *et al.*, 2018). In these cases, founding populations may bear variation that evolved in a range of environments and has subsequently been admixed by anthropogenic transport, giving selection a broader swath of adaptive diversity on which to act in new environments (Fig. 5.1a). Dispersal among regions in the invasive range will depend on vector activity, with extensive exchange (e.g. of small vessels, aquaculture gear) acting not only to spread species range initially but also potentially to maintain genetic connectivity among regions in the absence of larval dispersal (Ashton *et al.*, 2022).

Estuarine specialists may represent a special, intermediate case, where some species are highly successful invaders despite limited larval dispersal in the native range, due either to minimal time in the plankton or to physical or behavioral barriers to larval dispersal (Bilton *et al.*, 2002). In these cases, native range populations are often highly structured, suggesting a substantial role for local adaptation (Kawecki and Ebert, 2004; Sanford and Kelly, 2011). Such species, including *Eurytemora* spp. copepods and white-fingered mud crabs, have been found to harbor high standing diversity within estuaries, often for salinity tolerance (Stern and Lee, 2020; Blakeslee *et al.*, 2021). This variation allows rapid selection to occur along estuarine salinity gradients, facilitating invasion into new freshwater and brackish environments. In the case of *Eurytemora affinis*, repeated invasions of freshwater from brackish sources occurred in parallel across multiple lineages, drawing from many of the same candidate SNPs for salinity tolerance that have been maintained across salinity gradients by balancing selection (Stern and Lee, 2020).

5.7   Conclusions

Evolution in the native range shapes the adaptive diversity available to invaders and is strongly influenced by dispersal distance relative to selective gradients. High gene flow may promote

selection at a few large-effect alleles, which may be shared across much of the native range and provide a rich substrate for rapid adaptation after introduction. However, even in species with limited natural dispersal, introduction history – and notably repeated introduction events – can create opportunities for introduced populations to be genetically diversified. Selection can thus act on an enriched pool of standing genetic variation as well as novel genomic variants produced by recombination (admixture/hybridization).

The specific genomic underpinnings of adaptation during invasion are still not well understood, although we are beginning to characterize them through recent advances in population genomics based on RRS of the genome (e.g. RAD-seq, genotyping-by-sequencing, RNA sequencing) and WGS. Such genome-wide approaches have already provided good evidence that key adaptive traits may rely both on many loci of small effects and conversely on few large-effect structural variants. More sophisticated and generalizable conclusions will be greatly facilitated by high-quality reference genomes for species representing a wider variety of taxonomic and life-history variation, something that is so far very limited for marine NIS (Matheson and McGaughran, 2022). Such reference genomes will catalyze the next generation of breakthroughs, through approaches including haplotype-based analyses that allow the reconstruction of detailed ancestry tracts, and will help provide a better understanding of the relative roles of different types of genomic variation in rapid adaptation (e.g. Koch *et al.*, 2022; Stuart *et al.*, 2023). As these resources expand, we will be able to understand the nature and dynamics of adaptive diversity in a much wider range of NIS, particularly across the world's oceans.

Acknowledgments

C.K.T.'s work on this chapter was supported in part by the US National Science Foundation (NSF Award #1850996) and WHOI's Ocean Vision Program. F.V.'s work benefited from funding through the French National Research Agency (ANR) under the "Investissements d'Avenir" program with the reference ANR-16-IDEX-0006 (i-siteMUSE) and the DOCKEVOL project with the reference ANR-23-CE02-0020. Both authors are members of the Working Groups on Ballast and Other Ship Vectors and Marine Bioinvasions of the International Council for the Exploration of the Sea (ICES), and thank their fellow members for discussions contributing to this work.

References

Aires, T., Stuij, T.M., Muyzer, G., Serrao, E.A. and Engelen, A.H. (2021) Characterization and comparison of bacterial communities of an invasive and two native Caribbean seagrass species sheds light on the possible influence of the microbiome on invasive mechanisms. *Frontiers in Microbiology* 12: 653998. DOI: 10.3389/fmicb.2021.653998

Álvarez-Noriega, M., Burgess, S.C., Byers, J.E., Pringle, J.M., Wares, J.P. and Marshall, D.J. (2020) Global biogeography of marine dispersal potential. *Nature Ecology and Evolution* 4, 1196–1203. DOI: 10.1038/s41559-020-1238-y

Ashton, G.V, Zabin, C.J., Davidson, I.C. and Ruiz, G.M. (2022) Recreational boats routinely transfer organisms and promote marine bioinvasions. *Biological Invasions* 24, 1083–1096. DOI: 10.1007/s10530-021-02699-x

Azzurro, E., Nourigat, M., Cohn, F., Ben Souissi, J. and Bernardi, G. (2022) Right out of the gate: the genomics of Lessepsian invaders in the vicinity of the Suez Canal. *Biological Invasions* 24, 1117–1130. DOI: 10.1007/s10530-021-02704-3.

Bailey, S.A., Brown, L., Campbell, M.L., Canning-Clode, J., Carlton, J.T. *et al.* (2020) Trends in the detection of aquatic non-indigenous species across global marine, estuarine and freshwater ecosystems: a 50-year perspective. *Diversity and Distributions* 26, 1780–1797. DOI: 10.1111/ddi.13167

Barry, P., Broquet, T. and Gagnaire, P.-A. (2022) Age-specific survivorship and fecundity shape genetic diversity in marine fishes. *Evolution Letters* 6, 46–62. DOI: 10.1002/evl3.265

Barth, J.M.I., Berg, P.R., Jonsson, P.R., Bonanomi, S., Corell, H. *et al.* (2017) Genome architecture enables local adaptation of Atlantic cod despite high connectivity. *Molecular Ecology* 26, 4452–4466. DOI: 10.1111/mec.14207

Beaumont, M.A. and Nichols, R.A. (1996) Evaluating loci for use in the genetic analysis of population structure. *Proceedings of the Royal Society of London B: Biological Sciences* 263, 1619–1626. DOI: 10.1098/rspb.1996.0237

Bierne, N., Roze, D. and Welch, J.J. (2013) Pervasive selection or is it...? Why are F_{ST} outliers sometimes so frequent? *Molecular Ecology* 22, 2061–2064. DOI: 10.1111/mec.12241

Bilton, D.T., Paula, J. and Bishop, J.D.D. (2002) Dispersal, genetic differentiation and speciation in estuarine organisms. *Estuarine, Coastal and Shelf Science* 55, 937–952. DOI: 10.1006/ecss.2002.1037

Blakeslee, A.M.H., Pochtar, D.L., Fowler, A.E., Moore, C.S., Lee, T.S. *et al.* (2021) Invasion of the body snatchers: the role of parasite introduction in host distribution and response to salinity in invaded estuaries. *Proceedings of the Royal Society of London B: Biological Sciences* 288: 20210703. DOI: 10.1098/rspb.2021.0703

Bors, E.K., Herrera, S., Morris, J.A. and Shank, T.M. (2019) Population genomics of rapidly invading lionfish in the Caribbean reveals signals of range expansion in the absence of spatial population structure. *Ecology and Evolution* 9, 3306–3320. DOI: 10.1002/ece3.4952.

Bouchemousse, S., Liautard-Haag, C., Bierne, N. and Viard, F. (2016) Distinguishing contemporary hybridization from past introgression with postgenomic ancestry-informative SNPs in strongly differentiated *Ciona* species. *Molecular Ecology* 25, 5527–5542. DOI: 10.1111/mec.13854.

Briski, E., Chan, F.T., Darling, J.A., Lauringson, V., MacIsaac, H.J. *et al.* (2018) Beyond propagule pressure: importance of selection during the transport stage of biological invasions. *Frontiers in Ecology and the Environment* 16, 345–353. DOI: 10.1002/fee.1820

Campbell, M.A., Anderson, E.C., Garza, J.C. and Pearse, D.E. (2021) Polygenic basis and the role of genome duplication in adaptation to similar selective environments. *Journal of Heredity* 112, 614–625. DOI: 10.1093/jhered/esab049

Capinha, C., Essl, F., Seebens, H., Moser, D. and Pereira, H.M. (2015) The dispersal of alien species redefines biogeography in the Anthropocene. *Science* 348, 1248–1251. DOI: 10.1126/science.aaa8913

Carneiro, V.C. and Lyko, F. (2020) Rapid epigenetic adaptation in animals and its role in invasiveness. *Integrative and Comparative Biology* 60, 267–274. DOI: 10.1093/icb/icaa023

Chen, Y., Gao, Y., Huang, X., Li, S. and Zhan, A. (2021) Local environment-driven adaptive evolution in a marine invasive ascidian (*Molgula manhattensis*). *Ecology and Evolution* 11, 4252–4266. DOI: 10.1002/ece3.7322.

Colautti, R.I. and Lau, J.A. (2015) Contemporary evolution during invasion: evidence for differentiation, natural selection, and local adaptation. *Molecular Ecology* 24, 1999–2017. DOI: 10.1111/mec.13162

Exposito-Alonso, M., Becker, C., Schuenemann, V.J., Reiter, E., Setzer, C. *et al.* (2018) The rate and potential relevance of new mutations in a colonizing plant lineage. *PLOS Genetics* 14: e1007155. DOI: 10.1371/journal.pgen.1007155

Facon, B., Genton, B.J., Shykoff, J., Jarne, P., Estoup, A. and David, P. (2006) A general eco-evolutionary framework for understanding bioinvasions. *Trends in Ecology and Evolution* 21, 130–135. DOI: 10.1016/j.tree.2005.10.012

Forsström, T., Ahmad, F. and Vasemägi, A. (2017) Invasion genomics: genotyping-by-sequencing approach reveals regional genetic structure and signatures of temporal selection in an introduced mud crab. *Marine Biology* 164–186. DOI: 10.1007/s00227-017-3210-1.

Fraïsse, C., Le Moan, A., Roux, C., Dubois, G., Daguin-Thiébaut, C. *et al.* (2022) Introgression between highly divergent sea squirt genomes: an adaptive breakthrough? *Peer Community Journal* 2: e54. DOI: 10.24072/pcjournal.172

Frankham, R. (2005) Genetics and extinction. *Biological Conservation* 126, 131–140. DOI: 10.1016/j.biocon.2005.05.002

Gaither, M.R., Bowen, B.W. and Toonen, R.J. (2013) Population structure in the native range predicts the spread of introduced marine species. *Proceedings of the Royal Society of London B: Biological Sciences* 280: 20130409. DOI: 10.1098/rspb.2013.0409

Gao, Y., Chen, Y., Li, S., Huang, X., Hu, J. *et al.* (2022) Complementary genomic and epigenomic adaptation to environmental heterogeneity. *Molecular Ecology* 31, 3598–3612. DOI: 10.1111/mec.16500.

Gilbert, K.J. and Whitlock, M.C. (2017) The genetics of adaptation to discrete heterogeneous environments: frequent mutation or large-effect alleles can allow range expansion. *Journal of Evolutionary Biology* 30, 591–602. DOI: 10.1111/jeb.13029

Gribben, P.E. and Byers, J.E. (2020) Comparative biogeography of marine invaders across their native and introduced ranges. In: Hawkins, S.J, Allcock, A.L., Bates, A.E., Evans, A.J., Firth, L.B. *et al.* (eds) *Oceanography and Marine Biology – An Annual Review, Vol.* 58. CRC Press, Boca Raton, Florida, pp. 395–440.

Guzinski, J., Ballenghien, M., Daguin-Thiébaut, C., Lévêque, L. and Viard, F. (2018) Population genomics of the introduced and cultivated Pacific kelp *Undaria pinnatifida*: marinas – not farms – drive regional connectivity and establishment in natural rocky reefs. *Evolutionary Applications* 11, 1582–1597. DOI: 10.1111/eva.12647

Han, G.-D. and Dong, Y.-W. (2020) Rapid climate-driven evolution of the invasive species *Mytilus galloprovincialis* over the past century. *Anthropocene Coasts* 3, 14–29. DOI: 10.1139/anc-2019-0012.

Harper, K.E., Scheinberg, L.A., Boyer, K.E. and Sotka, E.E. (2022) Global distribution of cryptic native, introduced and hybrid lineages in the widespread estuarine amphipod *Ampithoe valida*. *Conservation Genetics* 23, 791–806. DOI: 10.1007/s10592-022-01452-8.

Hoban, S., Kelley, J.L., Lotterhos, K.E., Antolin, M.F., Bradburd, G. *et al.* (2016) Finding the genomic basis of local adaptation: pitfalls, practical solutions, and future directions. *American Naturalist* 188, 379–397. DOI: 10.1086/688018

Hodgins, K.A., Battlay, P. and Bock, D.G. (2025) The genomic secrets of invasive plants. *New Phytologist* 245, 1846–1863. DOI: 10.1111/nph.20368

Hudson, J., Viard, F., Roby, C. and Rius, M. (2016) Anthropogenic transport of species across native ranges: unpredictable genetic and evolutionary consequences. *Biology Letters* 12: 20160620. DOI: 10.1098/rsbl.2016.0620.

Hudson, J., Castilla, J.C., Teske, P.R., Beheregaray, L.B., Haigh, I.D. *et al.* (2021) Genomics-informed models reveal extensive stretches of coastline under threat by an ecologically dominant invasive species. *Proceedings of the National Academy of Sciences USA* 118: e2022169118. DOI: 10.1073/pnas.2022169118.

Hufbauer, R.A., Facon, B., Ravigné, V., Turgeon, J., Foucaud, J. *et al.* (2012) Anthropogenically induced adaptation to invade (AIAI): contemporary adaptation to human-altered habitats within the native range can promote invasions. *Evolutionary Applications* 5, 89–101. DOI: 10.1111/j.1752-4571.2011.00211.x

Jaspers, C., Ehrlich, M., Pujolar, J.M., Kunzel, S., Bayer, T. *et al.* (2021) Invasion genomics uncover contrasting scenarios of genetic diversity in a widespread marine invader. *Proceedings of the National Academy of Sciences USA* 118: e2116211118. DOI: 10.1073/pnas.2116211118

Jeffery, N.W., DiBacco, C., Wringe, B.F., Stanley, R.R.E, Hamilton, L.C. *et al.* (2017) Genomic evidence of hybridization between two independent invasions of European green crab (*Carcinus maenas*) in the Northwest Atlantic. *Heredity* 119, 154–165. DOI: 10.1038/hdy.2017.22.

Jeffery, N.W., Bradbury, I.R., Stanley, R.R.E, Wringe, B.F., Van Wyngaarden, M., Lowen, J.B., McKenzie, C.H., Matheson, K. *et al.* (2018) Genomewide evidence of environmentally mediated secondary contact of European green crab (*Carcinus maenas*) lineages in eastern North America. *Evolutionary Applications* 11, 869–882. DOI: 10.1111/eva.12601.

Johnston, E.L., Dafforn, K.A., Clark, G.F., Rius, M. and Floerl, O. (2017) How anthropogenic activities affect the establishment and spread of non-indigenous species post-arrival. In: Hawkins, S.J., Evans, A.J., Dale, A.C., Firth, L.B., Hughes, D.J. and Smith, I.P. (eds) *Oceanography and Marine Biology: An Annual Review, Vol.* 55. CRC Press, Boca Raton, Florida, pp. 389–419.

Kawecki, T.J. and Ebert, D. (2004) Conceptual issues in local adaptation. *Ecology Letters* 7, 1225–1241. DOI: 10.1111/j.1461-0248.2004.00684.x

Kimura, M. and Crow, J.F. (1964) The number of alleles that can be maintained in a finite population. *Genetics* 49, 725–738. DOI: 10.1093/genetics/49.4.725

Koch, E.L., Ravinet, M., Westram, A.M., Johannesson, K. and Butlin, R.K. (2022) Genetic architecture of repeated phenotypic divergence in *Littorina saxatilis* ecotype evolution. *Evolution* 76, 2332–2346. DOI: 10.1111/evo.14602

Kołodziejczyk, J., Fijarczyk, A., Porth, I., Robakowski, P., Vella, N. *et al.* (2025) Genomic investigations of successful invasions: the picture emerging from recent studies. *Biological Reviews of the Cambridge Philosophical Society* 100, 1396–1418. DOI: 10.1111/brv.70005

Lacoursière-Roussel, A., Bock, D.G., Cristescu, M.E., Guichard, F., Girard, P. *et al.* (2012) Disentangling invasion processes in a dynamic shipping–boating network. *Molecular Ecology* 21, 4227–4241. DOI: 10.1111/j.1365-294X.2012.05702.x

Le Cam, S., Daguin-Thiébaut, C., Bouchemousse, S., Engelen, A.H., Mieszkowska, N. and Viard, F. (2020) A genome-wide investigation of the worldwide invader *Sargassum muticum* shows high success albeit (almost) no genetic diversity. *Evolutionary Applications* 13, 500–514. DOI: 10.1111/eva.12837

Le Moan, A., Roby, C., Fraïsse, C., Daguin-Thiébaut, C., Bierne, N. and Viard, F. (2021) An introgression breakthrough left by an anthropogenic contact between two ascidians. *Molecular Ecology* 30, 6718–6732. DOI: 10.1111/mec.16189

Lotterhos, K.E. and Whitlock, M.C. (2014) Evaluation of demographic history and neutral parameterization on the performance of F_{ST} outlier tests. *Molecular Ecology* 23, 2178–2192. DOI: 10.1111/mec.12725

Lotterhos, K.E. and Whitlock, M.C. (2015) The relative power of genome scans to detect local adaptation depends on sampling design and statistical method. *Molecular Ecology* 24, 1031–1046. DOI: 10.1111/mec.13100

Matheson, P. and McGaughran, A. (2022) Genomic data is missing for many highly invasive species, restricting our preparedness for escalating incursion rates. *Scientific Reports* 12: 13987. DOI: 10.1038/s41598-022-17937-y

May, R.M. (1994) Biological diversity: differences between land and sea. *Philosophical Transactions of the Royal Society B: Biological Sciences* 343, 105–111. DOI: 10.1098/rstb.1994.0014

McFarlane, S.E. and Pemberton, J.M. (2019) Detecting the true extent of introgression during anthropogenic hybridization. *Trends in Ecology and Evolution* 34, 315–326. DOI: 10.1016/j.tree.2018.12.013.

Mérot, C., Oomen, R.A., Tigano, A. and Wellenreuther, M. (2020) A roadmap for understanding the evolutionary significance of structural genomic variation. *Trends in Ecology and Evolution* 35, 561–572. DOI: 10.1016/j.tree.2020.03.002

North, H.L., McGaughran, A. and Jiggins, C.D. (2021) Insights into invasive species from whole-genome resequencing. *Molecular Ecology* 30, 6289–6308. DOI: 10.1111/mec.15999

Oziolor, E.M., Reid, N.M., Yair, S., Lee, K.M., Guberman VerPloeg, S. *et al.* (2019) Adaptive introgression enables evolutionary rescue from extreme environmental pollution. *Science* 364, 455–457. DOI: 10.1126/science.aav4155

Palumbi, S.R. and Pinsky, M. (2014) Marine dispersal, ecology, and conservation. In: Bertness, M., Bruno, J., Silliman, B. and Stachowicz, J. (eds) *Marine Community Ecology and Conservation.* Sinauer Associates, Sunderland, Massachusettes, pp. 57–84.

Poh, Y.P., Domingues, V.S., Hoekstra, H.E. and Jensen, J.D. (2014) On the prospect of identifying adaptive loci in recently bottlenecked populations. *PLOS ONE* 9: e110579. DOI: 10.1371/journal.pone.0110579

Pokrovac, I. and Pezer, Z. (2022) Recent advances and current challenges in population genomics of structural variation in animals and plants. *Frontiers in Genetics* 13: 1060898. DOI: 10.3389/fgene.2022.1060898

Popovic, I., Bierne, N., Gaiti, F., Tanurdzic, M. and Riginos, C. (2021) Pre-introduction introgression contributes to parallel differentiation and contrasting hybridization outcomes between invasive and native marine mussels. *Journal of Evolutionary Biology* 34, 175–192. DOI: 10.1111/jeb.13746.

Prentis, P.J., Wilson, J.R.U., Dormontt, E.E., Richardson, D.M. and Lowe, A.J. (2008) Adaptive evolution in invasive species. *Trends in Plant Science* 13, 288–294. DOI: 10.1016/j.tplants.2008.03.004

Riquet, F., Daguin-Thiébaut, C., Ballenghien, M., Bierne, N. and Viard, F. (2013) Contrasting patterns of genome-wide polymorphism in the native and invasive range of the marine mollusc *Crepidula fornicata*. *Molecular Ecology* 22, 1003–1018. DOI: 10.1111/mec.12161.

Rius, M., Turon, X., Bernard, G., Volckaert, F. and Viard, F. (2015) Marine invasion genetics: from spatial and temporal patterns to evolutionary outcomes. *Biological Invasions* 17, 869–885. DOI: s10530-014-0792-0

Roesti, M., Gilbert, K.J. and Samuk, K. (2022) Chromosomal inversions can limit adaptation to new environments. *Molecular Ecology* 31, 4435–4439. DOI: 10.1111/mec.16609

Roman, J. and Darling, J.A. (2007) Paradox lost: genetic diversity and the success of aquatic invasions. *Trends in Ecology and Evolution* 22, 454–464. DOI: 10.1016/j.tree.2007.07.002

Sakai, A.K., Allendorf, F.W., Holt, J.S., Lodge, D.M., Molofsky, J. *et al.* (2001) The population biology of invasive species. *Annual Review of Ecology and Systematics* 32, 305–332. DOI: 10.1146/annurev.ecolsys.32.081501.114037

Sanford, E. and Kelly, M.W. (2011) Local adaptation in marine invertebrates. *Annual Review of Marine Science* 3, 509–535. DOI: 10.1146/annurev-marine-120709-142756

Sardain, A., Sardain, E. and Leung, B. (2019) Global forecasts of shipping traffic and biological invasions to 2050. *Nature Sustainability* 2, 274–282. DOI: 10.1038/s41893-019-0245-y

Satou, Y., Sato, A., Yasuo, H., Mihirogi, Y., Bishop, J. *et al.* (2021) Chromosomal inversion polymorphisms in two sympatric ascidian lineages. *Genome Biology and Evolution* 13: evab068. DOI: 10.1093/gbe/evab068

Schierenbeck, K. and Ellstrand, N. (2009) Hybridization and the evolution of invasiveness in plants and other organisms. *Biological Invasions* 11, 1093–1105. DOI:10.1007/s10530-008-9388-x

Shi, Y., Bouska, K.L., McKinney, G.J., Dokai, W., Bartels, A. *et al.* (2023) Gene flow influences the genomic architecture of local adaptation in six riverine fish species. *Molecular Ecology* 32, 1549–1566. DOI: 10.1111/mec.16317

Simon, A., Bierne, N. and Welch, J.J. (2018) Coadapted genomes and selection on hybrids: Fisher's geometric model explains a variety of empirical patterns. *Evolution Letters* 2, 472–498. DOI: 10.1002/evl3.66.

Simon, A., Arbiol, C., Nielsen, E.E., Couteau, J., Sussarellu, R. *et al.* (2020) Replicated anthropogenic hybridisations reveal parallel patterns of admixture in marine mussels. *Evolutionary Applications* 13, 575–599. DOI: 10.1111/eva.12879.

Smith, L.D. (2009) The role of phenotypic plasticity in marine biological invasions. In: Rilov, G. and Crooks, J.A. (eds) *Biological Invasions in Marine Ecosystems: Ecological, Management, and Geographic Perspectives*. Springer, Berlin, pp. 177–202.

Sotka, E.E., Baumgardner, A.W., Bippus, P.M., Destombe, C., Duermit, E.A. *et al.* (2018) Combining niche shift and population genetic analyses predicts rapid phenotypic evolution during invasion. *Evolutionary Applications* 11, 781–793 DOI: 10.1111/eva.12592

Stachowicz, J.J., Terwin, J.R., Whitlatch, R.B. and Osman, R.W. (2002) Linking climate change and biological invasion: ocean warming facilitates nonindigenous species invasions. *Proceedings of the National Academy of Sciences USA* 99, 15497–15500. DOI: 10.1073/pnas.242437499

Stern, D.B and Lee, C.E. (2020) Evolutionary origins of genomic adaptations in an invasive copepod. *Nature Ecology and Evolution* 4, 1084–1094. DOI: 10.1038/s41559-020-1201-y

Stuart, K.C., Edwards, R.J., Sherwin, W.B. and Rollins, L.A. (2023) Contrasting patterns of single nucleotide polymorphisms and structural variation across multiple invasions. *Molecular Biology and Evolution* 40: msad046. DOI: 10.1093/molbev/msad046

Sutherland, B.J.G., Rycroft, C., Ferchaud, A.L., Saunders, R., Li, L. *et al.* (2020) Relative genomic impacts of translocation history, hatchery practices, and farm selection in Pacific oyster *Crassostrea gigas* throughout the Northern Hemisphere. *Evolutionary Applications* 13, 1380–1399. DOI: 10.1111/eva.12965.

Tea, Y.K., Hobbs, J.A., Vitelli, F., DiBattista, J.D., Ho, S.Y.W. and Lo, N. (2020) Angels in disguise: sympatric hybridization in the marine angelfishes is widespread and occurs between deeply divergent lineages. *Proceedings of the Royal Society B: Biological Sciences* 287: 20201459. DOI: 10.1098/rspb.2020.1459

Tepolt, C.K. (2015) Adaptation in marine invasion: a genetic perspective. *Biological Invasions* 17, 887–903. DOI: 10.1007/s10530-014-0825-8

Tepolt, C.K. and Palumbi, S.R. (2015) Transcriptome sequencing reveals both neutral and adaptive genome dynamics in a marine invader. *Molecular Ecology* 24, 4145–4158. DOI: 10.1111/mec.13294

Tepolt, C.K. and Palumbi, S.R. (2020) Rapid adaptation to temperature via a potential genomic island of divergence in the invasive green crab, *Carcinus maenas*. *Frontiers in Ecology and Evolution* 8: 580701. DOI: 10.3389/fevo.2020.580701.

Tepolt, C.K., Grosholz, E.D., de Rivera, C.E. and Ruiz, G.M. (2022) Balanced polymorphism fuels rapid selection in an invasive crab despite high gene flow and low genetic diversity. *Molecular Ecology* 31, 55–69. DOI: 10.1111/mec.16143

Tigano, A. and Friesen, V.L. (2016) Genomics of local adaptation with gene flow. *Molecular Ecology* 25, 2144–2164. DOI: 10.1111/mec.13606

Touchard, F., Simon, A., Bierne, N. and Viard, F. (2023) Urban rendezvous along the seashore: ports as Darwinian field labs for studying marine evolution in the Anthropocene. *Evolutionary Applications* 16, 560–579. DOI: 10.1111/eva.13443

Touchard, F., Cerqueira, F., Bierne, N. and Viard, F. (2024) Adaptive alien genes are maintained amid a vanishing introgression footprint in a sea squirt. *Evolution Letters* 8, 600–609. DOI: 10.1093/evlett/qrae016

Viard, F., David, P. and Darling, J. (2016) Marine invasions enter the genomic era: three lessons from the past, and the way forward. *Current Zoology* 62, 629–642. DOI: 10.1093/cz/zow053

Viard, F., Riginos, C. and Bierne, N. (2020) Anthropogenic hybridization at sea: three evolutionary questions relevant to invasive species management. *Philosophical Transactions of the Royal Society B: Biological Sciences* 375: 20190547. DOI: 10.1098/rstb.2019.0547

Wilson, J.R.U., Dormontt, E.E., Prentis, P.J., Lowe, A.J. and Richardson, D.M. (2009) Something in the way you move: dispersal pathways affect invasion success. *Trends in Ecology and Evolution* 24, 136–144. DOI: 10.1016/j.tree.2008.10.007

Winston, J.E. (2012) Dispersal in marine organisms without a pelagic larval phase. *Integrative and Comparative Biology* 52, 447–457. DOI: 10.1093/icb/ics040

WoRMS Editorial Board (2025) World Register of Marine Species. Available at: www.marinespecies.org (accessed 23 July 2025).

Wright, S. (1931) Evolution in Mendelian populations. *Genetics* 16, 97–159.

Zenetos, A., Tsiamis, K., Galanidi, M., Carvalho, N., Bartilotti, C. *et al.* (2022) Status and trends in the rate of introduction of marine non-indigenous species in European seas. *Diversity* 14, 1077–1126. DOI: 10.3390/d14121077

6 From Genes to Ecosystems During Invasion: How Genomic Processes Reshape Functional Traits and Lead to Ecosystem Transformations

Jane Molofsky[1]* and Kattia Palacio-Lopez[2]

[1]*Department of Plant Biology, University of Vermont, Burlington, Vermont, USA;*
[2]*Natural Sciences Department, University of Houston-Downtown, Houston, Texas, USA*

Abstract

The high prevalence of invasive species around the world makes understanding the mechanisms behind their success of utmost importance. In recent years, the increased use of genomic techniques has opened the door for understanding how evolutionary processes such as the merging of genomes during admixture or selection on keystone genes have contributed to the invasiveness of species. While not all genetic changes are necessarily important for transforming a non-invasive species into an invasive one, there are multiple examples of how changes at the gene level result in changes in a functional trait profile or in the evolution of a new trait function, such that the species becomes a supercompetitor able to disrupt entire ecosystems. The identification of keystone genes is in its early stages. In contrast, admixture is well established as a mechanism leading to transgressive traits, some of which may contribute to ecosystem changes. This chapter explores links between genetic changes and functional trait changes that lead to the reorganization of ecosystems.

6.1 Introduction

Invasive species represent some of the most successful species on Earth. They have infiltrated every continent and have had devastating consequences on natural ecosystems and on economic output and human health (IPBES, 2023). Invasive species have the potential to become ecosystem engineers that transform multi-species communities into monocultures, resulting in the degradation of important ecosystem functions such as loss or transformation of the environment used by different species (Crooks, 2002; Emery-Butcher *et al.*, 2020). Indeed, invasive species are considered a major cause of habitat loss and extinctions (Clavero *et al.*, 2009; Dueñas *et al.*, 2021; Bellard *et al.*, 2022).

The rapid proliferation of invasive species has been greatly facilitated by the increased movement of taxa around the world through accelerating global trade and international travel (Hulme, 2015). This redistribution of biodiversity can also fundamentally alter evolutionary processes (Molofsky *et al.*, 2014; Smith *et al.*, 2020). The development of genome sequencing for non-model organisms has allowed a greater

---

*Corresponding author: Jane.Molofsky@uvm.edu

DOI: 10.1079/9781800626263.0006

exploration of how genomic processes shape evolution, including invasive species.

This chapter focuses on how genomic processes can lead to extreme phenotypic changes in introduced individuals and how these genetic changes can result in an ecosystem transformation. Ecosystem transformations are initiated when feedback loops amplify processes and create a new ecosystem state. For example, introduced individuals of the grass *Phalaris arundinacea* are larger in North America than in their native Europe. These larger individuals decompose more slowly, which creates a positive feedback loop favoring the spread of *P. arundinacea* at the expense of the native species (Molofsky *et al.*, 2014). The chapter first discusses genomic processes occurring during invasion, with a strong focus on admixture, which can lead to transgressive traits that have been shown to favor invasion success of introduced species including animals, plants and fungi. Finally, the chapter

details how functional traits alter ecosystems, providing a direct link between genomic changes and transformed environments. Specifically, it discusses how genomic processes can lead to the evolution of individuals with novel traits or extreme trait values that are adding new functions to communities and ecosystems, leaving them forever changed (Fig. 6.1).

6.2 Genomic Processes That Can Lead to Outsized Ecosystem-level Effects of Invasive Species

Invasive species that transform ecosystems are a special case of the concept of the "extended phenotype" whereby a genetic change in one species affects species at different trophic levels, ultimately leading to ecosystem change (Whitham *et al.*, 2006). The concept of the extended phenotype

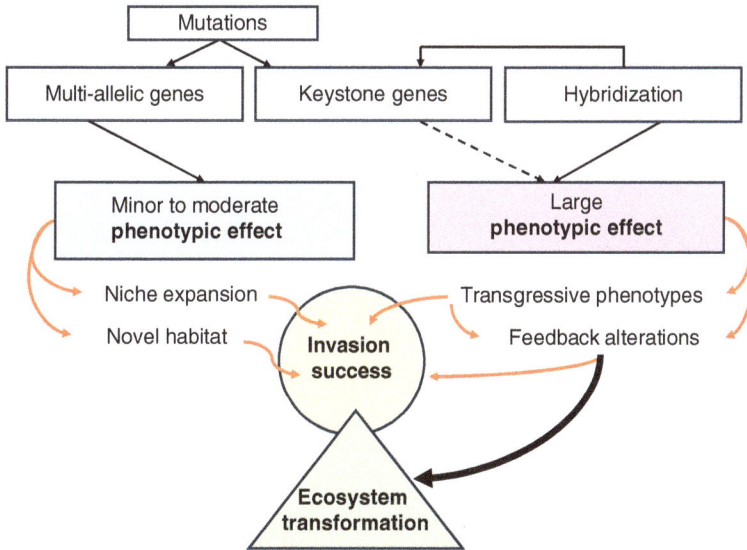

Fig. 6.1. Pathways that enhance invasion leading to ecosystem transformation. Invasion success results from different genomic-level processes including random mutations, hybridization, multi-allelic genes and keystone genes. Solid lines show the relationship among these processes. The dashed line indicates that the pathway is plausible, but evidence for its prevalence has not yet been demonstrated in an invasion context. Changes in the genome cause phenotypic changes that lead to shifts in population dynamics producing niche expansion and the colonization of new habitats (orange curved arrows). If phenotypic changes are large enough, they can lead directly to the alteration of the ecosystem feedback (bold black arrow). These effects are not mutually exclusive; they all contribute to the invasiveness of a species and, in some cases, can cause a regime shift to a new ecosystem state.

requires that specific genomic changes alter an organism's phenotype, ultimately rippling through the community and affecting other levels of organization. For introduced species, one route for initiating these effects is through selection on standing genetic variation or new mutations, which can result in the evolution of genes that have large effects on phenotypes as well as the broader ecosystem. These are known as keystone genes—genes that have a disproportionately large effect on a phenotype that consequently have a large effect on ecological processes in the ecosystem (Okuda et al., 2014; Skovmand et al., 2018). They are analogous to the "keystone species" term frequently used in ecology, which describes species that have outsized impacts on their ecosystems despite their relatively small numbers. A second route is through recombination occurring during the mixing of genomes from different species (hybridization) or from different populations (admixture) that causes extreme phenotypic changes (Fig. 6.1) (Rieseberg et al., 1999). These extreme phenotypic changes can add a new function to the ecosystem (Chapin et al., 1995), create new positive feedback relationships leading to the amplification of processes (Gaertner et al., 2014), or take on an extreme phenotypic value (Rieseberg et al., 1999). In all cases, this initiates irreversible dynamics that ultimately lead to an ecosystem change (Fig. 6.1). Because of the rapid advances in genomic technologies (Box 6.1), it is now possible to examine the role that genomic changes have in shaping the traits responsible for invasive behavior in introduced species. To date, a number of keystone genes have been identified (reviewed by Skovmand et al., 2018), and more recently, Barbour et al. (2022) identified a keystone gene in *Arabidopsis thaliana* that affects the food web structure. However, examples of keystone genes in invasive species have rarely been documented. One potential exception is the identification of herbicide-resistance polymorphisms, which represent some of the most convincing examples of large-effect genes and mutations that can have a critical role in invasion success (see also Chapter 12, this volume). In the invasive aquatic plant hydrilla (*Hydrilla verticillata*), for example, Michel et al. (2004) relied on a candidate gene approach to demonstrate that independent somatic mutations at a single nuclear gene (*pbs*) was responsible for the repeated evolution of resistance to fluridone, an important herbicide used to control this species. Given that hydrilla plants can reach densities of hundreds of millions of individuals per hectare (Michel et al., 2004), we consider the *pbs* gene to represent a candidate keystone gene acting during invasion. We expect similar findings to be made in other invasive species, especially with the development of genome-wide approaches to study the genetic basis of traits (Box 6.1), which provide more comprehensive information than candidate gene approaches. We emphasize, however, the critical need for ecological research that demonstrates the downstream ecosystem-level impact of genetic variation at candidate keystone genes.

Unlike the evidence for keystone genes and invasion, the evidence for admixture creating expanded phenotypes is well documented for plants, insects, amphibians, and birds, as well as mammals (Rieseberg et al., 2003). Indeed, the frequent admixture of populations that are brought within gene flow contact by human activity has been creating many opportunities for novel traits to arise, often with large ecosystem impacts (Molofsky et al., 2014). The mixing of genomes from different intra- and interspecific lineages leads to the breakdown of coadapted gene complexes and the expansion of trait expression, often resulting in transgressive traits that are outside the trait values of either of the parental species/lineages. Admixture can also lead to an increase in ploidy levels (polyploidy), which has been identified as a potential driver of invasiveness (te Beest et al., 2012; Hovick et al., 2023). Furthermore, admixture can break up genetic correlations and free populations from genetic constraints. When populations are locally adapted to a habitat over many generations, stabilizing selection—where intermediate trait values are favored at the expense of more extreme trait values—can constrain trait values to a narrow range (Palacio-Lopez et al., 2018). Admixture can then break up trait correlations and create individuals within populations that have trait values that are significantly different from ones that occur in the native range, favoring invasion (Barker et al., 2019).

6.3 Why Admixed Populations Often Lead to Invasive Behavior

One of the most important processes during biological invasions is the introduction of new

---

Box 6.1. Techniques used to study the genetic underpinnings of traits

The recent uptick in studies focusing on the genetic basis of traits in invasion ecology has been facilitated by quantitative trait locus (QTL) mapping, which can provide information on regions of the genome that affect phenotypes contributing to the ability of species to establish and spread into an introduced area and become invasive. This method has been successfully used in a few studies of invasive species (e.g. Paterson *et al.*, 1995; Prapas *et al.*, 2022). There are also a number of well-documented examples of the use of QTL mapping to find genes of large effect. However, this method can be challenging to implement for invasive species because it requires the establishment of experimental mapping populations, which are obtained by crossing two differentiated individuals. This can be time consuming and often difficult to achieve, particularly for invasive species, which often cannot be grown under natural conditions due to biosecurity concerns. In addition, because of the limited number of recombination events that occur in a QTL mapping population, the mapping resolution often does not allow the identification of specific genes. Rather, results may point to chromosomal regions likely to harbor one or more causal genes (Schielzeth and Husby, 2014). However, increased availability of genome-wide single-nucleotide polymorphism (SNP) markers for non-model species has paved the way for an alternative approach to trait mapping, known as genome-wide association studies (GWAS). GWAS use genome-wide sequencing across many individuals to identify genetic variants (typically SNPs) associated with specific traits and can be used to identify regions of the genome that are associated with a particular phenotype linked to invasive behavior using a number of different models that account for the confounding effect of population structure. Similar to QTL mapping, GWAS require the establishment of an experimental mapping population. However, these individuals are usually obtained from natural populations rather than from controlled crosses. Therefore, mapping resolution is substantially higher than for QTL mapping, as it exploits natural recombination events that have occurred in the focal species.

---

variants or genotypes into an area, which can alter the genetic diversity of the resident invasive population and shift its evolutionary trajectory. These newly introduced genotypes may become more invasive than the originally introduced ones (Lavergne and Molofsky, 2007). Early work on invasive species suggested that invasive populations would have lower genetic diversity than native populations because of the limited number of propagules or lower diversity of genotypes introduced (Tsutsui *et al.*, 2000; Hagenblad *et al.*, 2015; Buchholz *et al.*, 2023). As small populations are subject to extinction through both demographic stochasticity and genetic drift, the expansion of certain introduced species, despite these conditions, leads to the notion of the "paradox of invasion" (Pérez *et al.*, 2006; Estoup *et al.*, 2016). The success of invasive species is especially surprising because they often thrive in ecologically and evolutionarily novel conditions, which would typically result in low population fitness.

In introduced populations, individuals may originate from genetically distinct sources, resulting in a scenario in which the overall genetic diversity of the introduced population may be much higher than in conspecific populations in the native range (Lavergne and Molofsky, 2007). For example, many invasive plant species arise from planned introductions for horticulture, soil stabilization or other material benefits (Wootton *et al.*, 2005; Drew *et al.*, 2010). Seed contamination in shipping has also resulted in multiple unplanned introductions of plant species (Hulme, 2015; Wilson *et al.*, 2016). In these scenarios, the continual introduction of individuals from disparate locales into a common environment allows genetic mixing of divergent lineages, which automatically increases genetic diversity and thus evolutionary potential (Lavergne and Molofsky, 2007; Keller and Taylor, 2010).

Linkage disequilibrium—the correlation of different alleles within a chromosome leading them to be inherited together—will keep coadapted gene complexes together (Flint-Garcia *et al.*, 2003; Hohenlohe *et al.*, 2012). However, linkage disequilibrium can be broken up by the mixing of individuals from different populations or lineages (Edmands, 1999; Abbott *et al.*, 2013). The mixing of historically separate populations in the new range breaks up genetic correlations and allows new trait combinations while also increasing the overall genetic diversity in the newly formed populations. In addition, while we typically think of mutations as the source of novelty and innovation in evolution, we know that most mutations reduce fitness, even in new environments (Shaw *et al.*, 2002; Baer *et al.*, 2005). Nevertheless, the uncoupling of correlated traits that can accompany admixture means that some traits that could provide a fitness advantage are now free to be selected independently. Moreover, admixture can result in new trait values or trait values that occur outside of what is typically observed in the native range (i.e. novel phenotypes with extreme trait values not observed in either parental species), a process that has been termed "transgressive segregation" (Rieseberg *et al.*, 1999).

6.4　What Are Some Traits That Lead to Transgressive Segregation and Invasion?

Hybridization is generally associated with species establishment through transgressive segregation of phenotypes, which is manifested when hybrid progeny show more extreme phenotypes relative to those observed in either parental line (Mounger *et al.*, 2021). Transgressive phenotypes can be the result of novel trait values or novel trait interactions, and although transgressive phenotypes can appear *via* intraspecific crosses (admixture; cross-breeding of individuals from different locally adapted populations creating individuals with genomes that are a mix of both populations) as well as interspecific crosses (cross-breeding of individuals from different species), intraspecific transgression tends to be more frequent as a result of the regular mixing of individuals within the same species range (Rieseberg *et al.*, 1999).

Segregating hybrids can show extreme values for fitness-related traits, potentially leading to an adaptive phenotypic shift from the parental genotypes. Therefore, transgressive phenotypes could play an important role in accelerating the evolutionary trajectory of a population (Palacio-Lopez and Molofsky, 2021). Extreme phenotypes caused by transgressive segregation are heritably stable (Koide *et al.*, 2019) and increase the phenotypic variance that could be subject to selection under a particular environment (Molofsky *et al.*, 2014). An increasing number of studies

shows that population admixture plays an important role in the successful establishment of alien species across different taxa including plants (Keller and Taylor, 2010), mussels (Gillis *et al.*, 2009), fishes (Consuegra *et al.*, 2011), lizards (Kolbe *et al.*, 2008), fruit flies (Duchen *et al.*, 2013), and ladybirds (Lombaert *et al.*, 2011).

Transgressive segregation that follows from admixture can be underpinned by complementary gene action and epistasis (interaction between genes at different loci, where the expression of one gene is modified or completely masked by one or more other genes) when the hybridizing populations have different alleles at QTLs (Rieseberg *et al.*, 1999; Mao *et al.*, 2011). Recombination brings together novel allelic combinations that can produce extreme phenotypes, which may favor the invasiveness of species across different taxa (Table 6.1). In plants for example, purple loosestrife, *Lythrum salicaria*, is a European perennial that has been introduced multiple times into North America and is now reported as an invasive species (Lavoie, 2010). Hybrids of *L. salicaria* originating from crosses between populations in the native range, between populations in the introduced range, and between populations in the native and introduced ranges revealed a consistent transgressive plant allocation pattern through increased biomass and shoot:root ratio (Shi *et al.*, 2018). Similarly, radish (*Raphanus sativus*) is considered an invasive species in many regions including parts of North America. In this species, a study focusing on hybrids between *R. sativus* crop genotypes and their wild or weedy counterparts showed evidence of transgressive phenotypes in biomass and fecundity in ruderal and agrestal conditions (Vercellino *et al.*, 2023). This phenomenon is also common in fungal species that routinely hybridize and can produce new hybrid individuals that are more pathogenic than either of the parental species, as has been shown in hybrids between *Cryptococcus neoformans* and *Cryptococcus deneoformans* (Samarasinghe *et al.*, 2020).

In animals, transgressive traits have mainly been reported from interspecific hybridization, although examples of transgression following admixture are also known. Similar to the situation encountered in plants, fitness-related traits can drive phenotypic novelty (Table 6.1). For example, several species of predatory ladybird have been introduced in different countries as a

Table 6.1. Examples of transgressive segregation traits associated with invasive species. Transgressive segregation directionality (positive (+) or negative (−)) on traits across animals, fungi and plants reported as invasive species. Phenology traits include flowering time and bolting time. Growth traits include body mass, height, biomass and weight. Fitness traits include number of eggs, number of flowers and survival.

Type of hybridization	Classification	Parent taxa	Phenology	Growth	Fitness	Study
Positive transgressive segregation						
Admixture	Animal: fish	*Acipenser ruthenus*		+	+	Shivaramu *et al.* (2020)
	Animal: fish	*Clarias anguillaris*			+	Yisa *et al.* (2017)
	Animal: insect	*Cryptolaemus montrouzieri*			+	H.-S. Li *et al.* (2018)
	Animal: insect	*Harmonia axyridis*			+	Turgeon *et al.* (2011)
	Animal: salamander	*Ambystoma californiense* × *A. mavortium*		+	+	Fitzpatrick and Shaffer (2007)
	Animal: snail	*Melanoides tuberculata*		+		Facon *et al.* (2005)
	Fungi	*Cryptococcus neoformans*			+	Shahid *et al.* (2008)
	Plant	*Lythrum salicaria*		+		Shi *et al.* (2018)
	Plant	*Mimulus guttatus*		+		Y. Li *et al.* (2018)
	Plant	*Raphanus sativus*	+	+	+	Vercellino *et al.* (2023)
Interspecific hybridization	Animal: fish	*Catostomus commersonii* × *C. latipinnis*	+	+		Mandeville *et al.* (2022)
	Animal: fish	*Catostomus discobolus* × *C. commersonii*		+		Mandeville *et al.* (2022)
	Animal: fish	*Catostomus discobolus* × *C. latipinnis*		+		Mandeville *et al.* (2022)
	Plant	*Fallopia japonica* × *F. sachalinensis*		+		Parepa *et al.* (2014)
Negative transgressive segregation						
Admixture	Animal: fish	*Onchorhynchus gorbuscha* W.			−	Gharrett *et al.* (1999)
	Plants	*Centaurea solstitialis*			−	Irimia *et al.* (2021)
	Plants	*Chamaecrista fasciculata*		−	−	Fenster and Galloway (2000)

strategy to control mealybug pests. The Australian ladybird *Cryptolaemus montrouzieri*, for example, has been introduced multiple times in China as part of a biological control program, generating admixed genotypes between intraspecific lineages that outperform parental populations in terms of their fitness and potential for establishment under limiting conditions such as starvation and low temperatures (H.-S. Li *et al.*, 2018). Similarly, the Asian ladybird, *Harmonia axyridis*, was introduced in North America to control aphids. Comparison with parental populations showed that admixed genotypes of *H. axyridis* have a higher larval survival, an increment in fecundity and an earlier reproductive age (Turgeon *et al.*, 2011).

Observations of transgressive phenotypes are also found in aquatic habitats. Hybrids of the freshwater snail *Melanoides tuberculata* have shifted their life histories toward larger investment into biomass and a faster growth rate in juveniles (Facon *et al.*, 2005). Hybridization events in fish species are well documented and show that the long-term effect is complex (Chang *et al.*, 2022). For example, positive outcomes of hybridization have been reported in admixed individuals among three strains of the African catfish, *Clarias anguillaris*, which show superiority in growth performance under experimental conditions (Yisa *et al.*, 2017). Similarly, in one of the oldest groups of fish, the sterlet, *Acipenser ruthenus* (Du *et al.*, 2020), it has been reported that intraspecific hybrids perform better (higher body weight and cumulative survival) than purebreds, suggesting admixture as a strategy to overcome the extinction risk of the species (Shivaramu *et al.*, 2020). In addition, hybridization between two Lake Malawi cichlid populations of *Maylandia zebra* resulted in transgressive phenotypes with novel shapes beyond those of either parent. The findings suggest that complex genetic interactions, not just simple inheritance, drive these traits, and that hybridization may have contributed to cichlid diversification by generating new morphological variation (Husemann *et al.*, 2017).

Transgressive segregation *via* interspecific transgression, although less frequent than intraspecific transgression, can be more evolutionarily significant leading to adaptive radiation or ecological shifts (Stelkens *et al.*, 2009). Here, epistasis and other genetic interactions are the main mechanisms that drive transgressive phenotypes leading to novel phenotypic variations in hybrid populations (Rieseberg *et al.*, 2003; Mao *et al.*, 2011; Smith and Weigel, 2012). In animals, for example, interspecific hybrids of the *Catostomus* fish species showed novel ecological niches outside the range of their parental species. Hybrids show transgressive segregation in their feeding ecology that allows them to outperform their parents (Mandeville *et al.*, 2022). In plants, hybridization between two invasive knotweed species (*Fallopia japonica* × *F. sachalinensis*) showed transgressive segregation on competitive traits increasing the invasiveness of the hybrids (Parepa *et al.*, 2014; see also Chapter 9, this volume).

Transgressive phenotypes are not always beneficial, as mentioned previously (see also Table 6.1). Admixed genotypes can perform similar to or worse than the two parents. A phenotype that is similar to that of the parents could be explained by a similar degree of relatedness between progenitors (Palacio-Lopez *et al.*, 2018). Hybrids that perform worse than their parents show negative transgressive segregation, which could be the result of outbreeding depression (Verhoeven *et al.*, 2011). A genotype can experience outbreeding depression *via* two distinct mechanisms: (i) a disruption of allelic coadaptation (underdominance or complementary epistasis); or (ii) a disruption of local adaptation to environmental conditions (Waser and Price, 1989). Some examples of negative transgressive segregation include admixed individuals of the pink salmon (*Onchorhynchus gorbuscha* W.) that derived from two genetically distinct lines. Hybrids did not outperform their respective parents and showed evidence of outbreeding depression characterized by a reduced survival rate (Gharrett *et al.*, 1999). In plants, although hybridization events are commonly associated with rapid adaptation and reproductive success, as has been shown in interspecific hybrids of sunflowers (*Helianthus annuus* × *H. debilis*) (Mitchell *et al.*, 2019), this may not be the rule. Admixed genotypes of *Centaurea solstitialis* that resulted from crosses between native and non-native populations showed mixed fitness effects, including significant increases, decreases and no differences in reproductive success (number of seeds) compared with crosses within the population (Irimia *et al.*, 2021). In addition, in natural populations of the aggressive weed *Chamaecrista fasciculata*, the set of positive transgressive traits found in the F_1 hybrids reversed in subsequent generations as a result of dilution of epistatic interactions (Whitlock *et al.*, 1995; Fenster and Galloway, 2000).

6.5 Experimental Tests of Transgressive Segregation in Plants and Animals

Admixture is known to be important for expanding trait variation and for facilitating successful invasions through the transgressive segregation of traits (Mounger *et al.*, 2021). Population genomic studies have demonstrated that admixture is prevalent in invasive populations (e.g. Keller and Taylor, 2010). While admixture effects are widely discussed as being important for promoting invasive species evolution after introduction, most of these studies could be considered primarily speculative. Several studies report characteristics of hybrid genotypes that correlate with invasion. However, few experimental studies document the admixture events that immediately preceded invasion success (Keller and Taylor, 2010; Turgeon *et al.*, 2011; Shang *et al.*, 2019).

As transgressive phenotypes may depend on the environment, studies that involve both a crossing design to generate admixed genotypes and the exposure of parental and hybrid genotypes to limiting environmental conditions are very valuable. These types of studies can describe the entire process that led to the invasion. For example, with the understanding that intraspecific admixture is favored by multiple sequential introductions from different source populations into a single area (Lavergne and Molofsky, 2007), crosses of the herbaceous plant *Mimulus guttatus* between populations from its native and introduced range were performed to test the role of admixture in the invasion success of the species (Y. Li *et al.*, 2018). This study shows that admixture has a positive effect on total biomass and number of flowers produced, especially under drought conditions (Y. Li *et al.*, 2018). In a related experiment, Palacio-Lopez and Molofsky (2021) experimentally tested whether admixed individuals of *A. thaliana* could shift their phenotypes away from the parental phenotypes after one recombinant event (F_2 recombinant hybrids).

Analyzing a multi-dimensional trait profile, they found that approximately half of the admixed populations shifted their phenotype away from the parents and occupied a different phenotypic space than their parents, but whether the shift in phenotype resulted in higher fitness depended on the experimental conditions. Both experimental studies provide evidence that admixture can result in rapid evolution after introduction (Fitzpatrick and Shaffer, 2007). These studies also illustrate how experiments performed under controlled conditions can complement genomic analyses that typically rely on associations between traits or invasion success and genotypes (see also Chapter 10, this volume, as an example of how this integrative approach can be used in invasion science).

6.6 Transgressive Traits and Ecosystem Transformation

Not all traits that have a large phenotypic effect and/or result in transgressive segregation will have a large ecosystem impact (Fig. 6.1). For example, changes in plant size can lead to changes in competitive ability, but these shifts will not always result in changes in plant chemistry that can alter decomposition rates and nutrient cycles, ultimately bringing about substantial changes across the whole ecosystem. For animals, transgressive traits often lead to changes in color patterns (Mérot *et al.*, 2020), mating behaviors or physical traits, and expansion into new niches (Graham *et al.*, 2011). While changes in mating patterns and sometimes color patterns can result in speciation after hybridization, few definitive examples exist where the production of transgressive traits results in more invasive traits and finally an ecosystem shift.

Understanding functional traits has become an increasingly important area of ecological research and invasive species research in general (Divíšek *et al.*, 2018), but it can be difficult to determine which invasive traits are important (Strayer, 2012). Functional traits can be classified into two categories: (i) discrete traits: those that add a new function to the community; and (ii) continuous traits: those that increase or decrease an already existing trait value (Chapin *et al.*, 1995). In a classic paper, Chapin *et al.*

(1995) argued that introduced species that added a new feature such as nitrogen fixation in plants are more likely to have ecosystem consequences. However, these traits are present when the species is moved to the new range so can be considered pre-adaptation to invasive behavior (Schlaepfer *et al.*, 2010). In other plant examples, the increasing density of the invader can result in an ecosystem shift to a new state even if the trait was previously present in the ecosystem; such changes can lead to dramatic ecosystem shifts such as changes to the hydrology and fire cycle as invasive species begin to dominate landscapes (Dukes and Mooney, 2004). While most examples come from plant systems, a high population growth rate leading to a high density of invasive animal species can have similar effects. For example, invasive zebra mussels, *Dreissena polymorpha*, can reach high densities that dramatically alter nutrient cycling and water clarity in freshwater ecosystems (Strayer, 2009). Likewise, high densities of invasive Asian carp in North American rivers outcompete native fish species for food and alter the food web through shifts in the available zooplankton (Chick *et al.*, 2020). However, these examples focus on the increase in density of the introduced species and do not address the evolutionary modifications that may have underpinned these transitions. Indeed, enhanced evolutionary potential followed by selection for the more extreme trait values may precipitate the ecosystem regime shift, in which an ecosystem undergoes an abrupt change in function resulting in a completely different ecosystem (Molofsky *et al.*, 2014). Once a regime shift occurs, it is difficult to restore the ecosystem to its original state without intensive management intervention.

While transgressive traits have been documented extensively in plant hybrids, animal examples are less common, and when they do occur, they have more limited influence on ecosystems (Rieseberg *et al.*, 1999). In animals, one well-documented example is the California tiger salamander, *Ambystoma californiense*, which hybridized with the introduced barred tiger salamander, *Ambystoma tigrinum mavortium*, producing hybrid individuals that were larger than both the native and introduced species and thus could outcompete them (Ryan *et al.*, 2009).

Nevertheless, the connection to carryover effects into other trophic levels is less direct in these and other animal hybrids, limiting their ecosystem-level impacts despite their individual competitive advantages. Furthermore, the phenotypic expression in admixed or hybridizing populations exhibits greater variability than in non-admixed populations, as these individuals result from diverse backcrossing and hybridization events. This genetic complexity creates organisms with differential responses to environmental conditions, which complicates predictions about their potential impact on ecosystem processes (Stelkens *et al.*, 2009; Palacio-Lopez and Molofsky, 2021).

In contrast to animals, the high frequency of transgressive segregation in plants provides many examples where the more extreme phenotypic trait has resulted in an ecosystem-level change. Quantitative traits such as increased plant height, increased biomass or changes in leaf chemistry may have greater potential to evolve following introduction, as the genetic changes necessary only require a shift to a larger (or smaller) value. For example, introduced plants can have larger floral displays triggering changes in plant-pollinator activity and resulting in a change from one plant community to another (Muñoz and Cavieres, 2008; Gibson *et al.*, 2012; Goodell and Parker, 2017). Other possible changes in continuous traits that can affect entire ecosystems include shifts in the dominant rooting morphology, which can affect water availability, or changes in flammability when an annual grass starts to dominate an ecosystem previously occupied by shrubs (Dukes and Mooney, 2004). Changes in plant size either through biomass or height changes following establishment in the new community is another prevalent pattern (Blumenthal and Hufbauer, 2007). Blumenthal and Hufbauer (2007) documented the increase in size for individuals of 14 species taken from the native range (two populations per species) and from the invasive range (two populations per species) in a common garden. In all cases, size increases were found in the invasive populations but only in the absence of competition. Once introduced into the new range, plants are freed from their native enemies and thus can reallocate resources from defense to growth and reproduction via post-introduction evolution (Blossey and Notzold, 1995). In this context, the faster growth rate and larger size of introduced plants can result in ecosystem consequences for the invaded habitat through changes in leaf chemistry and the persistence of leaf litter (Molofsky *et al.*, 2014).

While the obvious change of being larger can provide a competitive advantage within the plant community, changes in leaf chemistry can also have dramatic effects on ecosystem processes by slowing down leaf decomposition (Eppinga and Molofsky, 2013). Plants with higher leaf carbon:nitrogen ratios, which can occur because of faster growth and greater biomass, may provide less food for microbes and hence will sustain a slower decomposition process (Aerts and de Caluwe, 1997; Güsewell and Gessner, 2009). Because of the greater biomass and slower decomposition, the leaf litter can build up in the ecosystem. Increased leaf litter may ultimately suppress germination and seedling growth for native taxa in the community, resulting in a change in the species composition (Molofsky and Augspurger, 1992; Kaproth *et al.*, 2013). Leaf litter can have profound impacts on plant communities through changes in seed germination ability under leaf litter leading to changes in plant community composition (Molofsky and Augspurger, 1992; Kortessis *et al.*, 2022). Moreover, litter can change the nature of the feedback relationships between plants and the leaf litter environment where the invasive plant is able to germinate under the thicker leaf litter while native species cannot (Eppinga *et al.*, 2011). This sets up a positive-feedback relationship between the invasive plant and the presence of litter and makes it more likely that the invasive species can continue to spread (Eppinga *et al.*, 2011; see also Fig. 6.1). Increased litter buildup after establishment is perhaps the most prevalent example of how genetic changes and selection for larger size can lead to a regime shift.

6.7 Conclusions

In this chapter, we discuss the direct effects of genomic processes on phenotypic expression and its subsequent effect on ecosystems. Two different genomic pathways have been identified that can disrupt ecosystem processes: keystone genes and hybridization followed by transgressive segregation (Fig. 6.1). The study of keystone genes is still in its infancy and will likely increase as whole-genome sequencing technology for non-model organisms becomes more accessible. Note, however, that even in model organisms and in non-invasive species, keystone genes are predicted to be relatively rare (Skovmand *et al.*, 2018). Indeed, only a small number of examples are currently available, the majority of which stem from species in their native range, or from crops (Skovmand *et al.*, 2018). We do not, however, exclude the possibility that future studies may reveal their importance for invasive species and their role in post-invasion ecosystem reorganization.

In contrast to keystone genes, intraspecific hybridization has been documented to cause transgressive traits and to lead to ecosystem regime shifts. Which invaders evolve on arrival in the new range and why are important questions to consider as we continue to move genetic material around the world. As the climate continues to change, how will these new evolutionary processes play out? The rapid change in environmental conditions results in populations living under conditions they may not have experienced before. Under such a scenario, what will be the fate of different species (Clements and Ditommaso, 2011)? As populations grow under increasingly novel conditions, we can expect to see a decline in population sizes leading to loss of biodiversity due to demographic stochasticity and genetic bottlenecks. However, some introduced species may harbor greater genetic potential than native species. Introduced species that undergo intraspecific hybridization and selection benefit from higher evolutionary potential on account of their greater genetic diversity, and may be more likely to adapt to these novel conditions and be able to track climate change (Molofsky and Collins, 2014). Although generalizations are difficult, the continual mixing and recombining of genetic material is fundamentally altering the evolutionary potential of many introduced species around the world with the ultimate consequences yet to be determined.

Acknowledgments

This work was supported in part by Hatch funds from the USDA National Institute of Food and Agriculture and by NIFA Award AWD00001453 to J.M. The authors thank the book editors for their insightful comments on the chapter.

References

Abbott, R., Albach, D., Ansell, S., Arntzen, J.W., Baird, S.J.E. *et al.* (2013) Hybridization and speciation. *Journal of Evolutionary Biology* 26, 229–246. DOI: 10.1111/j.1420-9101.2012.02599.x

Aerts, R. and de Caluwe, H. (1997) Nutritional and plant-mediated controls on leaf litter decomposition of *Carex* species. *Ecology* 78, 244–260. DOI: 10.2307/2265993.

Baer, C.F., Shaw, F., Steding, C., Baumgartner, M., Hawkins, A. *et al.* (2005) Comparative evolutionary genetics of spontaneous mutations affecting fitness in rhabditid nematodes. *Proceedings of the National Academy of Sciences USA* 102, 5785–5790. DOI: 10.1073/pnas.0406056102

Barbour, M.A., Kliebenstein, D.J. and Bascompte, J. (2022) A keystone gene underlies the persistence of an experimental food web. *Science* 376, 70–73. DOI: 10.1126/science.abf2232

Barker, B.S., Cocio, J.E., Anderson, S.R., Braasch, J.E., Cang, F.A. *et al.* (2019) Potential limits to the benefits of admixture during biological invasion. *Molecular Ecology* 28, 100–113. DOI: 10.1111/mec.14958

Bellard, C., Marino, C. and Courchamp, F. (2022) Ranking threats to biodiversity and why it doesn't matter. *Nature Communications* 13: 2616. DOI: 10.1038/s41467-022-30339-y

Blossey, B. and Notzold, R. (1995) Evolution of increased competitive ability in invasive nonindigenous plants: a hypothesis. *Journal of Ecology* 83, 887–889. DOI: 10.2307/2261425

Blumenthal, D.M. and Hufbauer, R.A. (2007) Increased plant size in exotic populations: a common-garden test with 14 invasive species. *Ecology* 88, 2758–2765. DOI: 10.1890/06-2115.1

Buchholz, M.J., Wright, E.A., Grisham, B.A., Bradley, R.D., Arsuffi, T.L. and Conway, W.C. (2023) Low genetic diversity among introduced axis deer: comments on the genetic paradox and invasive species. *Journal of Mammalogy* 104, 603–618. DOI: 10.1093/jmammal/gyad008

Chang, S.L., Ward, H.G.M., Elliott, L.D. and Russello, M.A. (2022) Genotyping-in-thousands by sequencing panel development and application for high-resolution monitoring of introgressive hybridization within sockeye salmon. *Scientific Reports* 12: 3441. DOI: 10.1038/s41598-022-07309-x

Chapin, F.S., Shaver, G.R., Giblin, A.E., Nadelhoffer, K.J. and Laundre, J.A. (1995) Responses of arctic tundra to experimental and observed changes in climate. *Ecology* 76, 694–711. DOI: 10.2307/1939337

Chick, J.H., Gibson-Reinemer, D.K., Soeken-Gittinger, L. and Casper, A.F. (2020) Invasive silver carp is empirically linked to declines of native sport fish in the Upper Mississippi river system. *Biological Invasions* 22, 723–734. DOI: 10.1007/s10530-019-02124-4

Clavero, M., Brotons, L., Pons, P. and Sol, D. (2009) Prominent role of invasive species in avian biodiversity loss. *Biological Conservation* 142, 2043–2049. DOI: 10.1016/j.biocon.2009.03.034

Clements, D.R. and Ditommaso, A. (2011) Climate change and weed adaptation: can evolution of invasive plants lead to greater range expansion than forecasted? *Weed Research* 51, 227–240. DOI: 10.1111/j.1365-3180.2011.00850.x

Consuegra, S., Phillips, N., Gajardo, G. and de Leaniz, C.G. (2011) Winning the invasion roulette: escapes from fish farms increase admixture and facilitate establishment of non-native rainbow trout. *Evolutionary Applications* 4, 660–671. DOI: 10.1111/j.1752-4571.2011.00189.x

Crooks, J.A. (2002) Characterizing ecosystem-level consequences of biological invasions: the role of ecosystem engineers. *Oikos* 97, 153–166. DOI: 10.1034/j.1600-0706.2002.970201.x

Divíšek, J., Chytrý, M., Beckage, B., Gotelli, N.J., Lososová, Z. *et al.* (2018) Similarity of introduced plant species to native ones facilitates naturalization, but differences enhance invasion success. *Nature Communications* 9: 4631. DOI: 10.1038/s41467-018-06995-4

Drew, J., Anderson, N. and Andow, D. (2010) Conundrums of a complex vector for invasive species control: a detailed examination of the horticultural industry. *Biological Invasions* 12, 2837–2851. DOI: 10.1007/s10530-010-9689-8.

Du, K., Stöck, M., Kneitz, S., Klopp, C., Woltering, J.M. *et al.* (2020) The sterlet sturgeon genome sequence and the mechanisms of segmental rediploidization. *Nature Ecology and Evolution* 4, 841–852. DOI: 10.1038/s41559-020-1166-x

Duchen, P., Živković, D., Hutter, S., Stephan, W. and Laurent, S. (2013) Demographic inference reveals African and European admixture in the North American *Drosophila melanogaster* population. *Genetics* 193, 291–301. DOI: 10.1534/genetics.112.145912

Dueñas, M.-A., Hemming, D.J., Roberts, A. and Diaz-Soltero, H. (2021) The threat of invasive species to IUCN-listed critically endangered species: a systematic review. *Global Ecology and Conservation* 26: e01476. DOI: 10.1016/j.gecco.2021.e01476

Dukes, J.S. and Mooney, H.A. (2004) Disruption of ecosystem processes in western North America by invasive species. *Revista Chilena de Historia Natural* 77, 411–437. DOI: 10.4067/S0716-078X2004000300003

Edmands, S. (1999) heterosis and outbreeding depression in interpopulation crosses spanning a wide range of divergence. *Evolution* 53: 1757. DOI: 10.2307/2640438

Emery-Butcher, H.E., Beatty, S.J. and Robson, B.J. (2020) The impacts of invasive ecosystem engineers in freshwaters: a review. *Freshwater Biology* 65, 999–1015. DOI: 10.1111/fwb.13479

Eppinga, M.B. and Molofsky, J. (2013) Eco-evolutionary litter feedback as a driver of exotic plant invasion. *Perspectives in Plant Ecology, Evolution and Systematics* 15, 20–31. DOI: 10.1016/j.ppees.2012.10.006

Eppinga, M.B., Kaproth, M.A., Collins, A.R. and Molofsky, J. (2011) Litter feedbacks, evolutionary change and exotic plant invasion. *Journal of Ecology* 99, 503–514. DOI: 10.1111/j.1365-2745.2010.01781.x

Estoup, A., Ravigné, V., Hufbauer, R., Vitalis, R., Gautier, M. and Facon, B. (2016) Is there a genetic paradox of biological invasion? *Annual Review of Ecology, Evolution, and Systematics* 47, 51–72. DOI: 10.1146/annurev-ecolsys-121415-032116

Facon, B., Jarne, P., Pointier, J.P. and David, P. (2005) Hybridization and invasiveness in the freshwater snail *Melanoides tuberculata*: hybrid vigour is more important than increase in genetic variance. *Journal of Evolutionary Biology* 18, 524–535. DOI: 10.1111/j.1420-9101.2005.00887.x

Fenster, C.B. and Galloway, L.F. (2000) Population differentiation in an annual legume: genetic architecture. *Evolution* 54, 1157–1172. DOI: 10.1111/j.0014-3820.2000.tb00551.x

Fitzpatrick, B.M. and Shaffer, H.B. (2007) Hybrid vigor between native and introduced salamanders raises new challenges for conservation. *Proceedings of the National Academy of Sciences USA* 104, 15793–15798. DOI: 10.1073/pnas.0704791104

Flint-Garcia, S.A., Thornsberry, J.M. and Buckler, E.S. (2003) Structure of linkage disequilibrium in plants. *Annual Review of Plant Biology* 54, 357–374. DOI: 10.1146/annurev.arplant.54.031902.134907

Gaertner, M., Biggs, R., Te Beest, M., Hui, C., Molofsky, J. and Richardson, D.M. (2014) Invasive plants as drivers of regime shifts: identifying high-priority invaders that alter feedback relationships. *Diversity and Distributions* 20, 733–744. DOI: 10.1111/ddi.12182

Gharrett, A.J., Smoker, W.W., Reisenbichler, R.R. and Taylor, S.G. (1999) Outbreeding depression in hybrids between odd- and even-broodyear pink salmon. *Aquaculture* 173, 117–129. DOI: 10.1016/S0044-8486(98)00480-3

Gibson, M.R., Richardson, D.M. and Pauw, A. (2012) Can floral traits predict an invasive plants impact on native plant–pollinator communities? *Journal of Ecology* 100, 1216–1223. DOI: 10.1111/j.1365-2745.2012.02004.x

Gillis, N.K., Walters, L.J., Fernandes, F.C. and Hoffman, E.A. (2009) Higher genetic diversity in introduced than in native populations of the mussel *Mytella charruana*: evidence of population admixture at introduction sites. *Diversity and Distributions* 15, 784–795. DOI: 10.1111/j.1472-4642.2009.00591.x

Goodell, K. and Parker, I.M. (2017) Invasion of a dominant floral resource: effects on the floral community and pollination of native plants. *Ecology* 98, 57–69. DOI: 10.1002/ecy.1639

Graham, A.M., Munday, M.D., Kaftanoglu, O., Page, R.E., Amdam, G.V. and Rueppell, O. (2011) Support for the reproductive ground plan hypothesis of social evolution and major QTL for ovary traits of Africanized worker honey bees (*Apis mellifera* L.) *BMC Evolutionary Biology* 11: 95. DOI: 10.1186/1471-2148-11-95

Güsewell, S. and Gessner, M.O. (2009) N:P ratios influence litter decomposition and colonization by fungi and bacteria in microcosms. *Functional Ecology* 23, 211–219. DOI: 10.1111/j.1365-2435.2008.01478.x

Hagenblad, J., Hülskötter, J., Acharya, K.P., Brunet, J., Chabrerie, O. *et al.* (2015) Low genetic diversity despite multiple introductions of the invasive plant species *Impatiens glandulifera* in Europe. *BMC Genetics* 16: 103. DOI: 10.1186/s12863-015-0242-8

Hohenlohe, P.A., Bassham, S., Currey, M. and Cresko, W.A. (2012) Extensive linkage disequilibrium and parallel adaptive divergence across three spine stickleback genomes. *Philosophical Transactions of the Royal Society B: Biological Sciences* 367, 395–408. DOI: 10.1098/rstb.2011.0245

Hovick, S.M., Adams, C.R., Anderson, N.O. and Kettenring, K.M. (2023) Progress on mechanisms and im-
    pacts of wetland plant invasions: a twenty-year retrospective analysis and priorities for the next twenty.
    Critical Reviews in Plant Sciences 42, 239–282. DOI: 10.1080/07352689.2023.2233232
Hulme, P.E. (2015) Invasion pathways at a crossroad: policy and research challenges for managing alien
    species introductions. *Journal of Applied Ecology* 52, 1418–1424. DOI: 10.1111/1365-2664.12470
Husemann, M., Tobler, M., McCauley, C., Ding, B. and Danley, P.D. (2017) Body shape differences in a pair
    of closely related Malawi cichlids and their hybrids: effects of genetic variation, phenotypic plasticity,
    and transgressive segregation. *Ecology and Evolution* 7, 4336–4346. DOI: 10.1002/ece3.2823
IPBES (2023) *Thematic Assessment Report on Invasive Alien Species and Their Control*. IPBES Secretar-
    iat, Bonn, Germany.
Irimia, R.E., Hierro, J.L., Branco, S., Sotes, G., Cavieres, L.A. *et al.* (2021) Experimental admixture among
    geographically disjunct populations of an invasive plant yields a global mosaic of reproductive incom-
    patibility and heterosis. *Journal of Ecology* 109, 2152–2162. DOI: 10.1111/1365-2745.13628
Kaproth, M.A., Eppinga, M.B. and Molofsky, J. (2013) Leaf litter variation influences invasion dynamics in
    the invasive wetland grass *Phalaris arundinacea*. *Biological Invasions* 15, 1819–1832. DOI: 10.1007/
    s10530-013-0411-5
Keller, S.R. and Taylor, D.R. (2010) Genomic admixture increases fitness during a biological invasion: ad-
    mixture increases fitness during invasion. *Journal of Evolutionary Biology* 23, 1720–1731. DOI:
    10.1111/j.1420-9101.2010.02037.x
Koide, Y., Sakaguchi, S., Uchiyama, T., Ota, Y., Tezuka, A. *et al.* (2019) Genetic properties responsible for
    the transgressive segregation of days to heading in rice. *G3: Genes, Genomes, Genetics* 9, 1655–
    1662. DOI: 10.1534/g3.119.201011
Kolbe, J.J., Larson, A., Losos, J.B. and de Queiroz, K. (2008) Admixture determines genetic diversity and
    population differentiation in the biological invasion of a lizard species. *Biology Letters* 4, 434–437. DOI:
    10.1098/rsbl.2008.0205
Kortessis, N., Kendig, A.E., Barfield, M., Flory, S.L., Simon, M.W. and Holt, R.D. (2022) Litter, plant compe-
    tition, and ecosystem dynamics: a theoretical perspective. *American Naturalist* 200, 739–754. DOI:
    10.1086/721438
Lavergne, S. and Molofsky, J. (2007) Increased genetic variation and evolutionary potential drive the suc-
    cess of an invasive grass. *Proceedings of the National Academy of Sciences USA* 104, 3883–3888.
    DOI: 10.1073/pnas.0607324104
Lavoie, C. (2010) Should we care about purple loosestrife? The history of an invasive plant in North America.
    Biological Invasions 12, 1967–1999. DOI: 10.1007/s10530-009-9600-7
Li, H.-S., Zou, S.-J., de Clercq, P. and Pang, H. (2018) Population admixture can enhance establishment
    success of the introduced biological control agent *Cryptolaemus montrouzieri*. *BMC Evolutionary
    Biology* 18: 36. DOI: 10.1186/s12862-018-1158-5
Li, Y., Stift, M. and van Kleunen, M. (2018) Admixture increases performance of an invasive plant beyond
    first-generation heterosis, *Journal of Ecology* 106, 1595–1606. DOI: 10.1111/1365-2745.12926
Lombaert, E., Guillemaud, T., Thomas, C.E., Lawson Handley, L.J., Li, J. *et al.* (2011) Inferring the origin of
    populations introduced from a genetically structured native range by approximate Bayesian computa-
    tion: case study of the invasive ladybird *Harmonia axyridis*. *Molecular Ecology* 20, 4654–4670. DOI:
    10.1111/j.1365-294X.2011.05322.x
Mandeville, E.G., Hall, R.O. and Buerkle, C.A. (2022) Ecological outcomes of hybridization vary extensively
    in Catostomus fishes. *Evolution* 76, 2697–2711. DOI: 10.1111/evo.14624
Mao, D., Liu, T., Xu, C., Li, X. and Xing, Y. (2011) Epistasis and complementary gene action adequately ac-
    count for the genetic bases of transgressive segregation of kilo-grain weight in rice. *Euphytica* 180,
    261–271. DOI: 10.1007/s10681-011-0395-0
Mérot, C., Debat, V., Le Poul, Y., Merrill, R.M., Naisbit, R.E. *et al.* (2020) Hybridization and transgressive
    exploration of colour pattern and wing morphology in *Heliconius* butterflies. *Journal of Evolutionary
    Biology* 33, 942–956. DOI: 10.1111/jeb.13626
Michel, A., Arias, R.S., Scheffler, B.E., Duke, S.O., Netherland, M. and Dayan, F.E. (2004) Somatic mutation-
    mediated evolution of herbicide resistance in the nonindigenous invasive plant hydrilla (*Hydrilla
    verticillata*). *Molecular Ecology* 13, 3229–3237. DOI: 10.1111/j.1365-294X.2004.02280.x
Mitchell, N., Owens, G.L., Hovick, S.M., Rieseberg, L.H. and Whitney, K.D. (2019) Hybridization speeds
    adaptive evolution in an eight-year field experiment. *Scientific Reports* 9: 6746. DOI: 10.1038/s41598-
    019-43119-4

Molofsky, J. and Augspurger, C.K. (1992) The effect of leaf litter on early seedling establishment in a tropical forest. *Ecology* 73, 68–77. DOI: 10.2307/1938721

Molofsky, J. and Collins, A.R. (2014) Using native and invasive populations as surrogate 'species' to predict the potential for native and invasive populations to shift their range. *Evolutionary Ecology Research* 16, 505–516.

Molofsky, J., Keller, S.R., Lavergne, S., Kaproth, M.A. and Eppinga, M.B. (2014) Human-aided admixture may fuel ecosystem transformation during biological invasions: theoretical and experimental evidence. *Ecology and Evolution* 4, 899–910. DOI: 10.1002/ece3.966

Mounger, J., Ainouche, M.L., Bossdorf, O., Cavé-Radet, A., Li, B. *et al.* (2021) Epigenetics and the success of invasive plants. *Philosophical Transactions of the Royal Society B: Biological Sciences* 376: 20200117. DOI: 10.1098/rstb.2020.0117

Muñoz, A.A. and Cavieres, L.A. (2008) The presence of a showy invasive plant disrupts pollinator service and reproductive output in native alpine species only at high densities. *Journal of Ecology* 96, 459–467. DOI: 10.1111/j.1365-2745.2008.01361.x.

Okuda, N., Watanabe, K., Fukumori, K., Nakano, S.I. and Nakazawa, T. (2014) Predator diversity changes the world: from gene to ecosystem. In: Okuda, N., Watanabe, K., Fukumori, K., Nakano, S.I. and Nakazawa, T. (eds) *Biodiversity in Aquatic Systems and Environments: Lake Biwa*. Springer, Tokyo, pp. 21–49.

Palacio-Lopez, K. and Molofsky, J. (2021) Phenotypic shifts following admixture in recombinant offspring of *Arabidopsis thaliana*. *Evolutionary Ecology* 35, 575–593. DOI: 10.1007/s10682-021-10118-9

Palacio-Lopez, K., Keller, S.R. and Molofsky, J. (2018) Genomic admixture between locally adapted populations of *Arabidopsis thaliana* (mouse ear cress): evidence of optimal genetic outcrossing distance. *Journal of Heredity* 109, 38–46. DOI: 10.1093/jhered/esx079

Parepa, M., Fischer, M., Krebs, C. and Bossdorf, O. (2014) Hybridization increases invasive knotweed success. *Evolutionary Applications* 7, 413–420. DOI: 10.1111/eva.12139

Paterson, A.H., Schertz, K.F., Lin, Y.R., Liu, S.C. and Chang, Y.L. (1995) The weediness of wild plants: molecular analysis of genes influencing dispersal and persistence of johnsongrass, *Sorghum halepense* (L.) Pers. *Proceedings of the National Academy of Sciences USA* 92, 6127–6131. DOI: 10.1073/pnas.92.13.6127

Pérez, J.E., Nirchio, M., Alfonsi, C. and Muñoz, C. (2006) The biology of invasions: the genetic adaptation paradox. *Biological Invasions* 8, 1115–1121. DOI: 10.1007/s10530-005-8281-0

Prapas, D., Scalone, R., Lee, J., Nurkowski, K.A., Bou-Assi, S. *et al.* (2022) Quantitative trait loci mapping reveals an oligogenic architecture of a rapidly adapting trait during the European invasion of common ragweed. *Evolutionary Applications* 15, 1249–1263. DOI: 10.1111/eva.13453

Rieseberg, L.H., Archer, M.A. and Wayne, R.K. (1999) Transgressive segregation, adaptation and speciation. *Heredity* 83, 363–372. DOI: 10.1038/sj.hdy.6886170

Rieseberg, L.H., Widmer, A., Arntz, A.M. and Burke, B. (2003) The genetic architecture necessary for transgressive segregation is common in both natural and domesticated populations. *Philosophical Transactions of the Royal Society B: Biological Sciences* 358, 1141–1147. DOI: 10.1098/rstb.2003.1283

Ryan, M.E., Johnson, J.R. and Fitzpatrick, B.M. (2009) Invasive hybrid tiger salamander genotypes impact native amphibians. *Proceedings of the National Academy of Sciences USA* 106, 11166–11171. DOI: 10.1073/pnas.0902252106

Samarasinghe, H., You, M., Jenkinson, T.S., Xu, J. and James, T.Y. (2020) Hybridization facilitates adaptive evolution in two major fungal pathogens. *Genes* 11: 101. DOI: 10.3390/genes11010101

Schielzeth, H. and Husby, A. (2014) Challenges and prospects in genome-wide quantitative trait loci mapping of standing genetic variation in natural populations. *Annals of the New York Academy of Sciences* 1320, 35–57. DOI: 10.1111/nyas.12397

Schlaepfer, D.R., Glättli, M., Fischer, M. and van Kleunen, M. (2010) A multi-species experiment in their native range indicates pre-adaptation of invasive alien plant species. *New Phytologist* 185, 1087–1099. DOI: 10.1111/j.1469-8137.2009.03114.x

Shahid, M., Han, S., Yoell, H. and Xu, J. (2008) Fitness distribution and transgressive segregation across 40 environments in a hybrid progeny population of the human-pathogenic yeast *Cryptococcus neoformans*. *Genome* 51, 272–281. DOI: 10.1139/G08-004

Shang, L., Li, L.-F., Song, Z.-P., Wang, Y., Yang, J. *et al.* (2019) High genetic diversity with weak phylogeographic structure of the invasive *Spartina alterniflora* (Poaceae) in China. *Frontiers in Plant Science* 10: 1467. DOI: 10.3389/fpls.2019.01467

Shaw, F.H., Geyer, C.J. and Shaw, R.G. (2002) A comprehensive model of mutations affecting fitness and inferences for *Arabidopsis thaliana*. *Evolution* 56, 453–463. DOI: 10.1111/j.0014-3820.2002.tb01358.x

Shi, J., Macel, M., Tielbörger, K. and Verhoeven, K.J.F. (2018) Effects of admixture in native and invasive populations of *Lythrum salicaria*. *Biological Invasions* 20, 2381–2393. DOI: 10.1007/s10530-018-1707-2

Shivaramu, S., Lebeda, I., Kašpar, V. and Flajšhans, M. (2020) Intraspecific hybrids versus purebred: a study of hatchery-reared populations of sterlet *Acipenser ruthenus*. *Animals* 10: 1149. DOI: 10.3390/ani10071149

Skovmand, L.H., Xu, C.C.Y., Servedio, M.R., Nosil, P., Barrett, R.D.H. and Hendry, A.P. (2018) Keystone genes. *Trends in Ecology and Evolution* 33, 689–700. DOI: 10.1016/j.tree.2018.07.00

Smith, A.L., Hodkinson, T.R., Villellas, J., Catford, J.A., Csergő, A.M. *et al.* (2020) Global gene flow releases invasive plants from environmental constraints on genetic diversity. *Proceedings of the National Academy of Sciences USA* 117, 4218–4227. DOI: 10.1073/pnas.1915848117

Smith, L.M. and Weigel, D. (2012) On epigenetics and epistasis: hybrids and their non-additive interactions. *EMBO Journal* 31, 249–250. DOI: 10.1038/emboj.2011.473

Stelkens, R.B., Schmid, C., Selz, O. and Seehausen, O. (2009) Phenotypic novelty in experimental hybrids is predicted by the genetic distance between species of cichlid fish. *BMC Evolutionary Biology* 9: 283. DOI: 10.1186/1471-2148-9-283

Strayer, D.L. (2009) Twenty years of zebra mussels: lessons from the mollusk that made headlines. *Frontiers in Ecology and the Environment* 7, 135–141. DOI: 10.1890/080020

Strayer, D.L. (2012) Eight questions about invasions and ecosystem functioning. *Ecology Letters* 15, 1199–11210. DOI: 10.1111/j.1461-0248.2012.01817.x

te Beest, M., Le Roux, J.J., Richardson, D.M., Brysting, A.K., Suda, J. *et al.* (2012) The more the better? The role of polyploidy in facilitating plant invasions. *Annals of Botany* 109, 19–45. DOI: 10.1093/aob/mcr277

Tsutsui, N.D., Suarez, A.V., Holway, D.A. and Case, T.J. (2000) Reduced genetic variation and the success of an invasive species. *Proceedings of the National Academy of Sciences USA* 97, 5948–5953. DOI: 10.1073/pnas.100110397

Turgeon, J., Tayeh, A., Facon, B., Lombaert, E., de Clercq, P. *et al.* (2011) Experimental evidence for the phenotypic impact of admixture between wild and biocontrol Asian ladybird (*Harmonia axyridis*) involved in the European invasion. *Journal of Evolutionary Biology* 24, 1044–1052. DOI: 10.1111/j.1420-9101.2011.02234.x

Vercellino, R.B., Hernández, F. and Presotto, A. (2023) The role of intraspecific crop–weed hybridization in the evolution of weediness and invasiveness: cultivated and weedy radish (*Raphanus sativus*) as a case study. *American Journal of Botany* 110: e16217. DOI: 10.1002/ajb2.16217

Verhoeven, K.J.F., Macel, M., Wolfe, L.M. and Biere, A. (2011) Population admixture, biological invasions and the balance between local adaptation and inbreeding depression. *Proceedings of the Royal Society B: Biological Sciences* 278, 2–8. DOI: 10.1098/rspb.2010.1272

Waser, N.M. and Price, M.V. (1989) Optimal outcrossing in *Ipomopsis aggregata*: seed set and offspring fitness. *Evolution* 43: 1097. DOI: 10.2307/2409589

Whitham, T.G., Bailey, J.K., Schweitzer, J.A., Shuster, S.M., Bangert, R.K. *et al.* (2006) A framework for community and ecosystem genetics: from genes to ecosystems. *Nature Reviews Genetics* 7, 510–523. DOI: 10.1038/nrg1877

Whitlock, M.C., Phillips, P.C., Moore, Francisco, B.G. Moore and Tonsor, S.J. (1995) Multiple fitness peaks and epistasis. *Annual Review of Ecology and Systematics* 26, 601–629.

Wilson, C.E., Castro, K.L., Thurston, G.B. and Sissons, A. (2016) Pathway risk analysis of weed seeds in imported grain: a Canadian perspective. *NeoBiota* 30, 49–74. DOI: 10.3897/neobiota.30.7502

Wootton, L.S., Halsey, S.D., Bevaart, K., McGough, A., Ondreicka, J. and Patel, P. (2005) When invasive species have benefits as well as costs: managing *Carex kobomugi* (Asiatic sand sedge) in New Jersey's coastal dunes. *Biological Invasions* 7, 1017–1027. DOI: 10.1007/s10530-004-3124-y

Yisa, M., Olufeagba, S.O., Iwalewa, M., Gabriel, S.S., Olowosegun, O.M. *et al.* (2017) Improving growth performance of fingerlings of *Clarias anguillaris* through intraspecific hybridization. *Agronomie Africaine* 29, 83–89.

7 Epigenetic Contributions to Biological Invasions

Juntao Hu[1], Xuena Huang[2], Man Luo[1], Yiyong Chen[2], Bing Chen[1], and Aibin Zhan[2,3]*

[1]*Department of Ecology and Evolutionary Biology, School of Life Sciences, Fudan University, Shanghai, China;* [2]*Research Center for Eco-Environmental Sciences, Chinese Academy of Sciences, Beijing, China;* [3]*University of Chinese Academy of Sciences, Chinese Academy of Sciences, Beijing, China*

Abstract

Epigenetic mechanisms such as DNA methylation and non-coding RNAs have emerged as critical factors in shaping phenotypic plasticity and adaptive potential in invasive species. In contrast to genetic mutations, epigenetic modifications can occur more rapidly in response to environmental changes, thus largely facilitating invasion success. This chapter explores the role of epigenetic variations in biological invasions, focusing on environment-driven rapid epigenetic changes, their potential contributions to local adaptation, and the interplay between genetic and epigenetic variations. Evidence suggests that epigenetic modifications can either compensate for reduced genetic diversity or produce novel traits that improve fitness in new environments or in response to rapid environmental challenges. However, distinguishing adaptive from maladaptive changes remains challenging, and the evolutionary significance of epigenetic inheritance is still debated. Integrating epigenomic data into ecological and evolutionary studies is crucial for understanding the role of epigenetics in invasion success, particularly in the context of climate change.

7.1 Introduction

The introduction of invasive species into novel habitats often imposes a transitory decrease in population size (Sakai *et al.*, 2001). During the invasion process, invasive species often exhibit reduced genetic diversity but a high potential for adapting to changing environments (Briski *et al.*, 2018). Thus, epigenetic mechanisms have been extensively proposed as a compensatory mechanism for genetic variation to facilitate biological invasions by rapidly reprogramming plastic phenotypes under changing environmental conditions

to survive and successfully establish (Liebl *et al.*, 2013; Artemov *et al.*, 2017). Epigenetic modifications can be sensitive to environmental stimuli, and these environmentally induced epigenetic changes can rapidly generate physiological or phenotypic variations by regulating gene expression networks. Therefore, the extent to which environmental changes can trigger epigenetic variations or how fast the epigenetic modifications can be reshaped by environmental changes is particularly important for understanding the role of epigenetics in rapid adaptation of invasive species (Marin *et al.*, 2020).

---

*Corresponding author: zhanaibin@hotmail.com or azhan@rcees.ac.cn

© CAB International 2025. *Invasion Genomics* (Eds D. Bock and M. Rius)
DOI: 10.1079/9781800626263.0007

7.2   Epigenetic Patterns and Associated Phenotypic Effects

7.2.1   Environment-driven (rapid) epigenetic changes during biological invasions

The broad definition of epigenetics encompasses any non-genetic molecular modifications of the genome, transcriptome or proteome that alter gene expression. Well-characterized epigenetic modifications fall mainly into three groups: DNA methylation, histone modification, and non-coding RNAs (Carneiro and Lyko, 2020). DNA methylation, as the most well-studied epigenetic mechanism, has been shown to be rapidly modified by various abiotic stresses and is often associated with gene expression silencing (Bossdorf *et al.*, 2008). However, accumulating studies have shown that its effects are complex and depend on the genomic position of DNA methylation, such as gene promoters, gene bodies or DNA repeats (Schubeler, 2015). As a result, the explicit relationship between epigenetic mechanisms and phenotypic variation on the success of biological invasions is complex and challenging to unveil, with little supporting scientific evidence (Marin *et al.*, 2020; Fu *et al.*, 2021a; Mounger *et al.*, 2021). Nevertheless, insights into the contributions of epigenetics to biological invasions can be gained by studying environment-driven rapid epigenetic changes in invasive species (Pu and Zhan, 2017; Gao *et al.*, 2022; Chen *et al.*, 2024a).

Environmental changes can induce extensive DNA methylation variations in invasive species, occurring rapidly within hours to days. In a study by Huang *et al.* (2017), the marine invasive ascidian *Ciona savignyi* was exposed to salinity and temperature stresses for 3 days. Interestingly, the authors observed rapid DNA methylation reprogramming after just 1 h of high-temperature stress. Furthermore, the majority of environmentally induced DNA methylation changes recovered to normal conditions after 48 h, indicating that these changes can be transient and reversible during environmental responses (i.e. epigenetic resilience; Huang *et al.*, 2017). Additionally, the study noted that DNA methylation variations among individuals significantly increased under environmental stresses. This increased variability might provide raw materials for natural selection and potentially enhance survival possibilities during the invasion process. Studies have shown that such dynamic DNA methylation responses are attributed to DNA methylation and demethylation processes. For example, Fu *et al.* (2021b) identified a complete set of key genes involved in DNA methylation processes in invasive *Ciona* spp., including DNA methylation writers (DNA methyltransferase 1 (*DNMT1*) and DNA methyltransferase 3 (*DNMT3*)), and a DNA methylation reader (methyl-CpG-binding domain gene (*MBD*)), and eraser (ten-eleven-translocation gene (*TET*)). They analyzed the gene expression patterns in response to recurrent salinity stresses and found time-dependent and stress-specific expression patterns. The stress-induced changes in the expression of (de)methylation-related genes could partially explain the dynamic DNA methylation response patterns observed in invasive species when facing environmental challenges.

Whether these environmentally induced DNA methylation changes are random or directed remains under debate, and their persistence varies depending on the species and environmental context. Hu *et al.* (2019) experimentally transplanted invasive brown anole lizards, *Anolis sagrei*, on to small islands with distinct habitats for at least 4 days. They used the reduced-representation bisulfite sequencing method (Meissner *et al.*, 2005) to assess epigenetic responses to new environments at the early stage of island colonization. Significant DNA methylation changes were observed following colonization, especially in low-quality environments. However, not all environmental changes induced sufficient stress to trigger an epigenetic response. Notably, differentially methylated cytosines were found in genes functionally associated with phenotypic plasticity during habitat change, such as those involved in signal transduction, the immune response and circadian rhythm.

Another source of insightful information on epigenetics comes from studies on the three-spined stickleback, *Gasterosteus aculeatus*, a crucial model organism for studying ecological adaptation. This species has repeatedly colonized freshwater habitats from ancestral marine habitats and can even switch between sea and freshwater habitats within one generation (Artemov *et al.*, 2017). Several representative studies have demonstrated the significant role of epigenetic mechanisms in the sticklebacks' environmental adaptation. For instance, Artemov *et al.* (2017) conducted an experiment in which marine sticklebacks were placed into a freshwater environment, and freshwater sticklebacks were placed into seawater for 4 days. The study investigated the immediate DNA methylation responses to osmotic conditions. The results indicated that the DNA methylation profiles of marine sticklebacks placed into freshwater partially converged with those of freshwater sticklebacks. Moreover, DNA methylation changes in several genes related to skeletal ossification in short-term responses were similar to those observed in evolutionary adaptation. This suggests that the immediate DNA methylation response to freshwater environments can be maintained in freshwater populations. Heckwolf *et al.* (2020) complemented a field survey with a two-generation salinity acclimation experiment, comparing DNA methylation profiles between wild-caught and laboratory-bred sticklebacks. This study aimed to study the roles of DNA methylation mechanisms in both the short-term rapid response and long-term adaptation. The study observed that 63% of differentially methylated sites between populations remained stable under experimental salinity changes. Transgenerational salinity acclimation shifted DNA methylation status to be more similar to the anticipated adaptive state than within-generation acclimation, indicating transgenerational DNA methylation plasticity in salinity adaptation. However, in a re-analysis of available published data sets on sticklebacks, Hu and Barrett (2023) investigated the role of the DNA methylation plastic response to salinity change. They found that most plastic and evolved DNA methylation differences were in non-concordant directions. This finding suggests that a significant proportion of plastic DNA methylation changes are likely to be maladaptive during colonization to new environments.

Based on the available evidence so far, elucidating the relationship between plastic DNA methylation changes and their phenotypic effects in invasive species remains challenging. However, a relevant case was provided by Xie *et al.* (2015) in the context of Crofton weed, *Ageratina adenophora*. Originating from Mexico, this plant species was successively introduced to south-western tropical regions, subtropical areas and finally

north-eastern China through cold tolerance evolution. Xie *et al.* (2015) investigated the DNA methylation state of the C-repeat/dehydration-responsive element binding factor (CBF) pathway, which plays a crucial role in the cold response by activating cold-responsive genes. Their results revealed a negative correlation between the DNA methylation level of the CBF inducer *ICE1* gene and cold tolerance among distinct geographical populations of *A. adenophora*. To further understand the relationship between DNA methylation and cold tolerance, the researchers subjected four populations to cold stress treatment for 24 h. They found that plants carrying slightly methylated *ICE1* genes rapidly and significantly upregulated *ICE1* and its downstream gene expression levels compared with plants with regularly methylated *ICE1*. This study established a clear relationship between the DNA methylation of a single gene, gene expression regulation and enhanced cold tolerance in invasive plants.

In summary, delving into how invasive species epigenetically respond to environmental changes offers deep insights into the mechanisms driving invasion success. Despite the rapid advancements in this research field, several critical questions regarding epigenetic mechanisms warrant further theoretical and experimental exploration: (i) Are environmentally induced epigenetic changes random or directed? (ii) To what extent do environmentally induced epigenetic changes persist, either transiently or across generations? (iii) What is the causal relationship between epigenetics and phenotypic traits, and how does this relationship influence the fitness of invasive populations and their ability to rapidly respond to environmental shifts during invasions? (v) What roles do environmentally induced epigenetic changes play in the long-term adaptation of invasive species?

7.2.2 Epigenetic variations and potential roles in local adaptation

During the biological invasion process, epigenetic variation is believed to play a crucial role in facilitating the establishment of exotic organisms in recipient ecosystems (Hu and Barrett, 2017; Hawes *et al.*, 2019). In contrast to traditional natural selection, which depends on genetic variation and is typically transgenerational, epigenetic changes offer a more rapid response to environmental challenges, occurring within a single generation (Banerjee *et al.*, 2019; Marin *et al.*, 2020). Successful invaders are anticipated to exhibit a propensity for epigenetic variation, potentially reflected in their epigenetic signatures, irrespective of the genetic diversity within invasive populations (Richards *et al.*, 2012; Schrey *et al.*, 2012; Vogt, 2023). Therefore, investigating epigenetic patterns between native and introduced populations is a crucial step for understanding the role of epigenetic variation in successful biological invasions.

A high level of intrapopulation DNA methylation diversity has been detected in numerous invasive species (Schrey *et al.*, 2012; Hawes *et al.*, 2019). For instance, a substantial percentage of methylation-susceptible loci were found to be polymorphic, such as those in the highly invasive ascidians *Ciona robusta* and *C. intestinalis* (Ni *et al.*, 2018). The increased intrapopulation methylation diversity was also evident in the high values of the Shannon index (a measure of diversity that accounts for both the number of methylation variants and their evenness within a population) for these two ascidian species (Ni *et al.*, 2018, 2019). Despite the challenges associated with reduced genetic variation in introduced or invasive species due to genetic bottlenecks or founder effects, introduced populations often exhibit increased DNA methylation diversity compared with native populations, potentially fostering phenotypic diversity/plasticity when it is most crucial (Schrey *et al.*, 2012; Hawes *et al.*, 2019). Recent studies on invasive ascidians, including *Didemnum vexillum* (Hawes *et al.*, 2019), *C. robusta* (Ni *et al.*, 2018), and *C. intestinalis* (Ni *et al.*, 2018, 2019; Chen *et al.*, 2024a), have indicated that high levels of DNA methylation diversity are maintained in introduced populations, serving as a compensatory mechanism under the scenario of lack of genetic variation in successful introduced species. Introduced populations often exhibit lower levels of global or genome-wide DNA methylation compared with populations in their native ranges (Ardura *et al.*, 2017; Sheldon *et al.*, 2018). For example, introduced populations of *D. vexillum* displayed lower levels of global DNA methylation compared with native populations within the original regions (Hawes *et al.*, 2019). Meanwhile, significant reductions in global methylation levels were observed during the range expansion of the pygmy mussel, *Xenostrobus securis*, in Europe (Ardura *et al.*, 2017). The decrease in global methylation is interpreted as a rapid method of increasing phenotypic plasticity. This, in turn, facilitates the thriving of invasive populations during range expansions, precisely when such adaptability is most crucial.

Investigating the epigenetic patterns of introduced populations from diverse recipient environments offers unique opportunities to assess the role of epigenetic variation in accommodation or adaptation during biological invasions (Schrey *et al.*, 2012; Hawes *et al.*, 2018). A study on introduced Australian house sparrow populations revealed higher epigenetic divergence among sampling sites than among invasive population clusters (Sheldon *et al.*, 2018). This observation suggests that patterns of epigenetic variation are more significantly influenced by local environments or founder events than the initial genetic diversity in introduction populations. However, potentially different genetic backgrounds among sampling sites at large geographical scales may make it difficult to isolate the effect of environmental DNA methylation patterns (Ni *et al.*, 2019). To mitigate the confounding effect of genetic heterogeneity in wild populations, fine/regional geographical scales relative to the overall distribution of a species have been suggested. Studies in *C. intestinalis* identified local environment-driven methylation patterns of introduced populations at fine geographical scales (Ni *et al.*, 2019; Chen *et al.*, 2024b). Another strategy to reduce the effects of genetic variation in investigating phenotypic variation through epigenetic mechanisms is to utilize genetically identical clones or asexually reproducing populations. The asexual snail *Potamopyrgus antipodarum* serves as a compelling example of wild animals utilizing epigenetic mechanisms for environmental adaptation (Thorson *et al.*, 2017). These snails manifested differences in shell shape between lakes and rivers, correlating with water current speed and associated with significant genome-wide DNA methylation divergence (Thorson *et al.*, 2017). Furthermore, a comparison of snail populations from a rural lake and two polluted urban lakes revealed differences in shell shape and allometric growth between lakes (Thorson *et al.*, 2019). These differences were associated with numerous differentially methylated DNA regions, indicating adaptation to diverse environments and stressors through epigenetic mechanisms (Thorson *et al.*, 2019).

The association between changes in methylation patterns and the process or status of biological invasions is intriguing. A study in the pygmy mussel, *X. securis*, and tubeworm *Ficopomatus enigmaticus* revealed that recently introduced populations appeared to be less methylated compared with older introduced populations (Ardura *et al.*, 2017). The interpretation of decreased methylation as a rapid means of increasing phenotypic plasticity, aiding invasive populations to thrive, suggests that generalized hypomethylation could be a signature of early introductions (Ardura *et al.*, 2017). The maximum alteration at the epigenomic level is expected to occur during the arrival through the early expansion phase, when the species needs to enhance its adaptive capacity to overcome existing environmental constraints and establish a successful population (Hu and Barrett, 2017; Mounger *et al.*, 2021). Thus, epigenetic marks can serve as indicators to recognize the invasive status, and changes in methylation patterns may provide insights into the dynamics of the biological invasion process.

DNA methylation is known to be responsive to environmental cues and has been linked to adaptation to various environmental conditions in invaded habitats (Ardura *et al.*, 2017, 2018). Increasing evidence indicates variations in DNA methylation patterns among populations in different environments (Ni *et al.*, 2018, 2019; Sheldon *et al.*, 2018). Significant population DNA methylation divergence has been demonstrated in recently colonized populations (Ni *et al.*, 2019; Chen *et al.*, 2024a). DNA methylation differences in targeted genes were also observed among individuals sampled from wild populations with varying temperature and salinity, with some of this variation correlated with environmental differences (Pu and Zhan, 2017).

Studies of invasive species have identified an association between DNA methylation variation and different environments, with a large number of environment-related methylated loci being identified (Ni *et al.*, 2018, 2019; Gao *et al.*, 2022; Chen *et al.*, 2024a). For example, in a recent study on the ascidian *Botryllus schlosseri*, the epigenetic structure and divergence in nine global populations were significantly influenced by local environmental variables, with over 33% of DNA methylation loci being significantly associated with these variables (Gao *et al.*, 2022). This local environment-driven population methylation pattern was also observed in invasive species at fine geographical scales (Chen *et al.*, 2024a).

More importantly, both studies (Gao *et al.*, 2022; Chen *et al.*, 2024a) identified epigenetic-related genes and biological pathways that play a role in adaptation to environmental heterogeneity. Gao *et al.* (2022) identified the receptor-type tyrosine-protein phosphatase delta (*PTPRD*) gene, annotated with one key DNA methylation locus. This gene, a member of the tyrosine-protein phosphatase family, is involved in the insulin receptor signaling pathway. An increasing methylation level of the *PTPRD* gene may inhibit gene expression, decrease insulin production, and increase glycogen storage, providing essential energy to cope with environmental stress. Enriched biological pathways such as the metabolic pathway in *B. schlosseri* (Gao *et al.*, 2022) and the spliceosome in *C. intestinalis* (Chen *et al.*, 2024a) have been demonstrated to be involved in responses to temperature and salinity stresses. These findings suggest that epigenetic variation, including DNA methylation variation, plays a key role in the adaptation of invasive species to different environments.

7.2.3 Adaptive and non-adaptive epigenetic effect on phenotypes: does epigenetic variation facilitate or hinder biological invasions?

Relationships between phenotypic plasticity and epigenetic variation in invasive species

Phenotypic plasticity, the ability of a genotype to produce multiple phenotypes in response to varying environmental conditions, is a vital mechanism for organisms to cope with environmental shifts under a rapidly changing climate (Leung *et al.*, 2022; Hackerott *et al.*, 2023). Phenotypic plasticity may be adaptive or maladaptive (non-adaptive) depending on whether it can move the traits closer to or further away from the new optimum in new environments (Campbell-Staton *et al.*, 2021; Wood *et al.*, 2023). Adaptive plasticity that leads to higher fitness (e.g. growth, survival or reproduction) can be considered an evolutionary strategy for organisms to adapt to new environments (Fox *et al.*, 2019; Hackerott *et al.*, 2023). Under a rapidly changing climate, adaptive phenotypic plasticity has been proposed as an important mechanism for organisms to persist in the new environments and enable evolutionary cues by "buying time" for adaptive evolution to occur (Kelly, 2019; Lafuente and Beldade, 2019). In contrast, maladaptive plasticity will result in lower fitness for individuals in new environments, which may ultimately lead to species extinction (Fox *et al.*, 2019; Hackerott *et al.*, 2023), although a number of recent studies have also suggested that maladaptive plasticity can also facilitate adaptation due to its role in generating greater phenotypic variability to accelerate the strength of directional selection (Grether, 2005).

Plasticity may be of more importance to invasive species because the initial invasive populations are usually small, and thus potentially suffer from elevated levels of inbreeding, reduced genetic diversity and lack of adaptation potential (Jaspers *et al.*, 2021). Recent studies have suggested that epigenetic variation, which regulates gene expression without altering DNA sequence, is an important mechanism for invasive species to mediate phenotypic variation in response to environmental change (Carneiro and Lyko, 2020). For example, Swaegers *et al.* (2023) conducted a common-garden experiment at rearing temperatures that match the ancestral and invaded thermal regimes, and tested the role of epigenetic mechanism (i.e. DNA methylation) in contributing to higher heat tolerance during the expansion of damselfly (*Ischnura elegans*) populations to warmer regions. The authors found a significant increase in heat tolerance in the invaded populations exposed to the hypermethylating chemical, suggesting the possibly important role of DNA methylation in regulating phenotypic plasticity for invasive species to respond to new environments after invasion (Swaegers *et al.*, 2023).

Due to the complex source of epigenetic variation, an increasing body of studies have focused on the interplay between environments, genetic variation, and epigenetic variation in inducing phenotypic plasticity (Heckwolf *et al.*, 2020; Dang *et al.*, 2023; Hackerott *et al.*, 2023; Sepers *et al.*, 2023; Silliman *et al.*, 2023; Usui *et al.*, 2023; Venney *et al.*, 2023), and have tried to understand the consequence of different interplay modes on biological invasions (Huang *et al.*, 2017; Carneiro and Lyko, 2020; Lamar *et al.*, 2020; Gao *et al.*, 2022; Chen *et al.*, 2024a). However, to what extent the epigenetic variation was driven by environmental changes or genetic variation is difficult to disentangle, and thus the extent to which epigenetic change contributes to variable phenotypes in natural populations remains controversial (Peaston and Whitelaw, 2006; Husby, 2022). While studies of invasive species are still in their infancy, noteworthy examples of strong associations between epigenetic variation and phenotypic plasticity under low genetic diversity include invasion of diverse habitats by monoclonal marbled crayfish (Gutekunst *et al.*, 2018) and a few Japanese knotweed genotypes (Richards *et al.*, 2012), suggesting the potential of autonomous epigenetic variation in producing phenotypic plasticity.

Potential role of epigenetic variation in facilitating invasion success

Phenotypic plasticity is considered a stepping stone for long-term adaptation from an evolutionary perspective following invasion into new environments. Both theoretical and empirical studies have suggested that epigenetic variation can also be under selection and can contribute to population divergence and adaptation (Estoup *et al.*, 2016; Verhoeven *et al.*, 2016; Heckwolf *et al.*, 2020). Below, we review the key findings from these studies, which provide some of the best examples for understanding the potential role of epigenetic variation for facilitating invasion success.

EXAMPLES OF INVASIVE VERTEBRATES.   Most studies to date have focused on the role of DNA methylation in invasion success of vertebrate animals. For instance, previous studies have found that introduced house sparrow (*Passer domesticus*) populations exhibited extensive methylation and phenotypic variation among and within populations despite limited genetic diversity (Schrey *et al.*, 2012; Liebl *et al.*, 2013). In addition, the authors found a negative correlation between DNA methylation diversity and genetic diversity in the introduced populations, suggesting that epigenetic variation might play a vital role in compensating for the reduced genetic diversity after invasion

(Schrey *et al.*, 2012; Liebl *et al.*, 2013). However, a similar correlation was not found between epigenetic and genetic differentiation in a different study of house sparrow populations (Sheldon *et al.*, 2018), suggesting the contribution of epigenetic variation to invasion success may be geographically heterogeneous. In a more recent study of invasive house sparrows, Hanson *et al.* (2022) found a positive correlation between the number of genome-wide CpG sites and the distance from the initially introduced site. Because gaining CpG sites usually corresponds to an increased epigenetic potential, the higher number of CpG sites may imply greater phenotypic plasticity and hence the role of epigenetic variation in facilitating the adaptation of invasive populations to new habitats (Hanson *et al.*, 2021, 2022). Similar observations have also been made in amphibians. For example, methylation change was found in the early stages of founder populations of the invasive brown anole lizard, *A. sagrei* (Hu *et al.*, 2019). Differentially methylated loci were identified in newly transplanted populations compared with their source populations only after a 4-day exposure to new habitats, with these loci annotated with genes relevant to signal transduction, the immune response and circadian rhythm. While the above studies showed that epigenetic variation can occur quickly and possibly compensate for the reduced genetic diversity after invasion, which allows invasive species to persist in new environment, the direct link between epigenetic variation and fitness is still lacking, and thus warrants further studies to provide a solid connection between epigenetic modifications and successful invasion.

EXAMPLES OF INVASIVE INVERTEBRATES.   While the methylation level in invertebrates is lower and the genomic distribution is sparser than in vertebrate animals (Roberts and Gavery, 2012), DNA methylation has also been suggested to facilitate the adaptation of invasive invertebrate species to new environments by influencing phenotypic variation, although the majority of results come from aquatic animals. For example, Ardura *et al.* (2017) discovered lower DNA methylation levels in the most recent populations compared with the older introduced populations of pygmy mussel, *X. securis*. The authors hypothesized this hypomethylation pattern as a mechanism that promotes

growth and therefore facilitates the spread of invasive populations. Other examples of possible associations between epigenetic variation and invasion success include the notorious marbled crayfish, *Procambarus virginalis*. Despite negligible genetic polymorphisms, considerable phenotypic plasticity and adaptability have been found in invasive populations across heterogeneous environments (Andriantsoa *et al.*, 2019; Carneiro and Lyko, 2020). Moreover, compared with its non-invasive ancestral species *Procambarus fallax*, marbled crayfish showed reduced gene body methylation and higher gene expression variability, indicating the potential role of epigenetic variation in promoting adaptability and invasiveness via increasing phenotypic variability (Gatzmann *et al.*, 2018; Vogt, 2022). Epigenetic mechanisms have also been implied in the adaptation of the New Zealand freshwater snail, *P. antipodarum*, an invasive species that exhibited significant changes in shell shape in fewer than 30 years (approximately 100 generations) after establishment in western USA from a single clonal lineage. Greater genome-wide DNA methylation differences were found between habitats than between replicate sites of the same habitat, suggesting the potential role of epigenetic variation in regulating the expression of adaptive shell shape differences between habitats (Thorson *et al.*, 2017). The role of epigenetic variation in facilitating rapid adaptive responses to different environments has also been detected in terrestrial invertebrates. Compared with native populations, DNA methylation changes in the invasive populations of Asian tiger mosquito, *Aedes albopictus*, were found to be associated with differences in tolerance to thermal stress and chemical compounds, suggesting that DNA methylation is likely to play an important role in regulating responses to novel environments (Oppold *et al.*, 2015; Kress *et al.*, 2017). While the above studies generally highlight the correlation between epigenetic variation and environmental variation, these studies have also failed to establish the causal link between epigenetic change and phenotype variation, and thus it is difficult to know whether the observed epigenetic divergence between populations can be translated into fitness difference. Functional validation of the discovered epigenetic variation could be particularly valuable for assessing the contribution of epigenetic mechanisms to adaptation.

EXAMPLES OF INVASIVE PLANTS. Epigenetic variation has also been found to facilitate rapid adaptation to new environments in invasive plants. For example, when comparing native and invasive populations of *Carpobrotus edulis*, Campoy *et al.* (2022) found that the invasive populations had higher levels of global DNA methylation than the native populations, and that genes with DNA methylation changes were involved in phenotypic changes in leaf traits, particularly the nitrogen isotope composition. When comparing the DNA methylation patterns between two invasive *Chenopodium ambrosioides* populations from heavy-metal-contaminated and uncontaminated sites, Zhang *et al.* (2022) found the global DNA methylation level of *C. ambrosioides* was lower in metalliferous habitats with a number of differentially methylated loci related to the trait of heavy-metal accumulation, suggesting that epigenetic variations can play an important role in adaptation to metalliferous environments. Although there was little genetic differentiation between populations, the invasive alligator weed (*Alternanthera philoxeroides*) still exhibited extensive phenotypic and DNA methylation variation in the introduced range (Gao *et al.*, 2010). Under common-garden environments, the authors further found a correlation between methylation reprogramming and the reversible phenotypic response to changing environmental stimuli, suggesting the role of epigenetic mechanisms in mediating the phenotypic change selected in new environments (Gao *et al.*, 2010). The pattern of higher levels of diversity in epigenetic variation than in genetic variation in invasive populations has also been found in other invasive plant species, such as bluegrass (*Poa annua*) (Chwedorzewska and Bednarek, 2012) and Japanese knotweed (*Fallopia japonica*) (Richards *et al.*, 2012; Zhang *et al.*, 2016). Therefore, phenotypic variation regulated by epigenetic mechanisms may be responsible for the establishment and spread of a wide range of invasive plant taxa and could be especially important in the absence of genetic diversity.

Collectively, the above examples in animals and plants suggest that both the initial entry and range expansion phases of invasive species are likely to be associated with low genetic diversity that is compensated by high epigenetic diversity, which contribute to phenotypic change in response to novel environments in invasive range.

Therefore, epigenetic variation could be crucial for the success of invasive species to establish and maintain populations in new environments (Marin *et al.*, 2020; Hanson *et al.*, 2022). Although most studies on epigenetic patterns have found significant epigenetic variation between invasive and native populations, or within invasive populations from different environments, results from these studies cannot fully reveal the ecological and evolutionary relevance of epigenetic variation in invasion without carefully partitioning the relationship between genetic and epigenetic variation, and validating the influence of epigenetic variation in fitness-related traits. Thus, this work will serve as an important reference for future studies to investigate short- and long-term epigenetic responses to the diverse stressors caused by environmental change in invasive species.

7.3 Relationship Between Genetic and Epigenetic Variation

7.3.1 Interactive roles between genetic and epigenetic variation

Investigating the roles of genetic and epigenetic variation underlying adaptation during biological invasions can significantly enhance our understanding of how introduced species rapidly adapt and successfully spread in diverse environments. However, an ongoing debate focuses on whether epigenetic variation contributes to environmental adaptation in wild populations independently or as an intermediate step under genetic control (Hu and Barrett, 2017; Marin *et al.*, 2020; Mounger *et al.*, 2021). According to Richards (2006), their relationships can be divided into three types: (i) obligatory relationship (epigenetic variation is entirely dependent on genetic variation); (ii) semi-dependency (genetic variation can partially explain epigenetic variation); and (iii) autonomous relationship (epigenetic variation is completely independent of genetic variation). Our understanding of these relationships comes primarily from studies of humans and laboratory-reared model species such as rats, demonstrating that epigenetic changes are largely under genetic control through methylation quantitative trait loci (mQTLs) (Taudt *et al.*, 2016). However, studies on genetic and epigenetic variation in invasive species have often examined them in isolation, and comparable investigations have yet to systematically explore their relationship and relative contributions in facilitating population-level adaptation to environmental variation (Hu and Barrett, 2017).

As highlighted earlier, epigenetic variation can complement genetic diversity in genetically limited populations, rapidly generating adaptive phenotypes to cope with environmental changes during invasions. According to Chen et al. (2024a), more methylated sites associated with the environment have been identified than environment-associated single-nucleotide polymorphisms (SNPs), indicating that DNA methylation is numerically more important as a source of variation than genetic variation. Researchers have also observed distinctive patterns in population-level genetic and epigenetic variation in invasive species, such as the invasive ascidians C. intestinalis (Chen et al., 2024a) and B. schlosseri (Gao et al., 2022). For example, clear epigenetic divergence was observed in five genetically similar populations in C. intestinalis, with a higher degree of population DNA methylation divergence than genetic divergence (Chen et al., 2024a). Furthermore, environmental effects are hypothesized to be more significant than those associated with genetic variation, as indicated by stronger correlations between epigenetic changes and environmental variables, along with a considerable number of environmentally associated methylation sites (Chen et al., 2024a). Differences in putatively adaptive genetic and epigenetic population structures suggest that local environments may affect different loci or regions in the genome and epigenome. In this case, environment-associated methylation sites were not proximal to putatively adaptive SNPs, indicating a partly independent effect of epigenetic variants from genetic variation in terms of the "positional perspective" (Chen et al., 2024a). The mQTL analysis revealed that genetic variation explained only 18.67% of methylation variation in C. intestinalis, supporting the idea that a substantial number of DNA methylation sites were autonomous from genetic variation (Chen et al., 2024a). An independent role of epigenetic variation from genetic variation has also been found in three-spine sticklebacks, where only a moderate proportion (~30%) of epigenetic variation

was accounted for by additive genetic variation (Hu et al., 2021).

Recent studies on invasive ascidians, specifically B. schlosseri (Gao et al., 2022) and C. intestinalis (Chen et al., 2024a), have delved into the relationship between genetic and epigenetic variation involved in environmental adaptation and invasion success. These studies adopted a strategy that focused on functional genes and biological pathways from a functional perspective. Both investigations identified shared and specific genes and biological pathways among candidate adaptive genetic and epigenetic loci (Gao et al., 2022; Chen et al., 2024a). This consistency suggests a complementary and independent role of genetic and epigenetic variations in contributing to environmental adaptation in these species (Gao et al., 2022; Chen et al., 2024a). The interplay of genetic and epigenetic mechanisms is likely to contribute to the adaptive capacity and geographical distribution of invasive species in different environments. Collectively, these studies offer a straightforward yet effective approach to discerning the respective contributions of genetic and epigenetic variations to the success of adaptation and biological invasion (Gao et al., 2022; Chen et al., 2024a).

7.3.2 Evolutionary potential of epigenetic variation in invasive species

While phenotypic changes can be induced by epigenetic variation within a single generation, the evolutionary relevance of epigenetic variation rests on whether these changes are heritable across generations while improving population persistence. Below, we review key findings related to heritable epigenetic variation in invasive species.

Due to the typically short generation time, the majority of inter- and transgenerational studies of epigenetic variation are performed in invasive invertebrates and plants. For example, Oppold et al. (2015) found that in the invasive Asian tiger mosquito, A. albopictus, a number of the DNA methylation alterations induced by insecticide in the F_0 generation could be stably transmitted until F_2 before returning to the normal state. Similarly, Brevik et al. (2021) studied DNA methylation changes in the Colorado potato beetle, Leptinotarsa decemlineata, exposed to an insecticide (neonicotinoid imidacloprid), and found a gradually disappearing methylation signature of insecticide exposure from the parental to the F_2 generations. These results suggest that although epigenetic variation can be altered under environmental stress and may contribute to insecticide resistance, DNA methylation retains its capacity to be reset once the environmental cues that initially induced the variation have disappeared. Another example comes from analysis of the muti-generational stability of DNA methylation in six invasive alligator weed populations sampled from distinct environments (Shi *et al.*, 2018). The authors found that some of the epigenetic differences between wild populations were maintained after ten generations of cultivation in a common environment, suggesting that part of the population epigenetic variation may be due to selection instead of stochastic transmission. Such stability of epigenetic modification could be beneficial in novel environments if these modifications are pre-adapted, particularly in invasive species with low genetic diversity.

Although an increasing number of studies have focused on the evolutionary relevance of epigenetic variation by analyzing the degree of its inheritance (Heckwolf *et al.*, 2020; Anastasiadi *et al.*, 2021; Hu *et al.*, 2021; Loison, 2021; Fitz-James and Cavalli, 2022), multi-generational studies of epigenetic variation in invasive species are still scarce, possibly due to the large amount of effort needed to breed, and the significant resources required to quantify epigenetic marks across generations (Anastasiadi *et al.*, 2021). In addition, it has been suggested that epigenetic variation can be directed by genetic variation, demonstrating that the autonomy of epigenetic variation in this context requires a sufficient number of samples to provide statistical power (Taudt *et al.*, 2016). Finally, as suggested by Anastasiadi *et al.* (2021), different reproductive modes (e.g. sexual versus asexual, oviparity versus viviparity) can have significant consequences on the persistence of epigenetic marks across generations. Thus, further empirical work built on solid experimental design and sufficient sample size while considering the reproductive mode of the focal species is warranted to provide a comprehensive understanding of the evolutionary relevance of epigenetic variation.

7.4 Epigenetic Prediction of (Mal) adaptation of Invasive Species Across Current and Future Climatic Scenarios

Due to the regulatory role of epigenetic variation in plasticity and adaptation, it is possible to also use epigenetic marks to predict climate (mal) adaptation of invasive populations, although such analysis is still lacking. Analogous to similar approaches used for genomic data, these methods first require identification of the epigenetic loci underlying local adaptation, and then use of these associations to predict the temporal and spatial outcomes of climate change. There are usually two approaches to unravel these candidate adaptive epigenetic loci: (i) analyzing the epigenetic basis of climate-adaptive phenotypes; and (ii) identifying epigenetic loci showing variation between populations and/or along environmental gradients. The technical details of the two approaches have been described elsewhere, and we refer readers to other reviews (Capblancq *et al.*, 2020; Aguirre-Liguori *et al.*, 2021). Below, we present advantages and caveats associated with epigenetic variation to support its value to predict (mal)adaptation in invasive species.

7.4.1 Why should we include epigenetic variation when predicting (mal)adaptation of invasive species in addition to genetic variation?

A number of recent reviews have demonstrated the power of using a variety of statistical models, and genomic and transcriptomic variation for monitoring and analyzing species' responses to environmental changes (Fitzpatrick and Keller, 2015; Gougherty *et al.*, 2021; Keagy *et al.*, 2023; Bernatchez *et al.*, 2024). We echo the insights from these valuable syntheses, and believe that epigenetic variation can also be used to predict the future expansion range and vulnerability of invasive species for the following three reasons:

1. Higher sensitivity of epigenetic variation to environmental change. Unlike genetic mutation, which typically occurs at the scale of generations, *de novo* methylation and demethylation processes could occurr as rapidly as 1 h after exposure to thermal stress in *Ciona savignyi* (Huang *et al.*, 2017), or within days after entering a new environment in *A. sagrei* (Hu *et al.*, 2019). In addition, the induced methylation variation can be associated with adaptive plasticity that facilitates population persistence. For example, the regulation of DNA methylation variation on genes relevant to heat-shock survival, cold coma recovery and development has been suggested as playing an important role in contributing to the wide range of thermal and salinity tolerance in invasive white fly (*Bemisia tabaci*) (Dai *et al.*, 2018; Ji *et al.*, 2020). Thus, including methylation variation when predicting (mal)adaptation of invasive species can allow us to integrate plasticity into predictions of climate (mal)adaptation.

2. Epigenetic variation can serve as a long-term memory of ancestral environments. One of the mechanisms for invasive species to successfully adapt to environments in the invasive range is pre-adaptation. A number of studies have suggested that epigenetic variation can serve as a long-term memory, easing adaptation to similar environmental scenarios (Yakovlev *et al.*, 2011; Skrøppa, 2022; Hu and Barrett, 2023; Viejo *et al.*, 2023). While such examples in invasive species are scarce, a recent study has found that DNA methylation-mediated stress memory could effectively affect the response to stress and enhance the overall performance under recurring environmental challenges in *Ciona robusta* (Fu *et al.*, 2021b). While the mechanism underlying epigenetic memory is still unclear, it is possibly due to the heritability of epigenetic variation across generations (Anastasiadi *et al.*, 2021). Although the extent to which epigenetic modifications can be heritable is species dependent, by including methylation variation when predicting (mal)adaptation of invasive species while testing the fitness difference between populations from native and invasive ranges in native versus invasive habitats, we can assess the possible contribution of pre-adaptation to persistence of invasive species under future climate scenarios.

3. Epigenetic variation may be more important than genetic variation in predicting (mal)adaptation in invasive species. This is especially true because higher epigenetic diversity has been suggested to compensate for lower genetic diversity in founding populations of invasive species

in a wide range of taxa such as house sparrow (*Passer domesticus*; Liebl *et al.*, 2013), bluegrass (*Poa annua*; Chwedorzewska and Bednarek, 2012), bentgrass (*Deyeuxia angustifolia*; Ni *et al.*, 2021), marbled crayfish (*Procambarus virginalis*; Carneiro and Lyko, 2020), ascidians (*Ciona intestinalis*; Ni *et al.*, 2019), pygmy mussel (*Xenostrobus securis*), and tubeworm (*Ficopomatus enigmaticus*) (Ardura *et al.*, 2017). Thus, epigenetic variation may be the major source of phenotypic variation that provides raw material for natural selection to act on in the invasive species.

7.4.2 Important considerations before using epigenetic variation for (mal) adaptation prediction in biological invasions

As mentioned above, epigenetic variation can regulate phenotypic variation and potentially play an important role in contributing to adaptation in invasive species. However, unlike adaptive genetic variation that is ultimately derived from mutation and shaped by natural selection, the existence of tissue-, sex-, age-, and ontogeny-specific epigenetic variation makes the identification of adaptive epigenetic loci complicated (Horvath, 2013; Weyrich *et al.*, 2020; Anastasiadi *et al.*, 2021; Vernaz *et al.*, 2021; Hu and Barrett, 2023). We thus emphasize the application of careful experimental design and sample collection plan when using epigenetic variation in predicting the vulnerability of invasive species. In addition, due to the various degrees of genetic control over epigenetic variation, it is necessary to first partition the relationship between epigenetic and genetic variation and separately analyze the role of autonomous epigenetic variation in predicting (mal)adaptation. However, it is important to note that there are distinct data structures between genetic and epigenetic variation, with DNA methylation levels being known to have a binomial distribution and to suffer from overdispersion (Akalin *et al.*, 2012). Thus, checking whether models can adapt to such a data structure is an important first step to minimize false positives and avoid incorrect predictions. A small number of pioneer studies (e.g. Chen *et al.*, 2024), including one in invasive species (Chen *et al.*, 2024b), has proven the applicability of these models to methylation data, making the wide use of methylation in predicting invasive success promising. Finally, whereever possible, the effect of adaptive epigenetic loci on fitness and the degree of transgenerational stability/heritability of identified loci should be assessed in common-garden environments. Only by doing so can we validate the evolutionary relevance of these loci and increase the robustness of the discovered relationship between epigenetic variation and biological invasions under climate change.

References

Aguirre-Liguori, J.A., Ramirez-Barahona, S. and Gaut, B.S. (2021) The evolutionary genomics of species' responses to climate change. *Nature Ecology and Evolution* 5, 1350–1360. DOI: 10.1038/s41559-021-01526-9

Akalin, A., Kormaksson, M., Li, S., Garrett-Bakelman, F., Figueroa, M. *et al.* (2012) methylKit: a comprehensive R package for the analysis of genome-wide DNA methylation profiles. *Genome Biology* 13: R87. DOI: 10.1186/gb-2012-13-10-r87

Anastasiadi, D., Venney, C.J., Bernatchez, L. and Wellenreuther, M. (2021) Epigenetic inheritance and reproductive mode in plants and animals. *Trends in Ecology and Evolution* 36, 1124–1140. DOI: 10.1016/j.tree.2021.08.006

Andriantsoa, R., Tonges, S., Panteleit, J., Theissinger, K., Carneiro, V.C. *et al.* (2019) Ecological plasticity and commercial impact of invasive marbled crayfish populations in Madagascar. *BMC Ecology* 19: 8. DOI: 10.1186/s12898-019-0224-1

Ardura, A., Zaiko, A., Morán, P., Planes, S. and Garcia-Vazquez, E. (2017) Epigenetic signatures of invasive status in populations of marine invertebrates. *Scientific Reports* 7: 42193. DOI: 10.1038/srep42193

Ardura, A., Clusa, L., Zaiko, A., Garcia-Vazquez, E. and Miralles, L. (2018) Stress-related epigenetic changes may explain opportunistic success in biological invasions in Antipode mussels. *Scientific Reports* 8: 10793. DOI: 10.1038/s41598-018-29181-4

Artemov, A.V., Mugue, N.S., Rastorguev, S.M., Zhenilo, S., Mazur, A.M. *et al.* (2017) Genome-wide DNA methylation profiling reveals epigenetic adaptation of stickleback to marine and freshwater conditions. *Molecular Biology and Evolution* 34, 2203–2213. DOI: 10.1093/molbev/msx156

Banerjee, A.K., Guo, W. and Huang, Y. (2019) Genetic and epigenetic regulation of phenotypic variation in invasive plants—linking research trends towards a unified framework. *NeoBiota* 49, 77–103. DOI: 10.3897/neobiota.49.33723

Bernatchez, L., Ferchaud, A.-L, Berger, C.S., Venney, C.J. and Xuereb, A. (2024) Genomics for monitoring and understanding species responses to global climate change. *Nature Reviews Genetics* 25, 165–183. DOI: 10.1038/s41576-023-00657-y

Bossdorf, O., Richards, C.L. and Pigliucci, M. (2008) Epigenetics for ecologists. *Ecology Letters* 11, 106–115. DOI: 10.1111/j.1461-0248.2007.01130.x

Brevik, K., Bueno, E.M., McKay, S., Schoville, S.D. and Chen, Y.H. (2021) Insecticide exposure affects intergenerational patterns of DNA methylation in the Colorado potato beetle, *Leptinotarsa decemlineata*. *Evolutionary Applications* 14, 746–757. DOI: 10.1111/eva.13153

Briski, E., Chan, F.T., Darling, J.A., Lauringson, V., MacIsaac, H.J. *et al.* (2018) Beyond propagule pressure: importance of selection during the transport stage of biological invasions. *Frontiers in Ecology and the Environment* 16, 345–353. DOI: 10.1002/fee.1820

Campbell-Staton, S.C., Velotta, J.P. and Winchell, K.M. (2021) Selection on adaptive and maladaptive gene expression plasticity during thermal adaptation to urban heat islands. *Nature Communications* 12: 6195. DOI: 10.1038/s41467-021-26334-4

Campoy, J.G., Sobral, M., Carro, B., Lema, M., Barreiro, R. and Retuerto, R. (2022) Epigenetic and phenotypic responses to experimental climate change of native and invasive *Carpobrotus edulis*. *Frontiers in Plant Science* 13: 888391. DOI: 10.3389/fpls.2022.888391

Capblancq, T., Fitzpatrick, M.C., Bay, R.A., Exposito-Alonso, M. and Keller, S.R. (2020) Genomic prediction of (mal)adaptation across current and future climatic landscapes. *Annual Review of Ecology, Evolution, and Systematics* 51, 245–269. DOI: 10.1146/annurev-ecolsys-020720-042553

Carneiro, V.C. and Lyko, F. (2020) Rapid epigenetic adaptation in animals and its role in invasiveness. *Integrative and Comparative Biology* 60, 267–274. DOI: 10.1093/icb/icaa023

Chen, B., Wang, M., Guo, Y., Zhang, Z., Zhou, W. *et al.* (2024) Climate-related naturally occurring epimuta-tion and their roles in plant adaptation in *A. thaliana*. *Molecular Ecology* 33: e17356. DOI: 10.1111/mec.17356

Chen, Y., Gao, Y., Zhang, Z. and Zhan, A. (2024b) Multi-omics inform invasion risks under global climate change. *Global Change Biology* 30: e17588. DOI: 10.1111/gcb.17588

Chen, Y., Ni, P., Fu, R., Murphy, K.J., Wyeth, R.C. *et al.* (2024a) (Epi)genomic adaptation driven by fine geographical scale environmental heterogeneity after recent biological invasions. *Ecological Applications* 34: e2772. DOI: 10.1002/eap.2772

Chwedorzewska, K. and Bednarek, P. (2012) Genetic and epigenetic variation in a cosmopolitan grass *Poa annua* from Antarctic and Polish populations. *Polish Polar Research* 33, 63–80. DOI: 10.2478/v10183-012-0004-5

Dai, T.M., Lü, Z.C., Wang, Y.S., Liu, W.X., Hong, X.Y. and Wan, F.H. (2018) Molecular characterizations of DNA methyltransferase 3 and its roles in temperature tolerance in the whitefly, *Bemisia tabaci* Mediterranean. *Insect Molecular Biology* 27, 123–132. DOI: 10.1111/imb.12354

Dang, X., Lim, Y.K., Li, Y., Roberts, S.B., Li, L. and Thiyagarajan, V. (2023) Epigenetic-associated phenotypic plasticity of the ocean acidification-acclimated edible oyster in the mariculture environment. *Molecular Ecology* 32, 412–427. DOI: 10.1111/mec.16751

Estoup, A., Ravigné, V., Hufbauer, R., Vitalis, R., Gautier, M. and Facon, B. (2016) Is there a genetic paradox of biological invasion? *Annual Review of Ecology, Evolution, and Systematics* 47, 51–72. DOI: 10.1146/annurev-ecolsys-121415-032116

Fitz-James, M.H. and Cavalli, G. (2022) Molecular mechanisms of transgenerational epigenetic inheritance. *Nature Reviews Genetics* 23, 325–341. DOI: 10.1038/s41576-021-00438-5

Fitzpatrick, M.C. and Keller, S.R. (2015) Ecological genomics meets community-level modelling of biodiversity, mapping the genomic landscape of current and future environmental adaptation. *Ecology Letters* 18, 1–16. DOI: 10.1111/ele.12376

Fox, R.J., Donelson, J.M., Schunter, C., Ravasi, T. and Gaitan-Espitia, J.D. (2019) Beyond buying time, the role of plasticity in phenotypic adaptation to rapid environmental change. *Philosophical Transactions of the Royal Society B: Biological Sciences* 374: 20180174. DOI: 10.1098/rstb.2018.0174

Fu, R., Huang, X., Chen, Y., Chen, Z. and Zhan, A. (2021a) Interactive regulations of dynamic methylation and transcriptional responses to recurring environmental stresses during biological invasions. *Frontiers in Marine Science* 8: 800745. DOI: 10.3389/fmars.2021.800745

Fu, R., Huang, X. and Zhan, A. (2021b) Identification of DNA (de)methylation-related genes and their transcriptional response to environmental challenges in an invasive model ascidian. *Gene* 768: 145331. DOI: 10.1016/j.gene.2020.145331

Gao, L., Geng, Y., Li, B., Chen, J. and Yang, J. (2010) Genome-wide DNA methylation alterations of *Alternanthera philoxeroides* in natural and manipulated habitats, implications for epigenetic regulation of rapid responses to environmental fluctuation and phenotypic variation. *Plant, Cell & Environment* 33, 1820–1827. DOI: 10.1111/j.1365-3040.2010.02186.x

Gao, Y., Chen, Y., Li, S., Huang, X., Hu, J. *et al.* (2022) Complementary genomic and epigenomic adaptation to environmental heterogeneity. *Molecular Ecology* 31, 3598–3612. DOI: 10.1111/mec.16500

Gatzmann, F., Falckenhayn, C., Gutekunst, J., Hanna, K., Raddatz, G. *et al.* (2018) The methylome of the marbled crayfish links gene body methylation to stable expression of poorly accessible genes. *Epigenetics and Chromatin* 11: 57. DOI: 10.1186/s13072-018-0229-6

Gougherty, A.V., Keller, S.R. and Fitzpatrick, M.C. (2021) Maladaptation, migration and extirpation fuel climate change risk in a forest tree species. *Nature Climate Change* 11, 166–171. DOI: 10.1038/s41558-020-00968-6

Grether, G.F. (2005) Environmental change, phenotypic plasticity, and genetic compensation. *American Naturalist* 166, E115–E123. DOI: 10.1086/432023

Gutekunst, J., Andriantsoa, R., Falckenhayn, C., Hanna, K., Stein, W. *et al.* (2018) Clonal genome evolution and rapid invasive spread of the marbled crayfish. *Nature Ecology and Evolution* 2, 567–573. DOI: 10.1038/s41559-018-0467-9

Hackerott, S., Virdis, F., Flood, P.J., Souto, D.G., Paez, W. and Eirin-Lopez, J.M. (2023) Relationships between phenotypic plasticity and epigenetic variation in two Caribbean *Acropora* corals. *Molecular Ecology* 32, 4814–4828. DOI: 10.1111/mec.17072

Hanson, H.E., Zimmer, C., Koussayer, B., Schrey, A.W., Maddox, J.D. and Martin, L.B. (2021) Epigenetic potential affects immune gene expression in house sparrows. *Journal of Experimental Biology* 224: jeb238451. DOI: 10.1242/jeb.238451

Hanson, H.E., Wang, C., Schrey, A.W., Liebl, A.L., Ravinet, M. *et al*. (2022) Epigenetic potential and DNA methylation in an ongoing house sparrow (*Passer domesticus*) range expansion. *American Naturalist* 200, 662–674. DOI: 10.1086/720950

Hawes, N.A., Fidler, A.E., Tremblay, L.A., Pochon, X., Dunphy, B.J. and Smith, K.F. (2018) Understanding the role of DNA methylation in successful biological invasions: a review. *Biological Invasions* 20, 2285–2300. DOI: 10.1007/s10530-018-1703-6

Hawes, N.A., Amadoru, A., Tremblay, L.A., Pochon, X., Dunphy, B. *et al*. (2019) Epigenetic patterns associated with an ascidian invasion, a comparison of closely related clades in their native and introduced ranges. *Scientific Reports* 9: 14275. DOI: 10.1038/s41598-019-49813-7

Heckwolf, M.J., Meyer, B.S., Häsler, R., Höppner, M.P., Eizaguirre, C. and Reusch, T.B.H. (2020) Two different epigenetic information channels in wild three-spined sticklebacks are involved in salinity adaptation. *Science Advances* 6: eaaz1138. DOI: 10.1126/sciadv.aaz1138

Horvath, S. (2013) DNA methylation age of human tissues and cell types. *Genome Biology* 14: 3156. DOI: 10.1186/gb-2013-14-10-r115

Hu, J. and Barrett, R.D.H. (2017) Epigenetics in natural animal populations. *Journal of Evolutionary Biology* 30, 1612–1632. DOI: 10.1111/jeb.13130

Hu, J. and Barrett, R.D.H. (2023) The role of plastic and evolved DNA methylation in parallel adaptation of threespine stickleback (*Gasterosteus aculeatus*). *Molecular Ecology* 32, 1581–1591. DOI: 10.1111/mec.16832

Hu, J., Askary, A.M., Thurman, T.J., Spiller, D.A., Palmer, T.M. *et al*. (2019) The epigenetic signature of colonizing new environments in Anolis lizards. *Molecular Biology and Evolution* 36, 2165–2170. DOI: 10.1093/molbev/msz133

Hu, J., Smith, S.J., Barry, T.N., Jamniczky, H.A., Rogers, S.M. and Barrett, R.D.H. (2021) Heritability of DNA methylation in threespine stickleback (*Gasterosteus aculeatus*). *Genetics* 217: iyab001. DOI: 10.1093/genetics/iyab001

Huang, X., Li, S., Ni, P., Gao, Y., Jiang, B. *et al*. (2017) Rapid response to changing environments during biological invasions: DNA methylation perspectives. *Molecular Ecology* 26, 6621–6633. DOI: 10.1111/mec.14382

Husby, A. (2022) Wild epigenetics, insights from epigenetic studies on natural populations. *Proceedings of the Royal Society B: Biological Sciences* 289: 20211633. DOI: 10.1098/rspb.2021.1633

Jaspers, C., Ehrlich, M., Pujolar, J.M., Künzel, S., Bayer, T. *et al*. (2021) Invasion genomics uncover contrasting scenarios of genetic diversity in a widespread marine invader. *Proceedings of the National Academy of Sciences USA* 118: e2116211118. DOI: 10.1073/pnas.2116211118

Ji, S.X., Wang, X.D., Shen, X.N., Liang, L., Liu, W.X. *et al*. (2020) Using RNA interference to reveal the function of chromatin remodeling factor ISWI in temperature tolerance in *Bemisia tabaci* Middle East–Asia Minor 1 Cryptic species. *Insects* 11: 113. DOI: 10.3390/insects11020113

Keagy, J., Drummond, C.P., Gilbert, K.J., Grozinger, C.M., Hamilton, J. *et al*. (2023) Landscape transcriptomics as a tool for addressing global change effects across diverse species. *Molecular Ecology Resources* 25: e13796. DOI: 10.1111/1755-0998.13796

Kelly, M. (2019) Adaptation to climate change through genetic accommodation and assimilation of plastic phenotypes. *Philosophical Transactions of the Royal Society B: Biological Sciences* 374: 20180176. DOI: 10.1098/rstb.2018.0176

Kress, A., Oppold, A.M., Kuch, U., Oehlmann, J. and Muller, R. (2017) Cold tolerance of the Asian tiger mosquito *Aedes albopictus* and its response to epigenetic alterations. *Journal of Insect Physiology* 99, 113–121. DOI: 10.1016/j.jinsphys.2017.04.003

Lafuente, E. and Beldade, P. (2019) Genomics of developmental plasticity in animals. *Frontiers in Genetics* 10: 720. DOI: 10.3389/fgene.2019.00720

Lamar, S.K., Beddows, I. and Partridge, C.G. (2020) Examining the molecular mechanisms contributing to the success of an invasive species across different ecosystems. *Ecology and Evolution* 10: 10254–10270. DOI: 10.1002/ece3.6688

Leung, C., Grulois, D. and Chevin, L.M. (2022) Plasticity across levels: relating epigenomic, transcriptomic, and phenotypic responses to osmotic stress in a halotolerant microalga. *Molecular Ecology* 31, 4672–4687. DOI: 10.1111/mec.16542

Liebl, A.L., Schrey, A.W., Richards, C.L. and Martin, L.B. (2013) Patterns of DNA methylation throughout a range expansion of an introduced songbird. *Integrative and Comparative Biology* 53, 351–358. DOI: 10.1093/icb/ict007

Loison, L. (2021) Epigenetic inheritance and evolution, a historian's perspective. *Philosophical Transactions of the Royal Society B: Biological Sciences* 376: 20200120. DOI: 10.1098/rstb.2020.0120

Marin, P., Genitoni, J., Barloy, D., Maury, S., Gibert, P. *et al.* (2020) Biological invasion: the influence of the hidden side of the (epi)genome. *Functional Ecology* 34, 385–400. DOI: 10.1111/1365-2435.13317

Meissner, A., Gnirke, A., Bell, G.W., Ramsahoye, B., Lander, E.S. and Jaenisch, R. (2005) Reduced representation bisulfite sequencing for comparative high-resolution DNA methylation analysis. *Nucleic Acids Research* 33, 5868–5877. DOI: 10.1093/nar/gki901

Mounger, J., Ainouche, M.L., Bossdorf, O., Cavé-Radet, A., Li, B. *et al.* (2021) Epigenetics and the success of invasive plants. *Philosophical Transactions of the Royal Society B: Biological Sciences* 376: 20200117. DOI: 10.1098/rstb.2020.0117

Ni, B., You, J., Li, J., Du, Y., Zhao, W. and Chen, X. (2021) Genetic and epigenetic changes during the upward expansion of *Deyeuxia angustifolia* Kom. in the Alpine Tundra of the Changbai Mountains, China. *Plants* 10: 291. DOI: 10.3390/plants10020291

Ni, P., Li, S., Lin, Y., Xiong, W., Huang, X. and Zhan, A. (2018) Methylation divergence of invasive *Ciona ascidians*: significant population structure and local environmental influence. *Ecology and Evolution* 8, 10272–10287. DOI: 10.1002/ece3.4504

Ni, P., Murphy, K.J., Wyeth, R.C., Bishop, C.D., Li, S. and Zhan, A. (2019) Significant population methylation divergence and local environmental influence in an invasive ascidian *Ciona intestinalis* at fine geographical scales. *Marine Biology* 166: 143. DOI: 10.1007/s00227-019-3592-3

Oppold, A., Kress, A., Vanden, B.J., Diogo, J.B., Kuch, U. *et al.* (2015) Epigenetic alterations and decreasing insecticide sensitivity of the Asian tiger mosquito *Aedes albopictus*. *Ecotoxicology and Environmental Safety* 122, 45–53. DOI: 10.1016/j.ecoenv.2015.06.036

Peaston, A.E. and Whitelaw, E. (2006) Epigenetics and phenotypic variation in mammals. *Mammalian Genome* 17, 365–374. DOI: 10.1007/s00335-005-0180-2

Pu, C. and Zhan, A. (2017) Epigenetic divergence of key genes associated with water temperature and salinity in a highly invasive model ascidian. *Biological Invasions* 19, 2015–2028. DOI: 10.1007/s10530-017-1409-1

Richards, C.L., Schrey, A.W. and Pigliucci, M. (2012) Invasion of diverse habitats by few Japanese knotweed genotypes is correlated with epigenetic differentiation. *Ecology Letters* 15, 1016–1025. DOI: 10.1111/j.1461-0248.2012.01824.x

Richards, E.J. (2006) Inherited epigenetic variation—revisiting soft inheritance. *Nature Reviews Genetics* 7, 395–401. DOI: 10.1038/nrg1834

Roberts, S.B. and Gavery, M.R. (2012) Is there a relationship between DNA methylation and phenotypic plasticity in invertebrates? *Frontiers in Physiology* 2: 116. DOI: 10.3389/fphys.2011.00116

Sakai, A.K., Allendorf, F.W., Holt, J.S., Lodge, D.M., Molofsky, J. *et al.* (2001) The population biology of invasive species. *Annual Review of Ecology and Systematics* 32, 305–332. DOI: 10.1146/annurev.ecolsys.32.081501.114037

Schrey, A.W., Coon, C.A., Grispo, M.T., Awad, M., Imboma, T. *et al.* (2012) Epigenetic variation may compensate for decreased genetic variation with introductions: a case study using house sparrows (*Passer domesticus*) on two continents. *Genetics Research International* 2012: 979751. DOI: 10.1155/2012/979751

Schubeler, D. (2015) Function and information content of DNA methylation. *Nature* 517, 321–326. DOI: 10.1038/nature14192

Sepers, B., Chen, R.S., Memelink, M., Verhoeven, K.J.F. and van Oers, K. (2023) Variation in DNA methylation in avian nestlings is largely determined by genetic effects. *Molecular Biology and Evolution* 40: msad086. DOI: 10.1093/molbev/msad086

Sheldon, E.L., Schrey, A., Andrew, S.C., Ragsdale, A. and Griffith, S.C. (2018) Epigenetic and genetic variation among three separate introductions of the house sparrow (*Passer domesticus*) into Australia. *Royal Society Open Science* 5: 172185. DOI: 0.1098/rsos.172185

Shi, W., Chen, X., Gao, L., Xu, C.Y., Ou, X. *et al.* (2018) Transient stability of epigenetic population differentiation in a clonal invader. *Frontiers in Plant Science* 9: 1851. DOI: 10.3389/fpls.2018.01851

Silliman, K., Spencer, L.H., White, S.J. and Roberts, S.B. (2023) Epigenetic and genetic population structure is coupled in a marine invertebrate. *Genome Biology and Evolution* 15: evad013. DOI: 10.1093/gbe/evad013

Skrøppa, T. (2022) Epigenetic memory effects in Norway spruce: are they present after the age of two years? *Scandinavian Journal of Forest Research* 37, 6–13. DOI: 10.1080/02827581.2022.2045349

Swaegers, J., de Cupere, S., Gaens, N., Lancaster, L.T., Carbonell, J.A. *et al.* (2023) Plasticity and associated epigenetic mechanisms play a role in thermal evolution during range expansion. *Evolution Letters* 8, 76–88. DOI: 10.1093/evlett/qrac007

Taudt, A., Colomé-Tatché, M. and Johannes, F. (2016) Genetic sources of population epigenomic variation. *Nature Reviews Genetics* 17, 319–332. DOI: 10.1038/nrg.2016.45

Thorson, J.L., Smithson, M., Beck, D., Sadler-Riggleman, I., Nilsson, E. *et al.* (2017) Epigenetics and adaptive phenotypic variation between habitats in an asexual snail. *Scientific Reports* 7: 14139. DOI: 10.1038/s41598-017-14673-6

Thorson, J.L., Smithson, M., Sadler-Riggleman, I., Beck, D., Dybdahl, M. and Skinner, M.K. (2019) Regional epigenetic variation in asexual snail populations among urban and rural lakes. *Environmental Epigenetics* 5: dvz020. DOI: 10.1093/eep/dvz020

Usui, T., Lerner, D., Eckert, I., Angert, A.L., Garroway, C.J. *et al.* (2023) The evolution of plasticity at geographic range edges. *Trends in Ecology and Evolution* 38, 831–842. DOI: 10.1016/j.tree.2023.04.004

Venney, C.J., Cayuela, H., Rougeux, C., Laporte, M., Mérot, C. *et al.* (2023) Genome-wide DNA methylation predicts environmentally driven life history variation in a marine fish. *Evolution* 77, 186–198. DOI: 10.1093/evolut/qpac028

Verhoeven, K.J.F., von Holdt, B.M. and Sork, V.L. (2016) Epigenetics in ecology and evolution: what we know and what we need to know. *Molecular Ecology* 25, 1631–1638. DOI: 10.1111/mec.13617

Vernaz, G., Malinsky, M., Svardal, H., Du, M., Tyers, A.M. *et al.* (2021) Mapping epigenetic divergence in the massive radiation of Lake Malawi cichlid fishes. *Nature Communications* 12: 5870. DOI: 10.1038/s41467-021-26166-2

Viejo, M., Tengs, T., Yakovlev, I., Cross, H., Krokene, P. *et al.* (2023) Epitype-inducing temperatures drive DNA methylation changes during somatic embryogenesis in the long-lived gymnosperm Norway spruce. *Frontiers in Plant Science* 14: 1196806. DOI: 10.3389/fpls.2023.1196806

Vogt, G. (2022) Phenotypic plasticity in the monoclonal marbled crayfish is associated with very low genetic diversity but pronounced epigenetic diversity. *Current Zoology* 69, 426–441. DOI: 10.1093/cz/zoac094

Vogt, G. (2023) Environmental adaptation of genetically uniform organisms with the help of epigenetic mechanisms—an insightful perspective on ecoepigenetics. *Epigenomes* 7: 1. DOI: 10.3390/epigenomes7010001

Weyrich, A., Yasar, S., Lenz, D. and Fickel, J. (2020) Tissue-specific epigenetic inheritance after paternal heat exposure in male wild guinea pigs. *Mammalian Genome* 31, 157–169. DOI: 10.1007/s00335-020-09832-6

Wood, D.P., Holmberg, J.A., Osborne, O.G., Helmstetter, A.J., Dunning, L.T. *et al.* (2023) Genetic assimilation of ancestral plasticity during parallel adaptation to zinc contamination in *Silene uniflora*. *Nature Ecology and Evolution* 7, 414–423. DOI: 10.1038/s41559-022-01975-w

Xie, H.J., Li, H., Liu, D., Dai, W.M., He, J.Y. *et al.* (2015) ICE1 demethylation drives the range expansion of a plant invader through cold tolerance divergence. *Molecular Ecology* 24, 835–850. DOI: 10.1111/mec.13067

Yakovlev, I.A., Asante, D.K., Fossdal, C.G., Junttila, O. and Johnsen, Ø. (2011) Differential gene expression related to an epigenetic memory affecting climatic adaptation in Norway spruce. *Plant Science* 180, 132–139. DOI: 10.1016/j.plantsci.2010.07.004

Zhang, H., Tang, Y., Li, Q., Zhao, S., Zhang, Z. *et al.* (2022) Genetic and epigenetic variation separately contribute to range expansion and local metalliferous habitat adaptation during invasions of *Chenopodium ambrosioides* into China. *Annals of Botany* 130, 1041–1056. DOI: 10.1093/aob/mcac139

Zhang, Y.Y., Parepa, M., Fischer, M. and Bossdorf, O. (2016) Epigenetics of colonizing species? A study of Japanese knotweed in Central Europe. In: Barrett, S.C.H., Colautti, R.I., Dlugosch, K.M. and Rieseberg, L.H. (eds) *Invasion Genetics: The Baker and Stebbins Legacy*. Wiley-Blackwell, Hoboken, New Jersey, pp. 328–340.

8 How DNA Sequencing of Herbarium Specimens Can Elucidate Our Understanding of Plant Invasions: Insights from Common Ragweed

Paul Battlay[1]*, Vanessa C. Bieker[2]*, Tianchun Wu[1], Michael D. Martin[2]*, and Kathryn A. Hodgins[1]*†

[1]School of Biological Sciences, Monash University, Clayton, Victoria, Australia;
[2]Department of Natural History, NTNU University Museum, Norwegian University of Science and Technology (NTNU), Trondheim, Norway

Abstract

Herbarium specimens, collected over centuries, offer a unique temporal resource for studying the ecological and evolutionary dynamics of invasive species. Advances in DNA extraction and sequencing technologies now enable researchers to analyze degraded DNA from these specimens, providing critical insights into genetic, phenotypic and ecological changes across time and space. These data have advanced our understanding of key processes in invasion biology, such as the timing of introductions, geographical range shifts and rapid adaptation to novel environments. Furthermore, preserved microbial DNA enables exploration of changes in plant–microbe interactions during invasion. For example, studies of invasive *Ambrosia artemisiifolia* have used herbarium samples to trace its origin, identify climate-driven selection, quantify selection strength on adaptive loci and characterize shifts in the microbial community during invasion. Despite challenges such as uneven sampling, integrating genomic, phenotypic, and ecological data from herbarium specimens has transformative potential for advancing our understanding of invasion dynamics.

8.1 Introduction

Invasive species are one of the major drivers of ecological change, threatening global biodiversity and ecosystems by outcompeting and causing the extinction of native species (Bellard *et al.*, 2016). They also have a large economic impact due to damage (e.g. agricultural yield loss) and control measures, with an estimated global cost of at least US$1.288 trillion between 1970 and 2017 (Diagne *et al.*, 2021). Some invasive species also impact human health, such as those that spread disease or cause allergies. Understanding the genetic basis of successful biological invasions may provide important tools to stymie the rate of new introductions worldwide, which we have so far been unable to control (Seebens *et al.*, 2017).

Invasive species provide an excellent model for studying ecological and evolutionary processes.

---

*These authors contributed equally.

†Corresponding author: kathryn.hodgins@monash.edu

© CAB International 2025. *Invasion Genomics* (Eds D. Bock and M. Rius)

DOI: 10.1079/9781800626263.0008

While this has been recognized for some time (Grinnell, 1919), interest in studying invasive species—as a means to understand adaptation in particular—has grown substantially for several key reasons. First, most biological invasions have occurred in the last few hundred years, and therefore abiotic and, in some cases, biotic factors (e.g. Alves *et al.*, 2019) that might influence selection in the introduced ranges are well documented. Second, the distributions and traits of invasive populations over this period are often recorded in scientific literature and through historic collections found in herbaria or museums (Fig. 8.1a), and now increasingly through human observation (e.g. van Horn *et al.*, 2017). Third, they are typically easily propagated, have short generation times and have other features amenable to experimental manipulation (Bock *et al.*, 2015). Finally, many invasive species have been introduced into multiple regions, where each invasion represents a replicate of a "natural experiment" unfolding over contemporary timescales, offering researchers the opportunity to study dynamic ecological and evolutionary processes in real time (Bock *et al.*, 2015).

Herbarium specimens—pressed and dried plants that were collected up to 400 years ago (Soltis, 2017)—offer a multi-faceted resource for the study of invasive plants. These specimens provide critical insights into the timing of species introductions and shifts in their distributions, helping to predict future range expansions (Hudson *et al.*, 2021; Putra *et al.*, 2023). They also facilitate the study of traits, including morphological and biochemical characteristics, as well as patterns of herbivory, across spatial and temporal scales, offering a unique historical perspective (e.g. Meineke *et al.*, 2019). Furthermore, herbarium specimens serve as a valuable source of genetic data, enabling genomic studies of both the host plant and its associated microbial communities (e.g. Bieker *et al.*, 2022). By integrating ecological, phenotypic, and genetic information, herbarium collections provide an important temporal dimension to the study of invasive species.

8.2 Herbarium Data: Challenges and Opportunities

Over 3400 herbaria across the world contain approximately 350 million specimens collected within the last approximately 400 years (Soltis, 2017). Due to efforts in digitization of herbarium metadata in recent years (Tulig *et al.*, 2012) and the availability of these through online repositories such as the Global Biodiversity Information Facility (GBIF), natural history collections are now easily accessible. Although the collection and preservation of plants was predominantly done for taxonomical and systematic purposes, advances in recent years have led to a remarkable expansion in the number and diversity of scientific applications for these records (Bieker and Martin, 2018; Graham *et al.*, 2004; Pyke and Ehrlich, 2010; Soltis, 2017; Kistler *et al.*, 2020; Burbano and Gutaker, 2023). However, as is the case for all natural history collections, herbaria suffer from several biases in their collections (Pyke and Ehrlich, 2010). These include non-random and uneven temporal sampling due to preferences of collectors and curators, higher sampling of easily accessible over remote locations (Pyke and Ehrlich, 2010; Willis *et al.*, 2017), and biases against unpleasantly colored, small, and introduced plants (Schmidt-Lebuhn *et al.*, 2013). In addition, the associated metadata may be absent, erroneous or inaccurate (e.g. labels containing incorrect information about taxonomic identification) (Pyke and Ehrlich, 2010).

Even without accessing the rich genetic contents of herbaria, our understanding of plant invasions has benefited from the metadata associated with individual herbarium specimens, specifically the plants' presence at particular geographical locations and times. For example, herbarium specimen metadata and GIS-based environmental data have been used together to improve predictions of the long-term spatial extent of invasions based on species-specific life history traits (Lavoie *et al.*, 2013). Species distribution models of *Ambrosia artemisiifolia* allowed the identification of specific genetic clusters that were more likely to spread in Australia (Putra *et al.*, 2023). Genetic information from present-day populations of *Gypsophila paniculata* was associated with time-stamped geographical coordinates of herbarium records to reconstruct the plants' invasion history in North America, concluding that there had been two distinct invasion events, each characterized by different amounts of genetic diversity (Lamar and Partridge, 2021). Analysis of such invasion curves inform us about individual species' temporal

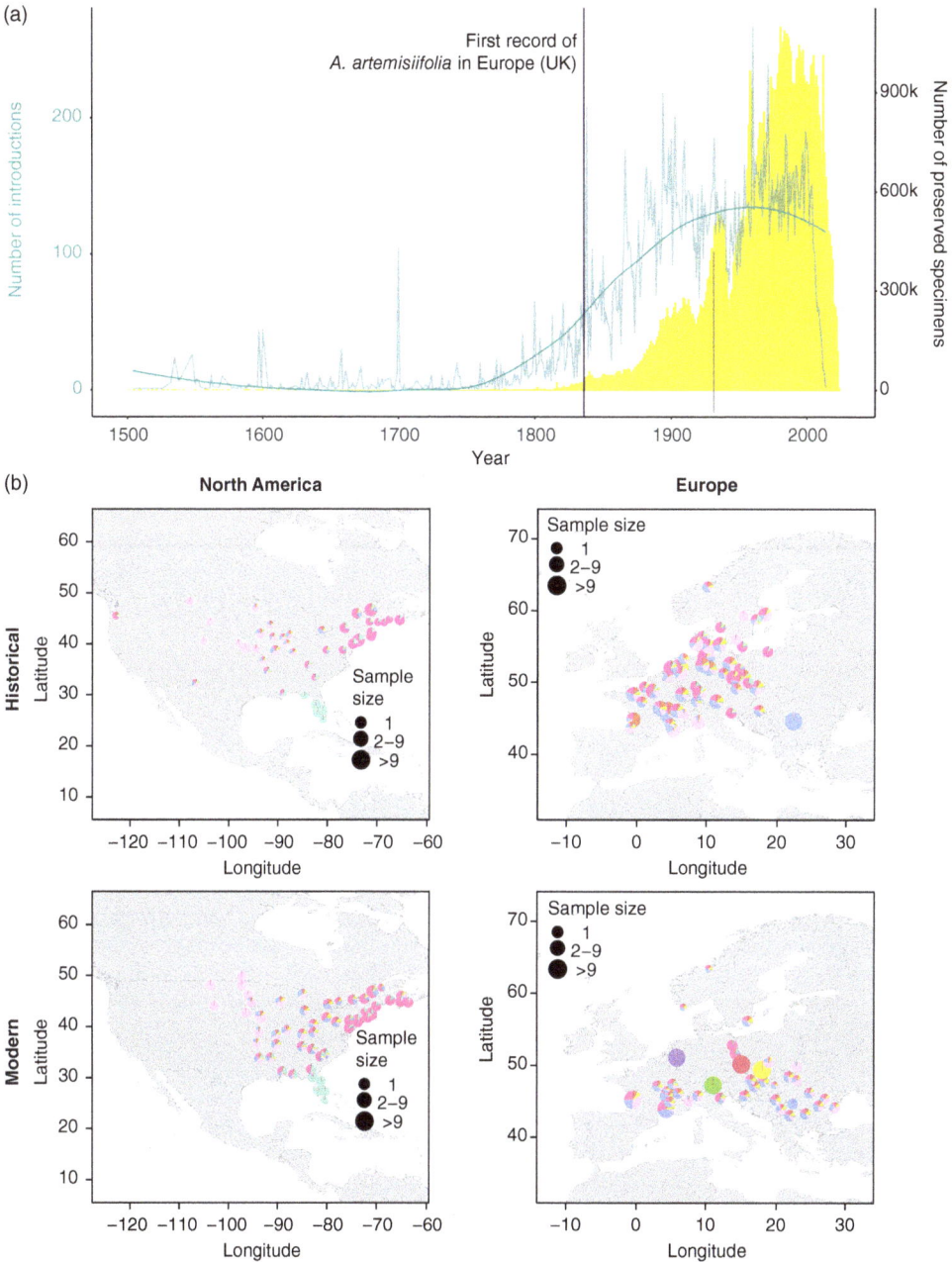

Fig. 8.1. (a) Introductions of invasive vascular plant species (green; regression line generated with geom_line (stat="smooth", method = "loess"); data from Global Alien Species First Record Database, DOI: 10.12761/sgn.2016.01.022) and the number of vascular plant herbarium specimens recorded per year (yellow bars; data from www.GBIF.org, accessed 22 May 2024, GBIF occurrence download https://doi.org/10.15468/dl.ah8ehu). The first record of *A. artemisiifolia* in Europe (purple line) is indicated according to Essl *et al.* (2015). (b) Genetic structure of historic and modern *A. artemisiifolia* populations (100 km groups) for the most likely number of ancestral populations (*K* = 9; after Bieker *et al.*, 2022). Figure used under license CC-BY-4.0.

progress through the three phases of invasion: introduction, establishment, and spread (Blackburn *et al.*, 2011). Such time-series data can also be used to investigate shifts in phenology (e.g. flowering time; Battlay *et al.*, 2023), as well as responses to anthropogenic activity, climate variation, and other biotic factors (Lister, 2011).

The utility of herbarium specimens for genetic studies has long been recognized (Rogers and Bendich, 1985). However, early attempts to amplify traditional barcode markers such as *rbcL* using the polymerase chain reaction (PCR) proved to be difficult, with extremely variable yields between specimens and variable success rates between markers (Savolainen *et al.*, 1995; Särkinen *et al.*, 2012; Korpelainen and Pietiläinen, 2019). The yield and quality of DNA from herbarium specimens is influenced by age and the preservation method used (Pyle and Adams, 1989; Adams and Sharma, 2010; Staats *et al.*, 2011; Weiß *et al.*, 2016). In addition, the DNA in dried leaf tissue decays about six times faster than in bones (Weiß *et al.*, 2016), resulting in highly fragmented and damaged DNA despite the relatively young age of the specimens (Adams and Sharma, 2010; Staats *et al.*, 2011; Weiß *et al.*, 2016). However, due to advances in ancient DNA extraction, sequencing, and bioinformatic tools over the last decades, including genotype likelihood-based methods that allow the analysis of low-depth sequencing data in population genomic studies, the accessibility of genetic material from herbarium specimens has increased immensely (Pääbo *et al.*, 2004; Jónsson *et al.*, 2013; Korneliussen *et al.*, 2014; Gutaker *et al.*, 2017). This allows the inclusion of specimens that cannot be obtained otherwise (or only through costly, time-intensive or difficult fieldwork), including endangered or extinct species, specimens from remote locations and type specimens (Shepherd and Perrie, 2014). For example, the phylogenetic position of a now-extinct monotypic genus (*Hesperelaea* A. Gray, Oleaceae) was inferred from genome skimming data of a 140-year-old specimen (Zedane *et al.*, 2016). This is particularly important in the context of invasive species studies, as many invaders cover vast regions of the world, making comprehensive collections challenging.

In addition to DNA from the specimen itself, herbarium specimens can also contain genetic material from associated pathogens, allowing the study of plant disease evolution and plant–pathogen interactions over time (Yoshida *et al.*, 2015). Studies of potato specimens infected with the oomycete *Phytophthora infestans* revealed that the population structure of the latter changed dramatically over time in Europe and that outbreaks were caused by different lineages (Martin *et al.*, 2013; Yoshida *et al.*, 2013). In another study, including *Citrus* spp. herbarium specimens in a phylogeny of *Xanthomonas citri* facilitated dating the diversification of this pathogenic bacterium, which causes Asiatic citrus canker in *Citrus* plants (Campos *et al.*, 2023). The high potential for herbarium specimens to be colonized by microbes during preservation, storage, and handling of the specimens hampers studies of the whole metagenomic community. Here, the degraded nature of DNA in herbarium specimens may be beneficial in distinguishing recent microbial contamination from microbes that were on the plant at the time of collection (Bieker *et al.*, 2020).

8.3 Demographic History of Biological Invasions

A main aim in most population genetic studies of invasive species is to understand invasion history, including the number of introduction events, their sources, and the demographic changes that coincide with species' introduction and spread. Such inferences are important as they can inform comparisons of introduced populations with their putative sources and aid in reconstructing evolutionary changes during the invasion (Hodgins *et al.*, 2018). In addition, if high-risk source populations can be identified, it may aid management and control efforts (Cristescu, 2015).

Species introductions are predicted to experience declines in genetic diversity during founding events, leading to a diminished capacity for adaptation and invasive spread (Dlugosch and Parker, 2008). However, evidence is mounting that invasions are frequently characterized by a complex demographic history, which can include founder events but also range expansion, admixture and hybridization (Kolbe *et al.*, 2007; Dlugosch and Parker, 2008; van Boheemen *et al.*, 2017; Rosinger *et al.*, 2021; Bieker *et al.*,

2022). The use of present-day genetic data alone may make it challenging to unravel this complexity because, for instance, the occurrence of genetic bottlenecks early on in an invasion can be obscured by subsequent introductions. Similarly, population bottlenecks combined with admixture may complicate identification of source populations. However, time-stratified genetic data from historical herbarium specimens can provide insight into the demographic history of complex invasions by allowing us to directly sample past diversity and thus infer temporal changes in genetic diversity and the presence, extent, and differential success of introduced lineages (Matsuhashi *et al.*, 2016).

In a population genomics study of the invasive weed *A. artemisiifolia*, the inclusion of over 300 historical herbarium specimen genomes, from both the North American native range east of the Rocky Mountains and Europe during relatively early stages of its invasion (collected 1844–1939), allowed identification of the source populations and admixture events in the introduced range (Bieker *et al.*, 2022). *A. artemisiifolia* populations today form three major ancestral genetic clusters that are weakly structured in eastern North America (Bieker *et al.*, 2022) and succinctly labeled according to their rough geographical distributions: Florida (South cluster); the Mississippi River watershed (West cluster); and the eastern third of the USA, from Louisiana to Maine, including Nova Scotia and New Brunswick in Canada, but excluding Florida (East cluster). There has been considerable mixing of the three clusters, especially in the "mid-east" region of the USA during the major expansion of agriculture in the 19th century (Martin *et al.*, 2014; van Boheemen *et al.*, 2017; Bieker *et al.*, 2022). This admixture in the native range largely pre-dated the global introduction of *A. artemisiifolia* populations, potentially with beneficial effects for the plants' invasion success (van Boheemen *et al.*, 2017). Indeed, using genomic data to estimate allele frequencies in various populations allowed the determination that the European invasive population was sourced mainly from the native range's admixed "mid-east" population, with little or no contribution from the South cluster population.

Beyond the identification of the European invasion's source, it was also found that mean genomic heterozygosity was lower in Europe, indicating a weak bottleneck effect, and that large temporal changes in population structure have occurred in Europe. The inclusion of genomic data from congeners facilitated testing for interspecific introgression in the introduced populations using the *D*-statistic (Bieker *et al.*, 2022). From this analysis, it was determined that the temporal changes in Europe were associated with introgression with other *Ambrosia* spp. While it was already known that *Ambrosia trifida* and *Ambrosia psilostachya* were similarly introduced to Europe during the 19th century (Montagnani *et al.*, 2017), it was not known that they had been hybridizing with *A. artemisiifolia*. These results suggest that interspecific hybridization may play an important role in the recruitment of novel diversity during evolution in invasive populations. Bieker *et al.* (2022) also compared present-day and historical genetic clusters at 16 specific geographical points throughout Europe and North America, which enabled tracking evolution in those regional populations. What was most striking about the temporal comparison was the substantial population structure turnover in Europe compared with the relative stasis of North America. Over more than 100 years of evolution in invasive populations across Europe, several distinct genetic groups have emerged, particularly in the Czech Republic, Austria, and France (Fig. 8.1b). These unique clusters were characterized by divergent allele frequencies, reductions in nucleotide diversity over time, and present-day low heterozygosity and effective population size. Introgression, strong selection and drift combined with multiple introductions and admixture likely contributed to these substantial temporal changes across Europe and the formation of these divergent genetic clusters.

8.4 Adaptive Evolution During Invasion

Invasive species can exhibit rapid adaptation to novel environmental conditions within their introduced ranges, leading to discernible adaptive evolutionary changes over short timeframes (Colautti and Barrett, 2013; Colautti and Lau, 2015; van Boheemen and Hodgins, 2020). A growing number of studies are using genomic data to identify recent signatures of selection in invasive populations. There are many well-established methods for elucidating regions of the genome under strong, recent selection using contemporary population genomic data, including extreme differentiation in allele frequencies across space (Akey *et al.*, 2002) and molecular footprints of selective sweeps (Kaplan *et al.*, 1989). However, the results of such "genome scans" are debatable because putative signatures of selection may be generated by neutral demographic events (Harris *et al.*, 2018); this is particularly pertinent in invasive species, which have often undergone complex, recent demographic shifts. Herbarium genomics allow the direct observation of changes in allele frequencies over time, providing an additional dimension for differentiating the effects of introduction history and selection.

Temporal genomic studies were once limited to microbial populations in laboratories and were largely off-limits to complex organisms evolving in nature due to logistical constraints. However, evolutionary genomics studies can now leverage natural history collections to explore "evolution in action" in populations experiencing recent environmental change or range expansion (Vandepitte *et al.*, 2014; Alves *et al.*, 2019; Bieker *et al.*, 2022; Kreiner *et al.*, 2022; Stuart *et al.*, 2022; Battlay *et al.*, 2023). These temporally resolved studies are providing a deeper understanding of the genetic architecture of adaptation, including the extent of genetic parallelism during adaptation to similar types of environmental change (Alves *et al.*, 2019; Kreiner *et al.*, 2022; Battlay *et al.*, 2023), as well as the source of the adaptive alleles (e.g. new mutations or standing variation) and their effect size (Battlay *et al.*, 2023; Vandepitte *et al.*, 2014).

Linking traits with environmental and genomic data offers a powerful method to dissect the genetic basis of local adaptation in invasive species. Analyses of contemporary *A. artemisiifolia* samples from common gardens have demonstrated strong patterns of parallel local climate adaptation in North America, Europe, and Australia at the phenotypic and genomic level (van Boheemen *et al.*, 2019; van Boheemen and Hodgins, 2020; Battlay *et al.*, 2023, 2025). Multiple latitudinal trait clines observed in the native range have re-evolved in invaded ranges of Europe and Australia (Fig. 8.2a). In particular, flowering-time and size-related traits show

strong clines with early flowering at a small size found at high latitudes and late flowering at a larger size at low latitudes. This pattern is likely due to selection for early flowering in shorter growing seasons at high latitudes, while delayed flowering is favored in longer growing seasons due to a trade-off between the timing of reproduction and size. Similarly, regions of the genome showing the strongest signals of climate adaptation are shared between North America and Europe more frequently than would be expected by chance (van Boheemen and Hodgins, 2020; Battlay *et al.*, 2023, 2025). These results provide strong evidence that rapid adaptation has occurred in invasive *A. artemisiifolia* since the species' introduction in response to environmental gradients in the invasive range that are similar to those in the native range.

The addition of temporally resolved data provides even greater support for recent local adaptation in Europe. These data allowed the identification of the specific local populations that experienced recent, strong selection to be identified, as the ancestral allele frequencies of each introduced population could be assessed. Here, temporal genomic analyses at the population scale were facilitated by herbarium samples fortuitously collected closely in both time and space, allowing direct comparisons with modern population samples from the same region. Selective sweeps—allele frequency shifts so rapid that they purge surrounding diversity—in these populations could be directly observed in regions of the genome that experienced extreme shifts in allele frequency over time coupled with extreme reductions in diversity in the modern samples relative to the historic samples (e.g. Fig. 8.2d). As expected, sweeps were more prevalent in European populations than North American populations, indicating more strong, recent selection in the invasive range than in the native range (Battlay *et al.*, 2023).

One of the most salient examples of recent, strong, climate-mediated selection in *A. artemisiifolia* is provided by the ortholog of *Arabidopsis thaliana*'s *EARLY FLOWERING 3* (*ELF3*) gene. This gene is involved in flowering-time adaptation in numerous other plant species, including wheat (Alvarez *et al.*, 2016) and barley (Zahn *et al.*, 2023). In *A. artemisiifolia*, variants in this region are strongly associated with the onset of flowering (Fig. 8.2b), a trait that shows strong latitudinal

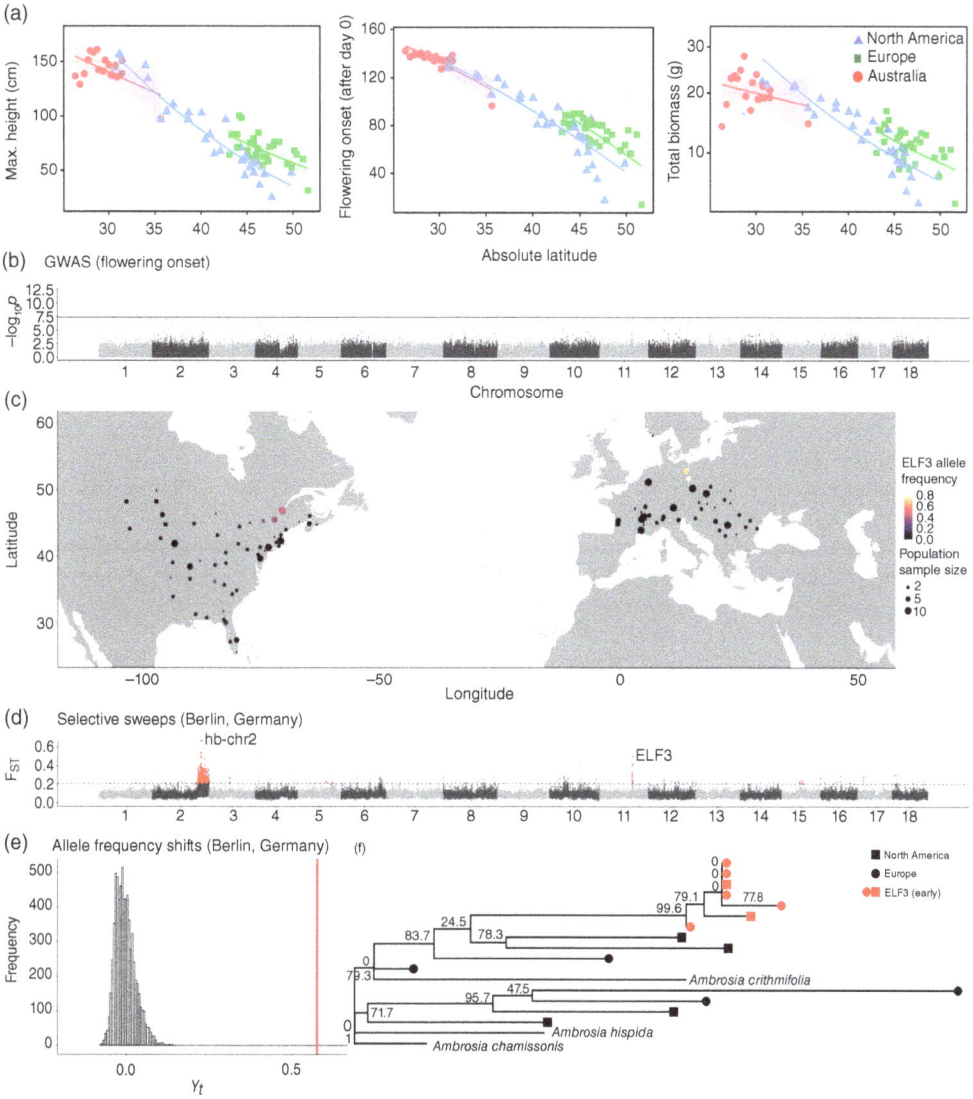

Fig. 8.2. Rapid climate adaptation in *A. artemisiifolia*. (a) Parallel latitudinal clines of putatively adaptive traits in the native and two invasive ranges. (b) Bonferroni-significant (above solid line) variants in *ELF3* are associated with early onset of flowering. (c) Early-flowering-associated single-nucleotide polymorphisms in *ELF3* are restricted to high-latitude populations in North America and Europe. (d) 10 kbp genomic windows showing signatures of selective sweeps (red points; extreme F_{ST} and extreme reductions in diversity between historic and modern samples) in temporally resolved samples from Berlin, Germany. (e) The temporal shift (y_t, a standardized measure of allele frequency change) in European *ELF3* allele frequency (red line) is more dramatic than 10,000 randomly sampled variants. (f) Maximum likelihood tree computed using IQ-Tree of the *ELF3* coding sequences suggests a single origin of early-flowering alleles (red) in North America (squares) and Europe (circles). Bootstrap scores are indicated on each node. Modified from Battlay *et al.* (2023) and van Boheemen *et al.* (2019). Figures used under license CC-BY-4.0.

clines in Europe, Australia, and North America (Fig. 8.2a). The geographical distribution of a non-synonymous single-nucleotide polymorphism (SNP) in this gene shows that the early-flowering allele is at high frequency in northern populations in both Europe and North America, specifically near Berlin and Quebec, respectively. This allele was rare in Berlin in the past but common in the modern samples (Fig. 8.2e), showing an allele frequency shift that is larger than expected based on 10,000 random genome-wide SNPs. Correspondingly, this region shows signatures of local selective sweeps in Berlin (Fig. 8.2d). A phylogenetic analysis of this gene region using North American and European samples with early- and late-flowering alleles clearly shows a recent single origin of the early-flowering allele, pointing to shared standing variation as the cause of the early-flowering variant in Europe rather than independent mutations in the same gene (Fig. 8.2f). This case study illustrates how the integration of temporal sampling into ecological genomics analysis can provide additional insight into the action of selection and the source of variation that underpins adaptation.

Many studies of invaders seek to identify traits or, more recently, regions of the genome that are under selection in the introduced range, in order to elucidate the types of evolutionary changes that might underpin invasion (Hodgins *et al.*, 2018). Most studies are based on comparisons of contemporary native and introduced populations, and evolutionary changes must be inferred. The difficulty with this approach is that the source populations may have been incorrectly identified, may remain unsampled or no longer exist, or evolutionary change may have occurred within the source populations. Temporal samples allow the ancestral state of the invasive and native populations to be estimated, providing a more accurate understanding of how evolution has proceeded following introduction. The use of historic samples allowed Bieker *et al.* (2022) to identify regions of the genome that showed divergence between Europe and North America, as well as divergence during the invasion of Europe. The analysis pointed to defense-response and flowering-time genes, which underpin traits that are likely important for adaptation during range expansion.

Temporal sampling can also capture information about the tempo of adaptation, allowing direct estimates of selection coefficients on individual loci (Alves *et al.*, 2019; Kreiner *et al.*, 2022). In *A. artemisiifolia*, the temporal analysis of selective sweeps in local populations in Europe identified a 14.5 Mbp region of chromosome 2 (*hb-chr2*; Fig. 8.2d) that had experienced

recent selection in most populations (Battlay *et al.*, 2023). Subsequent analyses revealed that this region showed population genomic signatures of a chromosomal inversion. Genotyping of historic and modern samples for the inversion haplotypes allowed the strength of selection acting on the putative inversion alleles to be estimated. Specifically, a 2.4% fitness advantage of one homozygous genotype over the other was estimated when directional selection was assumed. The extended timeframe covered by the herbarium samples enabled the detection of pronounced signals of selection for genetic loci, which might otherwise be overlooked in studies with limited temporal scope.

Beyond their utility as a source of historic DNA, museum and herbarium samples also provide important records of phenotypic variation that can be leveraged for studies of phenotypic change in invasive species (Flores-Moreno *et al.*, 2015). For example, nearly 1000 digitized images of European *A. artemisiifolia* herbarium samples collected as early as 1849 are freely available from GBIF (www.gbif.org; accessed 31 July 2025). Visual inspection of each image for the presence of flowers or fruit, coupled with sampling date and location metadata, allowed the inference of spatio-temporal patterns of trait variation across the European range. Phenological trait clines—specifically earlier flowering and fruit set in more northern latitudes that were also observed in common-garden experiments (Fig. 8.2a)—were recapitulated in these data but only in more recently collected herbarium samples (Fig. 8.3). This further supports the hypothesis that European *A. artemisiifolia* was generally poorly adapted to local environments upon introduction but has since rapidly evolved to better match the local growing conditions.

A major challenge to the utility of herbarium trait data for evolutionary inference is plasticity, where specimens have not experienced the same environments and thus trait differences through space and time do not necessarily reflect evolutionary change. Wu and Colautti (2022) were able to estimate flowering-time clines in purple loosestrife, *Lythrum salicaria*, during its invasion across North America using phenological measures estimated from herbarium sheets. They explicitly controlled for phenotypic plasticity using field experiments combined with a modeling approach, and demonstrated

(a)

(b)

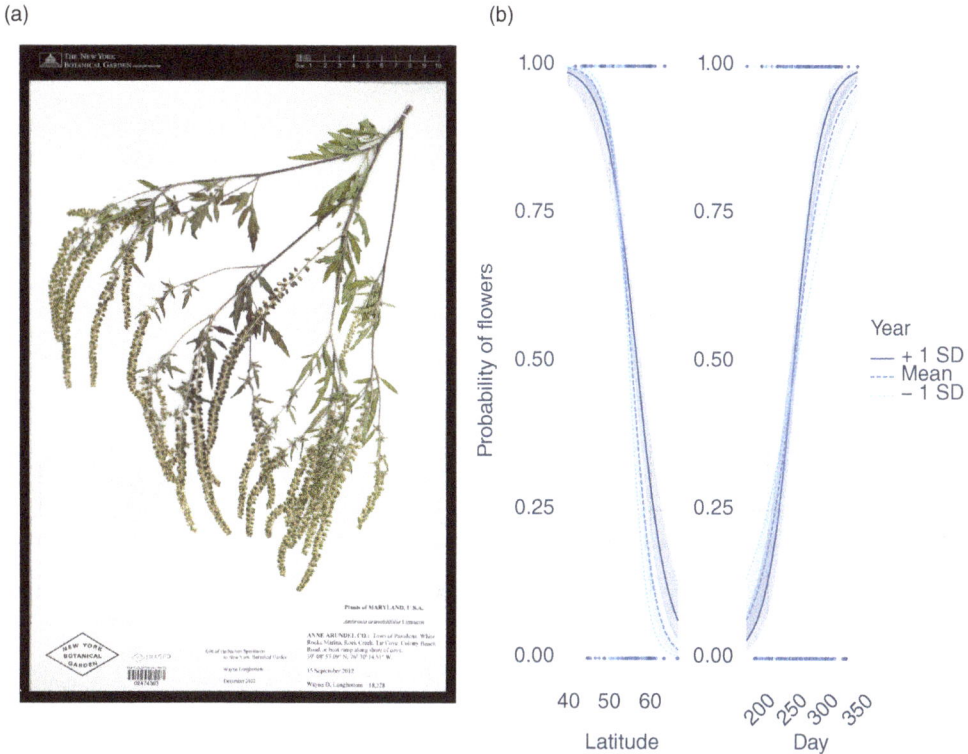

Fig. 8.3. Interpreting traits from herbarium specimens. (a) Example of a digitized herbarium specimen, *Ambrosia artemisiifolia* L., collected in the USA, by The New York Botanical Garden (licensed under http://creativecommons.org/licenses/by/4.0/). (b) Interaction plots illustrating the results of generalized linear models examining the presence of mature male inflorescences (probability of flowers) in digitized *A. artemisiifolia* herbarium specimens from Europe as a function of collection day, latitude of origin and collection year. The predicted probability of observing flowers is plotted as a function of latitude or collection day for different collection years (mean collection year ± 1 SD). Confidence intervals for the predictions and raw data are shown. Modified from Battlay *et al.* (2023).

the rapid evolution of multiple flowering-time clines with a predicted initial punctuated shift in the cline followed by evolutionary stasis. This approach has great promise for identifying evolutionary shifts in traits of other species in response to recent environmental change.

A hopeful avenue for discerning temporal evolutionary dynamics involves merging genomic and phenotypic data sets to reconstruct trait evolution (Swarts *et al.*, 2017; Lang *et al.*, 2024). For instance, Lang *et al.* (2024) used measures of stomatal density, genomic data and functional information to identify latitudinal clines in stomatal density across modern and historic samples of *A. thaliana*. They demonstrated a reduction in stomatal density over time, which may reflect

adaptation to climate change. Similarly, genomic estimation of complex traits in ancient maize provided evidence for adaptation to temperate North America (Swarts *et al.*, 2017). The further integration of phenotypic and temporally resolved genetic data will be key to future investigations of trait evolution in species expanding to novel ranges or adapting to changing climate. Such integration should not only reveal the tempo of trait evolution over time but also the dynamics of the genetic architecture of adaptation during invasion. For example, large-effect loci are predicted to be important during the initial stages of adaptation to sudden environmental change (Orr, 1998), and therefore might contribute disproportionately early on, while small-effect loci might be important later as the population approaches the optimum (Fisher, 1930; Orr, 1998). Temporal genomic studies in invaders could potentially provide an empirical test of these predictions in naturally evolving populations.

8.5   Metagenomics

A plant's microbiome can be a key factor in its fitness and therefore may be important to invasion success (Berg *et al.*, 2017). Within the native range, microbes and hosts have likely co-evolved for a long time, resulting in specialized interactions. However, when a host species is introduced to a new range, the associated microbes may not get introduced at the same time. In case of beneficial microbes, this may hamper the ability of the host to establish in the introduced range. However, the release from specialized enemies (e.g. pathogens) could facilitate invasion success (Blossey and Notzold, 1995; Williamson and Griffiths, 1996). In addition, introduced species are likely exposed to new microbes to which they need to adapt.

As herbarium specimens contain not only DNA from the specimen itself but also from associated microbes (Bieker *et al.*, 2020), they offer an opportunity to study the role of microbes on successful invasions over time. However, due to slow drying, long storage and handling of the specimens since their collection, herbarium samples are prone to microbial contamination (Bieker *et al.*, 2020), making it challenging to identify the natural microbial community. Thus, studies investigating microbes in herbarium specimens have focused on pathogens, often using plant material with signs of disease (Martin *et al.*, 2013; Yoshida *et al.*, 2015; Campos *et al.*, 2023). With recent advances in microbial community analyses (e.g. assembly-based methods (Nurk *et al.*, 2017), improved methods and databases for short-read classifications (Wood *et al.*, 2019; Lu *et al.*, 2022), and a better understanding of common herbarium contaminants), analyses of temporal changes in microbial communities may soon be tractable.

8.6   Future Directions

For centuries, botanists have diligently collected specimens to place into natural history collections with the goal of preserving and understanding plant diversity. It is unlikely that these early collectors could have predicted the wealth of information these samples could provide. Population genomic analysis of herbarium data is in its infancy, and we expect that the insights gleaned from it thus far only hint at what is to come for the study of invasion biology, and plant evolution more generally. Below, we outline some key future directions for this field.

Herbarium-derived sequences add a temporal dimension to population genomic data, which have already provided powerful insights into the ecological and evolutionary processes involved in biological invasions. More extensive herbarium sampling will increase the resolution of these data sets and allow more subtle evolutionary patterns to be detected. For example, the use of multiple historic time points could similarly be employed to better understand how any changes in adaptive variants might precipitate invasion following the "lag phase," which often occurs following introduction.

The utility of herbarium specimens may also be expanded by increasing the type of variation identifiable from their sequences. There is mounting evidence that large-effect structural variants play a key role in the range expansion of some invasive species (Tepolt and Palumbi, 2020; Santangelo *et al.*, 2022) including *A. artemisiifolia* (Battlay *et al.*, 2023, 2025). However, identification of many types of structural variation is intractable in most existing short-read sequence data, particularly the fragmented sequences derived from herbarium specimens. The combination of multiple reference-quality genome assemblies into a "pangenome" reference allows structural variants present in the pangenome to be genotyped across existing sequenced samples (Liao *et al.*, 2023). By this approach, previously overlooked forms of genetic variation may be tracked across invasive species' colonization histories. Epigenetic modifications occur at higher rates than genetic variants and can be heritable (Schmitz *et al.*, 2011); hence, they may play a crucial role in rapid adaptation (Mounger *et al.*, 2021), particularly for invasions with limited genetic diversity. Genomic patterns of DNA methylation and changes in chromatin architecture can be extracted from ancient or museum DNA (Gokhman *et al.*, 2016) and are an exciting direction for future studies of adaptation in invasive species.

A major goal in the study of invasive species is to understand the attributes that make them so successful. However, studies overwhelmingly focus on extant invasions, despite the fact that most species' introductions are expected to fail (Williamson and Fitter, 1996). Sampled among herbarium specimens are invasions that have subsequently died out, and therefore many of the population-scale and genomic analyses that have classically been applied to successful invasions may also be applied to those that have become extirpated. Such investigations may provide important insights into the requirements of successful invasions, particularly when failed and successful invasions exist in distinct ranges for the same species (Zenni and Nuñez, 2013).

As human-driven environmental upheaval continues to escalate the rate at which many species must adapt to avoid extinction, understanding the genetic basis of adaptation to sudden environmental shifts—a key question in evolutionary biology—becomes ever more important. Bolstered by the temporal aspect provided by herbarium specimens, invasive species are poised to become a powerful model for understanding the process of rapid adaptation.

References

Adams, R.P. and Sharma, L.N. (2010) DNA from herbarium specimens: I. correlation of DNA size with specimen age. *Phytologia* 92, 346–353.

Akey, J.M., Zhang, G., Zhang, K., Li, J. and Shriver, M.D. (2002) Interrogating a high-density SNP map for signatures of natural selection. *Genome Research* 12, 1805–1814. DOI: 10.1101/gr.631202

Alvarez, M.A., Tranquilli, G., Lewis, S., Kippes, N. and Dubcovsky, J. (2016) Genetic and physical mapping of the earliness *per se* locus *Eps-A^m 1* in *Triticum monococcum* identifies *EARLY FLOWERING 3* (*ELF3*) as a candidate gene. *Functional and Integrative Genomics* 16, 365–382. DOI: 10.1007/s10142-016-0490-3

Alves, J.M., Carneiro, M., Cheng, J.Y., de Matos, A.L., Rahman, M.M. *et al.* (2019) Parallel adaptation of rabbit populations to myxoma virus. *Science* 363, 1319–1326. DOI: 10.1126/science.aau7285

Battlay, P., Wilson, J., Bieker, V.C., Lee, C., Prapas, D., Petersen, B., Craig, S. *et al.* (2023) Large haploblocks underlie rapid adaptation in the invasive weed *Ambrosia artemisiifolia*. *Nature Communications* 14: 1717. DOI: 10.1038/s41467-023-37303-4

Battlay, P., Craig, S., Putra, A.R., Monro, K., de Silva, N.P. *et al.* (2025) Rapid parallel adaptation in distinct invasions of *Ambrosia artemisiifolia* is driven by large-effect structural variants. *Molecular Biology and Evolution* 42: msae270. DOI: 10.1093/molbev/msae270

Bellard, C., Cassey, P. and Blackburn, T.M. (2016) Alien species as a driver of recent extinctions. *Biology Letters* 12: 20150623. DOI: 10.1098/rsbl.2015.0623

Berg, G., Köberl, M., Rybakova, D., Müller, H., Grosch, R. and Smalla, K. (2017) Plant microbial diversity is suggested as the key to future biocontrol and health trends. *FEMS Microbiology Ecology* 93. DOI: 10.1093/femsec/fix050

Bieker, V.C. and Martin, M.D. (2018) Implications and future prospects for evolutionary analyses of DNA in historical herbarium collections. *Botany Letters* 165, 409–418. DOI: 10.1080/23818107.2018.1458651

Bieker, V.C., Sánchez Barreiro, F., Rasmussen, J.A., Brunier, M., Wales, N. and Martin, M.D. (2020) Metagenomic analysis of historical herbarium specimens reveals a postmortem microbial community. *Molecular Ecology Resources* 20, 1206–1219. DOI: 10.1111/1755-0998.13174

Bieker, V.C., Battlay, P., Petersen, B., Sun, X., Wilson, J. *et al.* (2022) Uncovering the genomic basis of an extraordinary plant invasion. *Science Advances* 8: eabo5115. DOI: 10.1126/sciadv.abo5115

Blackburn, T.M., Pyšek, P., Bacher, S., Carlton, J.T., Duncan, R.P. *et al.* (2011) A proposed unified framework for biological invasions. *Trends in Ecology and Evolution* 26, 333–339. DOI: 10.1016/j.tree.2011.03.023

Blossey, B. and Notzold, R. (1995) Evolution of increased competitive ability in invasive nonindigenous plants: a hypothesis. *Journal of Ecology* 83, 887–889. DOI: 10.2307/2261425

Bock, D.G., Caseys, C., Cousens, R.D., Hahn, M.A., Heredia, S.M. *et al.* (2015) What we still don't know about invasion genetics. *Molecular Ecology* 24, 2277–2297. DOI: 10.1111/mec.13032

Burbano, H.A. and Gutaker, R.M. (2023) Ancient DNA genomics and the renaissance of herbaria. *Science* 382, 59–63. DOI: 10.1126/science.adi1180

Campos, P.E., Pruvost, O., Boyer, K., Chiroleu, F., Cao, T.T. *et al.* (2023) Herbarium specimen sequencing allows precise dating of *Xanthomonas citri* pv. *citri* diversification history. *Nature Communications* 14: 4306. DOI: 10.1038/s41467-023-39950-z

Colautti, R.I. and Barrett, S.C.H. (2013) Rapid adaptation to climate facilitates range expansion of an invasive plant. *Science* 342, 364–366. DOI: 10.1126/science.1242121

Colautti, R.I. and Lau, J.A. (2015) Contemporary evolution during invasion: evidence for differentiation, natural selection, and local adaptation. *Molecular Ecology* 24, 1999–2017. DOI: 10.1111/mec.13162

Cristescu, M.E. (2015) Genetic reconstructions of invasion history. *Molecular Ecology* 24, 2212–2225. DOI: 10.1111/mec.13117

Diagne, C., Leroy, B., Vaissière, A.-C, Gozlan, R.E., Roiz, D. *et al.* (2021) High and rising economic costs of biological invasions worldwide. *Nature* 592, 571–576. DOI: 10.1038/s41586-021-03405-6

Dlugosch, K.M. and Parker, I.M. (2008) Founding events in species invasions: genetic variation, adaptive evolution, and the role of multiple introductions. *Molecular Ecology* 17, 431–449. DOI: 10.1111/j.1365-294X.2007.03538.x

Essl, F., Biró, K., Brandes, D., Broennimann, O., Bullock, J.M. *et al.* (2015) Biological flora of the British Isles: *Ambrosia artemisiifolia*. *Journal of Ecology* 103, 1069–1098. DOI: 10.1111/1365-2745.12424

Fisher, R.A. (1930) *The Genetical Theory of Natural Selection*. Clarendon Press, Oxford, UK.

Flores-Moreno, H., García-Treviño, E.S., Letten, A.D. and Moles, A.T. (2015) In the beginning: phenotypic change in three invasive species through their first two centuries since introduction. *Biological Invasions* 17, 1215–1225. DOI: 10.1007/s10530-014-0789-8

Gokhman, D., Meshorer, E. and Carmel, L. (2016) Epigenetics: it's getting old. Past meets future in paleoepigenetics. *Trends in Ecology and Evolution* 31, 290–300. DOI: 10.1016/j.tree.2016.01.010

Graham, C.H., Ferrier, S., Huettman, F., Moritz, C. and Peterson, A.T. (2004) New developments in museum-based informatics and applications in biodiversity analysis. *Trends in Ecology and Evolution* 19, 497–503. DOI: 10.1016/j.tree.2004.07.006

Grinnell, J. (1919) The English sparrow has arrived in Death Valley: an experiment in nature. *American Naturalist* 53, 468–472. DOI: 10.1086/279725.

Gutaker, R.M., Reiter, E., Furtwängler, A., Schuenemann, V.J. and Burbano, H.A. (2017) Extraction of ultra-short DNA molecules from herbarium specimens. *BioTechniques* 62, 76–79. DOI: 10.2144/000114517

Harris, R.B., Sackman, A. and Jensen, J.D. (2018) On the unfounded enthusiasm for soft selective sweeps II: examining recent evidence from humans, flies, and viruses. *PLOS Genetics* 14: e1007859. DOI: 10.1371/journal.pgen.1007859

Hodgins, K.A., Bock, D.G. and Rieseberg, L.H. (2018) Trait evolution in invasive species. *Annual Plant Reviews Online* 1, 459–496. DOI: 10.1002/9781119312994.apr0643

Hudson, J., Castilla, J.C., Teske, P.R., Beheregaray, L.B., Haigh, I.D. *et al.* (2021) Genomics-informed models reveal extensive stretches of coastline under threat by an ecologically dominant invasive species. *Proceedings of the National Academy of Sciences USA* 118: e2022169118. DOI: 10.1073/pnas.2022169118

Jónsson, H., Ginolhac, A., Schubert, M., Johnson, P.L.F. and Orlando, L. (2013) mapDamage2.0: fast approximate Bayesian estimates of ancient DNA damage parameters. *Bioinformatics* 29, 1682–1684. DOI: 10.1093/bioinformatics/btt193.

Kaplan, N.L., Hudson, R.R. and Langley, C.H. (1989) The 'hitchhiking effect' revisited. *Genetics* 123, 887–899. DOI: 10.1093/genetics/123.4.887

Kistler, L., Bieker, V.C., Martin, M.D., Pedersen, M.W., Madrigal, J.R. and Wales, N. (2020) Ancient plant genomics in archaeology, herbaria, and the environment. *Annual Review of Plant Biology* 71, 605–629. DOI: 10.1146/annurev-arplant-081519-035837

Kolbe, J.J., Larson, A. and Losos, J.B. (2007) Differential admixture shapes morphological variation among invasive populations of the lizard *Anolis sagrei*. *Molecular Ecology* 16, 1579–1591. DOI: 10.1111/j.1365-294X.2006.03135.x

Korneliussen, T.S., Albrechtsen, A. and Nielsen, R. (2014) ANGSD: analysis of next generation sequencing data. *BMC Bioinformatics* 15: 356. DOI: 10.1186/s12859-014-0356-4

Korpelainen, H. and Pietiläinen, M. (2019) The effects of sample age and taxonomic origin on the success rate of DNA barcoding when using herbarium material. *Plant Systematics and Evolution* 305, 319–324. DOI: 10.1007/s00606-019-01568-4

Kreiner, J.M., Latorre, S.M., Burbano, H.A., Stinchcombe, J.R., Otto, S.P. *et al.* (2022) Rapid weed adaptation and range expansion in response to agriculture over the past two centuries. *Science* 378, 1079–1085. DOI: 10.1126/science.abo7293

Lamar, S.K. and Partridge, C.G. (2021) Combining herbarium databases and genetic methods to evaluate the invasion of a popular horticultural species, baby's breath (*Gypsophila paniculata*), in the United States. *Biological Invasions* 23, 37–52. DOI: 10.1007/s10530-020-02354-x

Lang, P.L.M., Erberich, J.M., Lopez, L., Weiß, C.L., Amador, A. *et al.* (2024) Century-long timelines of herbarium genomes predict plant stomatal response to climate change. *Nature Ecology and Evolution* 8, 1641–1653. DOI: 10.1038/s41559-024-02481-x

Lavoie, C., Shah, M.A., Bergeron, A. and Villeneuve, P. (2013) Explaining invasiveness from the extent of native range: new insights from plant atlases and herbarium specimens. *Diversity and Distributions* 19, 98–105. DOI: 10.1111/ddi.12014

Liao, W.-W., Asri, M., Ebler, J., Doerr, D., Haukness, M. *et al.* (2023) A draft human pangenome reference. *Nature* 617, 312–324. DOI: 10.1038/s41586-023-05896-x

Lister, A.M. (2011) Natural history collections as sources of long-term datasets. *Trends in Ecology and Evolution* 26, 153–154. DOI: 10.1016/J.TREE.2010.12.009

Lu, J., Rincon, N., Wood, D.E., Breitwieser, F.P., Pockrandt, C. *et al.* (2022) Metagenome analysis using the Kraken software suite. *Nature Protocols* 17, 2815–2839. DOI: 10.1038/s41596-022-00738-y

Martin, M.D., Cappellini, E., Samaniego, J.A., Lisandra Zepeda, M., Campos, P.F. *et al.* (2013) Reconstructing genome evolution in historic samples of the Irish potato famine pathogen. *Nature Communications* 4: 2172. DOI: 10.1038/ncomms3172

Martin, M.D., Zimmer, E.A., Olsen, M.T., Foote, A.D., Gilbert, M.T.P. and Brush, G.S. (2014) Herbarium specimens reveal a historical shift in phylogeographic structure of common ragweed during native range disturbance. *Molecular Ecology* 23, 1701–1716. DOI: 10.1111/mec.12675

Matsuhashi, S., Kudoh, H., Maki, M., Cartolano, M., Tsiantis, M. *et al.* (2016) Invasion history of *Cardamine hirsuta* in Japan inferred from genetic analyses of herbarium specimens and current populations. *Biological Invasions* 18, 1939–1951. DOI: 10.1007/s10530-016-1139-9

Meineke, E.K., Classen, A.T., Sanders, N.J. and Davies, T.J. (2019) Herbarium specimens reveal increasing herbivory over the past century. *Journal of Ecology* 107, 105–117. DOI: 10.1111/1365-2745.13057

Montagnani, C., Gentili, R., Smith, M., Guarino, M.F. and Citterio, S. (2017) The worldwide spread, success, and impact of ragweed (*Ambrosia* spp.). *Critical Reviews in Plant Sciences* 36, 139–178. DOI: 10.1080/07352689.2017.1360112

Mounger, J., Ainouche, M.L., Bossdorf, O., Cavé-Radet, A., Li, B. *et al.* (2021) Epigenetics and the success of invasive plants. *Philosophical Transactions of the Royal Society B: Biological Sciences* 376: 20200117. DOI: 10.1098/rstb.2020.0117

Nurk, S., Meleshko, D., Korobeynikov, A. and Pevzner, P.A. (2017) MetaSPAdes: a new versatile metagenomic assembler. *Genome Research* 27, 824–834. DOI: 10.1101/gr.213959.116

Orr, H.A. (1998) The population genetics of adaptation: the distribution of factors fixed during adaptive evolution. *Evolution* 52, 935–949. DOI: 10.1111/j.1558-5646.1998.tb01823.x

Pääbo, S., Poinar, H., Serre, D., Jaenicke-Després, V., Hebler, J. *et al.* (2004) Genetic analyses from ancient DNA. *Annual Review of Genetics* 38, 645–679. DOI: 10.1146/annurev.genet.37.110801.143214

Putra, A.R., Hodgins, K.A. and Fournier-Level, A. (2023) Assessing the invasive potential of different source populations of ragweed (*Ambrosia artemisiifolia* L.) through genomically informed species distribution modelling. *Evolutionary Applications* 17: e13632. DOI: 10.1111/eva.13632

Pyke, G.H. and Ehrlich, P.R. (2010) Biological collections and ecological/environmental research: a review, some observations and a look to the future. *Biological Reviews* 85, 247–266. DOI: 10.1111/j.1469-185X.2009.00098.x

Pyle, M.M. and Adams, R.P. (1989) *In situ* preservation of DNA in plant specimens. *Taxon* 38: 576. DOI: 10.2307/1222632.

Rogers, S.O. and Bendich, A.J. (1985) Extraction of DNA from milligram amounts of fresh, herbarium and mummified plant tissues. *Plant Molecular Biology* 5, 69–76. DOI: 10.1007/BF00020088

Rosinger, H.S., Geraldes, A., Nurkowski, K.A., Battlay, P., Cousens, R.D. *et al.* (2021) The tip of the iceberg: genome wide marker analysis reveals hidden hybridization during invasion. *Molecular Ecology* 30, 810–825. DOI: 10.1111/mec.15768

Santangelo, J.S., Ness, R.W., Cohan, B., Fitzpatrick, C.R., Innes, S.G. *et al.* (2022) Global urban environmental change drives adaptation in white clover. *Science* 375, 1275–1281. DOI: 10.1126/science.abk0989

Särkinen, T., Staats, M., Richardson, J.E., Cowan, R.S. and Bakker, F.T. (2012) How to open the treasure chest? Optimising DNA extraction from herbarium specimens. *PLOS ONE* 7: e43808. DOI: 10.1371/journal.pone.0043808

Savolainen, V., Cuenoud, P., Spichiger, R., Martinez, M.D.P., Crevecoeur, M.M. and Manen, J.F. (1995) The use of herbarium specimens in DNA phylogenetics—evaluation and improvement. *Plant Systematics and Evolution* 97, 87–98. DOI: 10.1007/BF00984634

Schmidt-Lebuhn, A.N., Knerr, N.J. and Kessler, M. (2013) Non-geographic collecting biases in herbarium specimens of Australian daisies (*Asteraceae*). *Biodiversity and Conservation* 22, 905–919. DOI: 10.1007/s10531-013-0457-9

Schmitz, R.J., Schultz, M.D., Lewsey, M.G., O'Malley, R.C., Urich, M.A. *et al.* (2011) Transgenerational epigenetic instability is a source of novel methylation variants. *Science* 334, 369–373. DOI: 10.1126/science.1212959

Seebens, H., Blackburn, T.M., Dyer, E.E., Genovesi, P., Hulme, P.E. *et al.* (2017) No saturation in the accumulation of alien species worldwide. *Nature Communications* 8: 14435. DOI: 10.1038/ncomms14435

Shepherd, L. and Perrie, L. (2014) Genetic analyses of herbarium material: is more care required? *Taxon* 63, 972–973. DOI: 10.12705/635.2

Soltis, P.S. (2017) Digitization of herbaria enables novel research. *American Journal of Botany* 104, 1281–1284. DOI: 10.3732/ajb.1700281

Staats, M., Cuenca, A., Richardson, J.E., Vrielink-van Ginkel, R., Petersen, G. *et al.* (2011) DNA damage in plant herbarium tissue. *PLOS ONE* 6: e28448. DOI: 10.1371/journal.pone.0028448

Stuart, K.C., Sherwin, W.B., Austin, J.J., Bateson, M., Eens, M. *et al.* (2022) Historical museum samples enable the examination of divergent and parallel evolution during invasion. *Molecular Ecology* 31, 1836–1852. DOI: 10.1111/mec.16353

Swarts, K., Gutaker, R.M., Benz, B., Blake, M., Bukowski, R.B. *et al.* (2017) Genomic estimation of complex traits reveals ancient maize adaptation to temperate North America. *Science* 357, 512–515. DOI: 10.1126/science.aam9425

Tepolt, C.K. and Palumbi, S.R. (2020) Rapid adaptation to temperature via a potential genomic island of divergence in the invasive green crab, *Carcinus maenas*. *Frontiers in Ecology and Evolution* 8: 580701. DOI: 10.3389/fevo.2020.580701

Tulig, M., Tarnowsky, N., Bevans, M., Kirchgessner, A. and Thiers, B.M. (2012) Increasing the efficiency of digitization workflows for herbarium specimens. *ZooKeys* 209, 103–113. DOI: 10.3897/zookeys.209.3125

van Boheemen, L.A. and Hodgins, K.A. (2020) Rapid repeatable phenotypic and genomic adaptation following multiple introductions. *Molecular Ecology* 29, 4102–4117. DOI: 10.1111/mec.15429.

van Boheemen, L.A., Lombaert, E., Nurkowski, K.A., Gauffre, B., Rieseberg, L.H. and Hodgins, K.A. (2017) Multiple introductions, admixture and bridgehead invasion characterize the introduction history of *Ambrosia artemisiifolia* in Europe and Australia. *Molecular Ecology* 26, 5421–5434. DOI: 10.1111/mec.14293

van Boheemen, L.A., Atwater, D.Z. and Hodgins, K.A. (2019) Rapid and repeated local adaptation to climate in an invasive plant. *New Phytologist* 222, 614–627. DOI: 10.1111/nph.15564

van Horn, G., Aodha, O.M., Song, Y., Cui, Y., Sun, C. *et al.* (2017) The iNaturalist species classification and detection dataset. In: *2018 IEEE/CVF Conference on Computer Vision and Pattern Recognition, Salt Lake City, Utah*. IEEE, New York, pp. 8769–8778. DOI: 10.1109/CVPR.2018.00914

Vandepitte, K., de Meyer, T., Helsen, K., van Acker, K., Roldán-Ruiz, I. *et al.* (2014) Rapid genetic adaptation precedes the spread of an exotic plant species. *Molecular Ecology* 23, 2157–2164. DOI: 10.1111/mec.12683

Weiß, C.L., Schuenemann, V.J., Devos, J., Shirsekar, G., Reiter, E. *et al.* (2016) Temporal patterns of damage and decay kinetics of DNA retrieved from plant herbarium specimens. *Royal Society Open Science* 3: 160239. DOI: 10.1098/rsos.160239

Williamson, M. and Fitter, A. (1996) The varying success of invaders. *Ecology* 77, 1661–1666. DOI: 10.2307/2265769

Williamson, M. and Griffiths, B. (1996) *Biological Invasions*. Springer Science and Business Media, Dordrecht, Netherlands.

Willis, C.G., Ellwood, E.R., Primack, R.B., Davis, C.C., Pearson, K.D. *et al.* (2017) Old plants, new tricks: phenological research using herbarium specimens. *Trends in Ecology and Evolution* 32, 531–546. DOI: 10.1016/j.tree.2017.03.015

Wood, D.E., Lu, J. and Langmead, B. (2019) Improved metagenomic analysis with Kraken 2. *Genome Biology* 20: 257. DOI: 10.1186/s13059-019-1891-0

Wu, Y. and Colautti, R.I. (2022) Evidence for continent-wide convergent evolution and stasis throughout 150 y of a biological invasion. *Proceedings of the National Academy of Sciences USA* 119: e2107584119. DOI: 10.1073/pnas.2107584119

Yoshida, K., Schuenemann, V.J., Cano, L.M., Pais, M., Mishra, B. *et al.* (2013) The rise and fall of the *Phytophthora infestans* lineage that triggered the Irish potato famine. *eLife* 2: e00731. DOI: 10.7554/eLife.00731

Yoshida, K., Sasaki, E. and Kamoun, S. (2015) Computational analyses of ancient pathogen DNA from herbarium samples: challenges and prospects. *Frontiers in Plant Science* 6: 771. DOI: 10.3389/fpls.2015.00771

Zahn, T., Zhu, Z., Ritoff, N., Krapf, J., Junker, A. *et al.* (2023) Novel exotic alleles of *EARLY FLOWERING 3* determine plant development in barley. *Journal of Experimental Botany* 74, 3630–3650. DOI: 10.1093/jxb/erad127

Zedane, L., Hong-Wa, C., Murienne, J., Jeziorski, C., Baldwin, B.G. and Besnard, G. (2016) Museomics illuminate the history of an extinct, paleoendemic plant lineage (*Hesperelaea*, Oleaceae) known from an 1875 collection from Guadalupe Island, Mexico. *Biological Journal of the Linnean Society* 117, 44–57. DOI: 10.1111/bij.12509

Zenni, R.D. and Nuñez, M.A. (2013) The elephant in the room: the role of failed invasions in understanding invasion biology. *Oikos* 122, 801–815. DOI: 10.1111/j.1600-0706.2012.00254.x

9 The Ecological Genomic Processes of the Iconic Japanese Knotweed Invasion

Bethany Burns[1], Malika L. Ainouche[2], Elena Barni[3], Jingwen Bi[5], Bernd Blossey[6], Peipei Cao[4,5], Armand Cavé-Radet[2,7], Stacy B. Endriss[8,9], Elisa Giaccone[3], Uta Grunert[7], Yaolin Guo[5], Ramona E. Irimia[7], Ruiting Ju[5], Sophie Karrenberg[10], Kyle A. Keefer[1], Katie Lee[6], Zhiyong Liao[11], Madalin Parepa[7], Armel Salmon[2], Marc W. Schmid[12], Nicole Sebesta[3], Isolde Van Riemsdijk[7,13], Shengyu Wang[5,14], Jihua Wu[5], Wei Yuan[15], Lei Zhang[5], Weihan Zhao[5,16], Yujie Zhao[5], Xin Zhuang[5,17], Bo Li[5,18], Oliver Bossdorf[7], and Christina L. Richards[1,7,11]*

[1]University of South Florida, Tampa, FL, USA; [2]UMR CNRS 6553 ECOBIO, University of Rennes, France; [3]University of Turin, Turin, Italy; [4]Research Institute of Subtropical Forestry, Hangzhao, China; [5]Fudan University, Shanghai, China; [6]Department of Natural Resources, Cornell University, Ithaca, NY, USA; [7]University of Tübingen, Tübingen, Germany; [8]University of North Carolina Wilmington, Wilmington, NC, USA; [9]Virginia Tech, Blacksburg, VA, USA; [10]Uppsala University, Uppsala, Sweden; [11]Xishuangbanna Tropical Botanical Garden, Xishuangbanna, China; [12]MWSchmid GmbH, Glarus, Switzerland; [13]Lund University, Lund, Sweden; [14]Université Paris-Saclay, Paris, France; [15]Max Planck Institute, Tübingen, Germany; [16]University of Konstanz, Konstanz, Germany; [17]University of Helsinki, Helsinki, Finland; [18]Yunnan University, Kunming, China

Abstract

Many plant invasions establish from only a few individuals and exhibit clonal spread, providing an opportunity to examine the genomic mechanisms that underlie the success of particularly aggressive individuals. One of the world's most invasive species, the Japanese knotweed complex, is well suited for such investigations. Simultaneously developing ecological and genomics work in the native and introduced ranges of this species complex has provided insight into the evolutionary history of the species, and has revealed that traits such as high clonality, as well as shifts in trait combinations and genetic polymorphisms, may contribute to invasion success. These findings improve our understanding of potential future environmental impacts of this invasion (and invasions more broadly) in the context of changing climate. Ongoing work will continue to investigate the role of somatic mutations and genomic associations with increased plasticity and biotic interactions to assess eco-evolutionary feedback in the context of invasion and global change.

9.1 Introduction

Since Herbert Baker's seminal work (Baker, 1965) to characterize the "ideal weed," many studies of invasive species have focused on identifying the properties of specific traits that contribute to invasiveness. Although no single trait appears to universally explain why species become invasive

*Corresponding author: clr@usf.edu

© CAB International 2025. *Invasion Genomics* (Eds D. Bock and M. Rius)
DOI: 10.1079/9781800626263.0009

(van Kleunen *et al.*, 2018; Gioria *et al.*, 2023), the shift in performance of a given genotype from its native source location compared with its introduced location reflect changes in genome-level processes. Hence, genomics approaches offer powerful tools for unraveling how current phenotypic responses in introduced populations may result from the interactions between the history of the introduced species and the current environments that the species experience.

Genetic markers have shown that many invasive populations undergo only modest reductions in genetic variation following introduction to a new range thanks to multiple introductions or hybridization, among other mechanisms (Dlugosch *et al.*, 2015; Estoup *et al.*, 2016; Schierenbeck and Ellstrand, 2009). At the other extreme, some introduced clonal plant species reportedly have very little genetic diversity even after they become well established, arguing for the importance of the phenotypic plasticity of "general-purpose genotypes" (Ainouche *et al.*, 2004; Gao *et al.*, 2010; Zhang *et al.*, 2010; Richards *et al.*, 2012). In fact, some of the world's most successful invasive plants are thought to be genetically uniform in their introduced ranges (Hollingsworth and Bailey, 2000; Gao *et al.*, 2010; Zhang *et al.*, 2010). This uniformity results from vegetative or clonal reproduction, which was one of the characteristics of the ideal weed identified by Herbert Baker (1965). Recent work has identified that the majority of the plants (30/37) on the 100 most invasive species list of the International Union for Conservation of Nature (IUCN) reproduce at least partly clonally, with clonality being the main mode of reproduction for nine of these species (Mounger *et al.*, 2021).

In addition to the standing level of genetic or clonal diversity, the ability to respond to challenging environmental conditions is also constrained by the complex evolutionary history of whole-genome duplication followed by diploidization and genome fractionation, as well as hybridization (van de Peer *et al.*, 2017; Clark and Donoghue, 2018). These genome-level processes have important functional ramifications on the potential of a species to be invasive (reviewed by Mounger *et al.*, 2021). The dynamics of how these different sources of diversity will impinge on the process of introduction, establishment and spread will vary dramatically among species. However, evolutionary studies of introduced

plants that integrate such evolutionary history and molecular data with phenotypes, plant fitness and environmental interactions are still rare (van Kleunen *et al.*, 2018; McGaughran *et al.*, 2024). In general, we have only a limited understanding of how genomic processes translate into phenotypic diversity and performance in response to complex environmental conditions in invasive species (Bock *et al.*, 2015; van Kleunen *et al.*, 2018; Neinavaie *et al.*, 2021). Furthermore, to understand how these processes contribute to invasion, we need appropriate comparative studies of the native and introduced ranges (van Kleunen *et al.*, 2010; Levis and Pfennig, 2016).

Here, we focus on the well-known complex of introduced Japanese knotweed, which is listed on the IUCN's top 100 most invasive species list (Fig. 9.1). We have worked with a global network of researchers to develop and apply genomic resources to field surveys and ecological experiments in the introduced and native ranges to begin to disentangle the genomic basis of knotweed invasions. We present what is known about sources of variation in the genome structure of Japanese knotweed and how this variation interacts with novel abiotic and biotic factors to facilitate evolutionary processes across its global distribution. This work models the interdisciplinary effort needed to understand the complex process of invasion.

9.2 Study System

The invasive knotweed complex (Japanese knotweed *sensu lato (s.l.)*) is composed of three herbaceous polyploid taxa within the Polygonaceae: *Reynoutria japonica* (typically $2n = 8\times = 88$), *Reynoutria sachalinensis* (typically $2n = 4\times = 44$) and the hybrid of the two species, *Reynoutria* × *bohemica* (typically $2n = 6\times = 66$). All three taxa are known to be perennial dioecious or subdioecious, but we found some monoecious Reynoutria japonica in the north of China. They reproduce and disperse predominantly clonally in their introduced range through rhizome fragments (Mandák *et al.*, 2003), with some evidence of sexual reproduction in eastern USA (Grimsby *et al.*, 2007). *Reynoutria japonica* and *R. sachalinensis* are originally from East Asia. An early record of a Chinese accession of *R. japonica* was reported

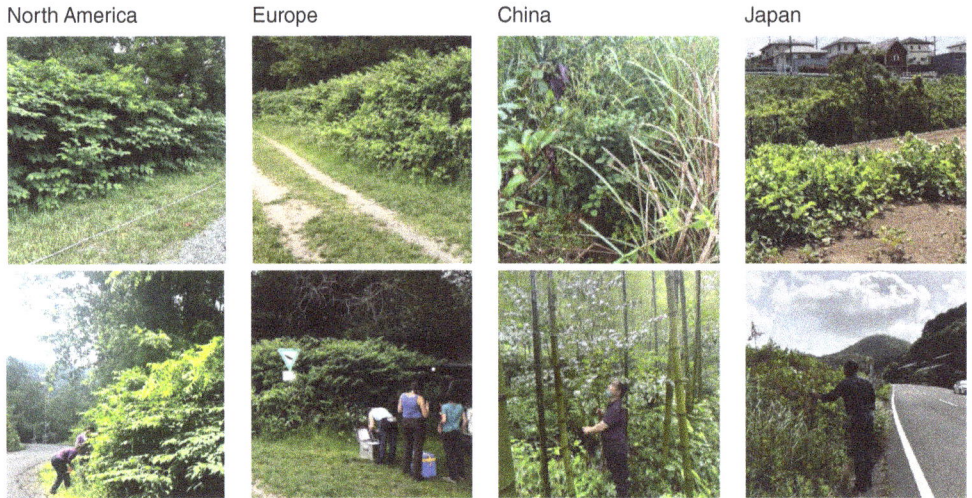

Fig. 9.1. Japanese knotweed growing in the wild in both introduced (North America and Europe) and native ranges (China and Japan). Populations in introduced ranges exhibit taller, thicker monocultures compared with populations in native ranges. Image credits: C.L. Richards (N. America), M. Parepa (Europe), Z. Liao (China) and W. Hu (Japan), used with permission.

in 1825, growing in the Chiswick Garden of the Horticultural Society of London, but it did not survive or contribute to the invasion (Bailey and Conolly, 2000). *Reynoutria japonica* was then brought to the Netherlands from Japan in the mid-1800s by Philipp Franz von Siebold, where it was sold as an ornamental plant and distributed around Europe (Bailey and Conolly, 2000). Three distinct introductions of *R. japonica* to North America in the 1860s have been reported, two from von Seibold's introduction and one directly from Japan (Del Tredici, 2017). *Reynoutria sachalinensis* was first reported at the Botanic Garden at St Petersburg, Russia, in 1864, and in 1876 at the Botanical Garden of Harvard University, USA. The first report of the hybrid *R. × bohemica* was at the Manchester Botanic Garden, UK, in 1872 (Bailey and Wisskirchen, 2004). Herbarium records from Europe and North America show a relatively short lag phase of 40–83 years before Japanese knotweed populations rapidly expanded and became invasive (Pyšek and Prach, 1993; Pyšek and Hulme, 2005; Del Tredici, 2017). The rapid growth of knotweed populations has resulted in large monocultures that have lower species diversity compared with adjacent uninvaded areas (Aguilera *et al.*, 2010).

Japanese knotweed *s.l.* are recognized as powerful ecosystem engineers that shape the physical and biological environment, with important impacts on nutrient cycling (Bímová *et al.*, 2004; Murrell *et al.*, 2011). Established populations of Japanese knotweed *s.l.* have colonized a variety of habitats in Europe (Murrell *et al.*, 2011; Parepa, *et al.*, 2013a, 2019; Zhang *et al.*, 2016; Garnica *et al.*, 2022) and the USA (Maerz *et al.*, 2005; Siemens and Blossey, 2007; Walls, 2010; Richards *et al.*, 2012), causing both ecological and economic disturbances (Grabar, 2019; Eschen *et al.*, 2023), including negative impacts on native biodiversity (Maerz *et al.*, 2005; Lecerf *et al.*, 2007; Siemens and Blossey, 2007; Topp *et al.*, 2008; Fogelman *et al.*, 2018; Renčo *et al.*, 2021). Riparian habitats have been especially affected by the spread of Japanese knotweeds, showing decreased abundance and species richness of plants and invertebrates, including endangered native species (Bailey and Conolly, 2000; Murrell *et al.*, 2011), with cascading effects on higher trophic levels (Gerber *et al.*, 2008). For example, introduced knotweed may affect ecosystem processes by changing the composition of leaf litter subsidies in riparian areas (Fogelman *et al.*, 2018). Genetic and phenotypic traits of these species have long been considered to contribute to establishment and impacts across large latitudinal ranges following both deliberate introductions and natural dispersal (Pyšek *et al.*, 2003; Bímová *et al.*, 2004; Mandák *et al.*, 2005; Bailey *et al.*, 2009). Furthermore, the hybrid *R. × bohemica* has greater genetic diversity than either parent in the introduced range, which is thought to contribute to more aggressive spread (Pyšek *et al.*, 2003; Bímová *et al.*, 2004; Mandák *et al.*, 2005; Bailey *et al.*, 2009). Given that *Reynoutria japonica* is listed among the 100 world's worst invasive alien species by the IUCN (Lowe *et al.*, 2000), and that the Polygonaceae includes other introduced species (Matesanz and Sultan, 2013; Matesanz *et al.*, 2015), investigating the genomics of these species can provide valuable insights into the mechanisms of invasion and future management of invasive species.

9.3   Early Studies of Knotweed Diversity

Many early studies in Europe and North America reported only one, male-sterile, octoploid ($2n = 8\times = 88$) genotype among the introduced *R. japonica* populations (Bailey and Conolly, 2000; Hollingsworth and Bailey, 2000; Richards *et al.*, 2012; Gaskin *et al.*, 2014; Zhang *et al.*, 2016). While some introduced populations include tetraploids ($2n = 44$), Japanese populations harbor both tetraploid and octoploid ($2n = 88$) *R. japonica* plants with high genetic and morphological diversity (Inamura *et al.*, 2000; Zhou *et al.*, 2003; Bailey, 2013). The introduction of a hermaphrodite *R. sachalinensis* ($2n = 4\times = 44$) to Europe in the 1860s provided a male parent for hybridization with *R. japonica* and the formation of the hybrid *R. × bohemica* on multiple occasions in different places (Hollingsworth and Bailey, 2000; Bailey *et al.*, 2009). Hybridization may contribute to invasion success by introducing new genetic diversity for natural selection to act on (te Beest *et al.*, 2012; Kagawa and Takimoto, 2018). Furthermore, interspecific hybridization, and exposure to stressful conditions, can reprogram gene expression and result in phenotypic novelty and functional plasticity even in the absence of inter-individual genetic variation (Chen and Yu, 2013; Wendel *et al.*, 2016). The hybrid species is thought to have spread in Europe mostly through vegetative propagation, but the initial hybridization events could be important in shaping the potential of those vegetative lineages. In the case of Japanese knotweed, as for many introduced species, it is difficult to separate the effects of hybridization and polyploidy experimentally as the two are confounded (Ainouche *et al.*, 2009; Bock *et al.*, 2015; Salmon and Ainouche, 2015). Both events increase genetic and epigenetic variation and therefore evolutionary potential (Ainouche *et al.*, 2009). It is therefore unclear how hybridization and/or genome duplication events may be contributing to traits that facilitate invasion (Gammon and Kesseli, 2010; te Beest *et al.*, 2012).

This genome complexity could facilitate knotweed's success, as the genomic processes of hybridization and polyploidization have been associated with increased phenotypic plasticity and invasion potential (Ainouche *et al.*, 2009; Mounger *et al.*, 2021). Controlled experimental studies have supported the idea that some potentially important knotweed traits are highly plastic (Pyšek *et al.*, 2003; Richards *et al.*, 2008; Van-Wallendael *et al.*, 2018; S. Wang *et al.*, 2025). However, many of these studies were limited in scope. For example, studies of Czech populations suggested that transgressive segregation may have contributed to even more aggressive spread in the hybrid species compared with *R. japonica* or *R. sachalinensis* (Pyšek *et al.*, 2003; Mandák *et al.*, 2004). In fact, some *R. × bohemica* genotypes regenerated quicker and grew faster, while others emerged earlier and produced more biomass compared to other genotypes (Pyšek *et al.*, 2003). We confirmed the quicker regeneration of *R. × bohemica* and showed that replicates of individuals of the hybrid species from Switzerland and Germany performed better in competition with native species compared with either of its parent species (Parepa, *et al.*, 2013b). Importantly, we also found phenotypic variation among populations of *R. japonica*, even though these individuals were genetically uniform (Parepa, *et al.*, 2013b).

In the North American range, we also found significant differences in most traits and trait plasticities within and among populations of both *R. japonica* and *R. × bohemica* in New York. We showed that plants of *R. japonica* from both roadside and salt marsh habitats had highly plastic responses to salt treatment, even though we had only one amplified fragment length polymorphism (AFLP) haplotype of *R. japonica* (Richards *et al.*, 2008). Using reciprocal transplants in the field, we showed that plants from salt marsh, beach and roadside habitats were differentiated. We also found some evidence for local adaptation of Japanese knotweeds: plants from the marsh habitat had greater biomass and plants from beaches and roadsides had greater survival in their "home" sites compared with other plants (Fig. 9.2; Yuan *et al*., 2024). Another study did not find support for local adaptation in *R. japonica* populations along a latitudinal gradi-

ent in the temperate deciduous forest habitats of Kentucky, New York, and New Hampshire, but did find among-population differences in height, basal stem diameter and biomass (VanWallendael

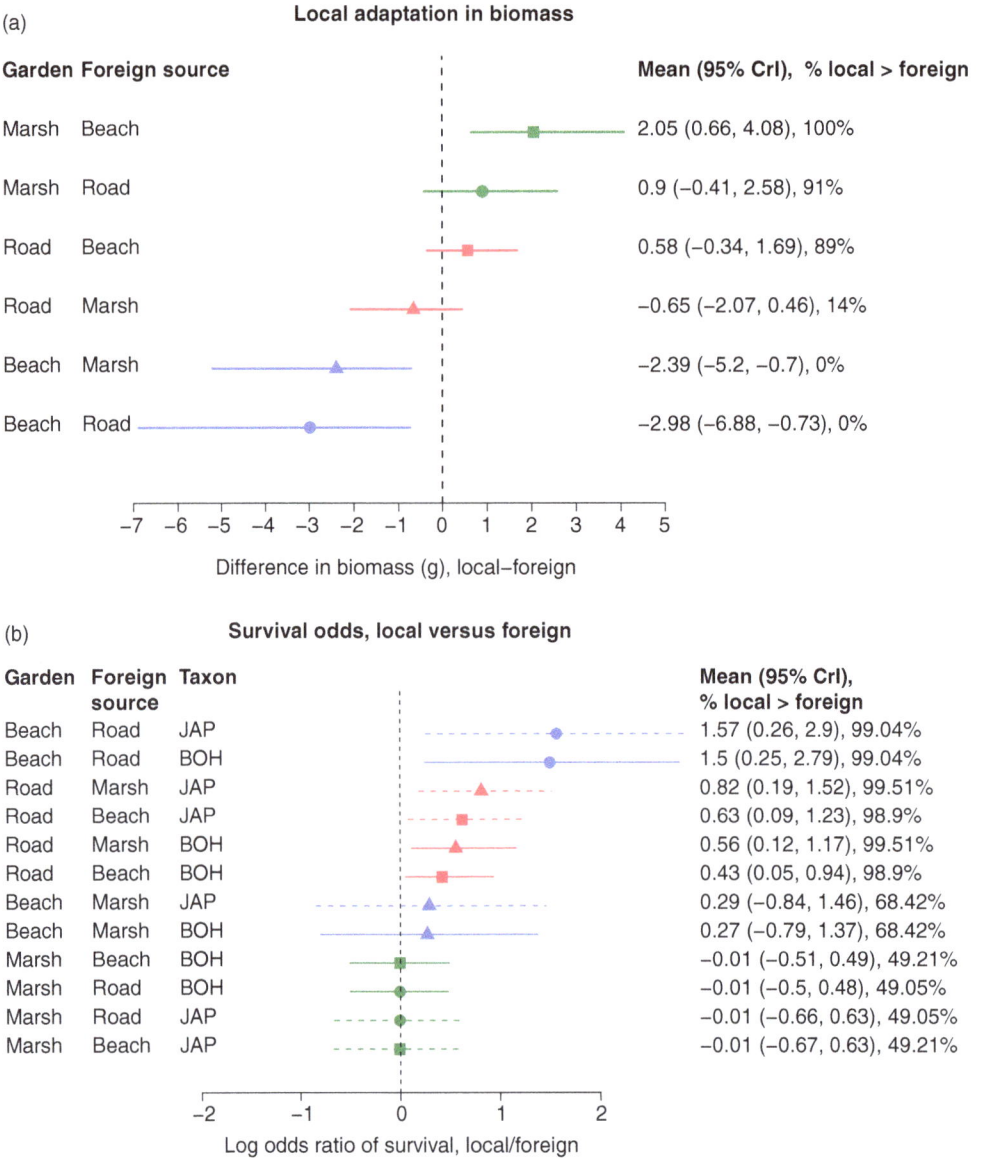

Fig. 9.2. (a) Evidence of local adaptation in marsh plants compared with beach and roadside plants grown in marsh habitats as measured by total dry biomass (g). (b) Survival of beach plants compared with roadside plants grown in beach habitats, and roadside plants compared with beach or marsh plants grown in roadside habitats. Survival in *Reynoutria japonica* (dashed lines) and *Reynoutria × bohemica* (solid lines) are indicated separately. Plants grown in beach sites are depicted in blue, marsh sites in green and roadside sites in red. Symbols and whiskers are differences of fitted estimates and credible intervals (CrI) estimated from Bayesian statistical models. Reprinted from Yuan *et al*. (2024) as per the open-access policy of Springer Nature, http://creativecommons.org/licenses/by/4.0/.

et al., 2018). These responses to complex environmental conditions suggest that the knotweed invasion has involved some level of trait evolution, but the genomic or epigenomic mechanisms that underlie these changes have yet to be resolved.

9.4   Recreating the Knotweed Invasion on a Global Scale

Identifying how responses and performance of introduced populations compare with those of native populations has been a research priority (Verlaque et al., 2011; van Kleunen et al., 2018). While R. japonica is well known in traditional medicine in the native range (Peng et al., 2013), until recently very little was known about the levels of genetic and phenotypic variation in natural populations (Inamura et al., 2000; Zhou et al., 2003; Bailey, 2013). To examine in situ and heritable trait variation across the native and introduced ranges of Japanese knotweed, we started the collaborative project "Genomics and Epigenomics of Plant Invasion," with collections of at least five individuals at each of 56 native (China and Japan) and 100 introduced (50 each in the USA and Europe) Japanese knotweed s.l. populations (Fig. 9.3). We collected data on trait variation in field conditions, and we established common gardens from rhizomes collected from USA populations at Cornell University, from European populations at University of Tübingen (Germany) and from Chinese populations at Fudan University in Shanghai (China). Subsequently, we collected rhizomes from the putative source of the introduction in six populations in Kyushu, Japan (however, we were unable to replicate our sampling scheme and did not obtain data from these Japanese plants in situ). We imported and grew the rhizome fragments under quarantine conditions in a glasshouse for one growing season at the Xishuangbanna Tropical Botanical Garden (XTBG) as required by the Chinese legislation and then established two common gardens that included the collection of all 156 populations at XTBG and Fudan University in China (see S. Wang et al., 2025 for full common garden set up).

9.4.1   Recreating past evolutionary history

Using the global knotweed collection and published data on an additional 27 individuals obtained from the NCBI database (www.ncbi.nlm.nih.gov; accessed 31 July 2025), we evaluated the evolutionary history of R. japonica with seven chloroplast DNA fragments (rbcL, matK, rbcL-accD, accD-psaI, ndhF, accD, and trnL-intron). We confirmed that R. japonica originated from Japan and spread westward to the Korean Peninsula and eastern China around 1.42-2.58 Mya (Fig. 9.4; Zhang et al., 2024). Subsequently, sea-level changes and land-bridge configurations isolated the Chinese populations from the Japanese ones. We confirmed that all of the introduced populations of R. japonica that we sampled in western Europe and most of those that we sampled in the eastern USA shared the same haplotype (H45) as several populations in Kyushu, Japan. We also confirmed that a second

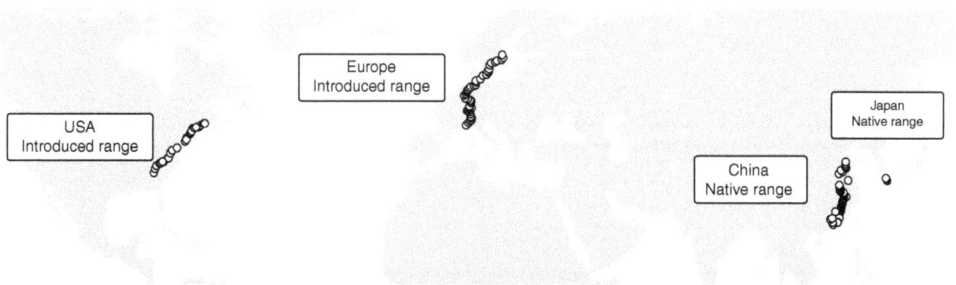

Fig. 9.3. The global collection of knotweed plants in native and introduced ranges. We collected plant data from 156 populations (indicated by circles) along 2000 km transects on a north-to-south gradient in the native ranges (50 populations in China, six populations in Japan) and introduced ranges (50 populations in the USA, 50 populations in Europe).

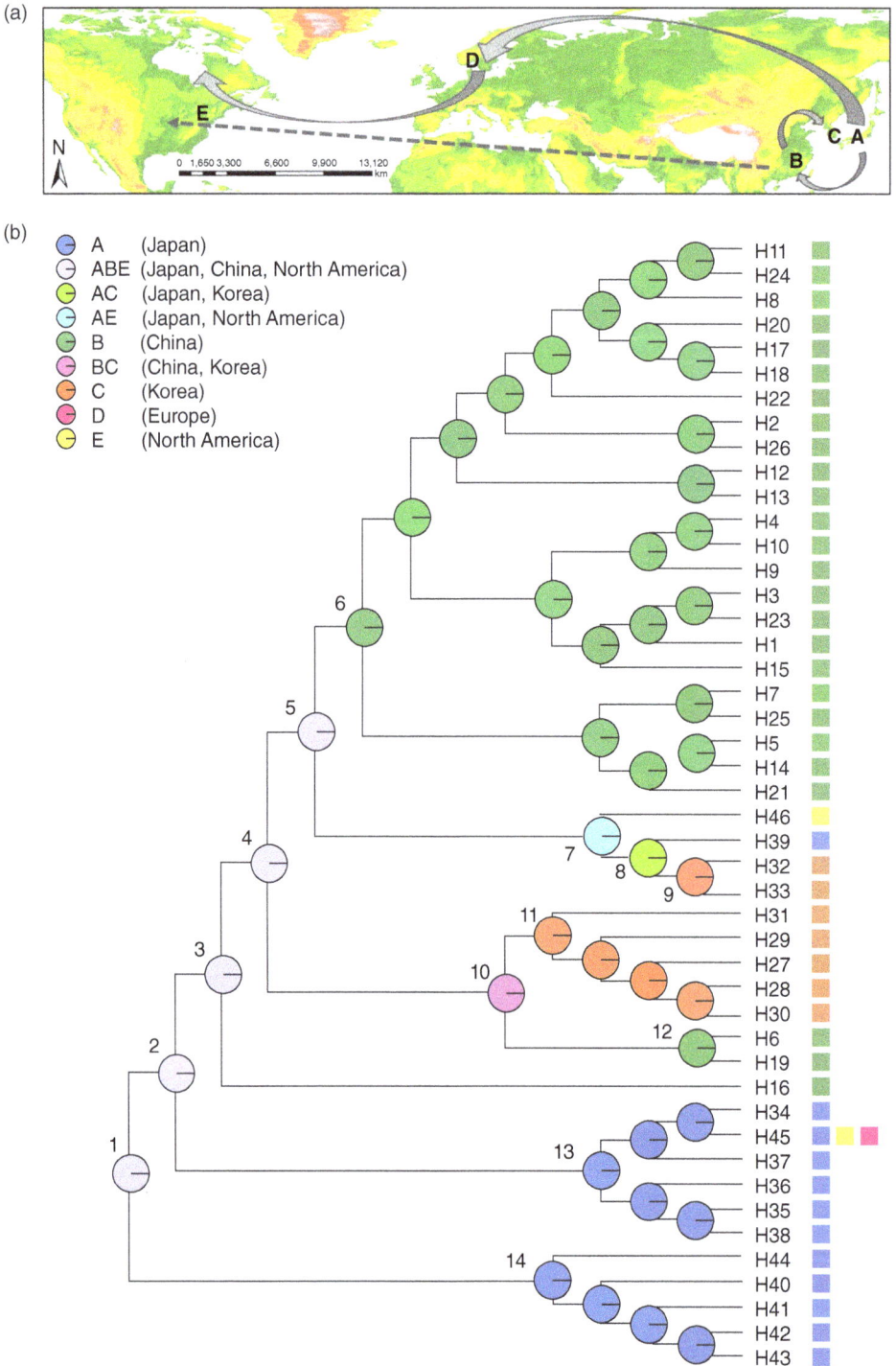

Fig. 9.4. (a) Inferred migration routes of *Reynoutria japonica* based on RASP analysis. (b) Ancestral area reconstruction of *R. japonica* based on RASP analysis. The haplotypes are labeled H1–H46. Squares in different colors represent the sampling locations in different areas. The areas A–E with different colors are defined according to the physical geographical regions in (a) and (b), with area A as the putative ancestral area. Reprinted from Zhang *et al.* (2024) as per the policy of John Wiley and Sons. Figure used with permission from Wiley.

lineage was introduced to North America (H46), which we found in a single population in the southern part of Virginia. However, we were unable to resolve whether the most likely source of that lineage was from Japan, Korea or China (Fig. 9.4; Zhang *et al.*, 2024).

In addition to phylogenetic approaches with contemporary samples, herbaria have provided important resources for tracking long-term changes between plants and environmental conditions dating back to the 16th century (Franck, 2018; Lang *et al.*, 2019). We obtained herbarium collections of Japanese knotweed *s.l.* spanning the entire 180-year history of global introductions of the species. We generated low-coverage whole-genome shotgun sequencing data for 150 herbarium accessions, temporally and spatially distributed across the species' native and introduced ranges (R.E. Irimia, unpublished data). Our sampling design included some of the oldest specimens available, which represent the focal introduction points of the species into Europe (e.g. *R. japonica*: London, UK, 1850; Leiden, Netherlands, 1851; Nancy, France: 1853; *R. sachalinensis*: Allgäu, Germany: 1878) and North America (e.g. *R. japonica*: New York, 1891; Delaware, 1893; Florida, 1894; North Carolina, 1897; *R. sachalinensis*: Massachusetts, 1895; Ontario, 1901), as well as from the putative source of origin of these introductions in the native ranges in Asia (e.g. *R. japonica*: Nagasaki, Japan, 1828 and 1862; Hubei, China, 1887; *R. sachalinensis*: Sakhalin island, Russia, 1860).

Our preliminary results indicated that the historical European and North American accessions of *R. japonica* showed high genetic affinity to accessions from different geographical areas across Japan, suggesting multiple independent introduction events. We detected reduced population genetic structure in introduced regions but no substantial reduction in genetic diversity and observed heterozygosity when compared with the native range, suggesting that the species did not undergo severe genetic bottlenecks following introduction (R.E. Irimia, unpublished data). However, more in-depth sequencing data are needed to elucidate these fine-scale genetic patterns. This type of herbarium work will expand our understanding of the history of the invasion process, providing information on specimens and diversity that are perhaps no longer present in current populations.

To link these genomics signatures to traits, we analyzed leaf traits of the herbarium specimens. Leaves and reproductive structures (seeds, flowers and fruit) in herbarium specimens provide information on important trait shifts over time (Heberling, 2022; Willems *et al.*, 2022; Hamma, 2023). Stomatal development depends on both environmental conditions and genetically-based differences. Stomatal densities have been shown to decrease with rising concentrations of carbon dioxide in several species (Woodward *et al.*, 2002; Lampard *et al.*, 2008; Casson and Hetherington, 2010; Zhang *et al.*, 2012). Stomatal characteristics are also affected by changes in genome size and ploidy (van de Peer *et al.*, 2021; Haworth *et al.*, 2023), but these relationships have not yet been investigated in knotweeds.

Specific leaf area can also be assessed on herbarium studies and has been studied across many contexts in ecology and serves as an easy measurement correlated to relative growth rate, photosynthetic capacity, leaf lifespan and leaf nitrogen content (Nicotra *et al.*, 2010). In addition, specific leaf area has been predicted to decrease with increasing carbon dioxide, while stomatal size may present an inverse relationship. Elevated specific leaf area is also a common predictor for invasiveness (Grotkopp and Rejmánek, 2007; Hodgins *et al.*, 2018). We investigated shifts in stomatal density and size and specific leaf area within *R. japonica* and *R. × bohemica* to identify evolutionary changes within the invasion over the past 180 years. We predicted that shifts in stomatal size and density, as well as specific leaf area, have responded to climate change (Fig. 9.5) (R.E. Irimia *et al.*, unpublished data; K. Keefer *et al.*, unpublished data; Haworth *et al.*, 2023). We will compare the changing patterns of leaf traits in the introduced and native ranges and associate trait variation with shifts in climate and genomic changes that have arisen over time.

9.4.2 Association between trait variation and genomics processes

Besides tracking the history of populations moving across the landscape, genomics may provide information about the mechanisms that underlie changes in traits and trait variation that contribute to invasion. In our field survey of natural

Reynoutria japonica   Reynoutria × bohemica

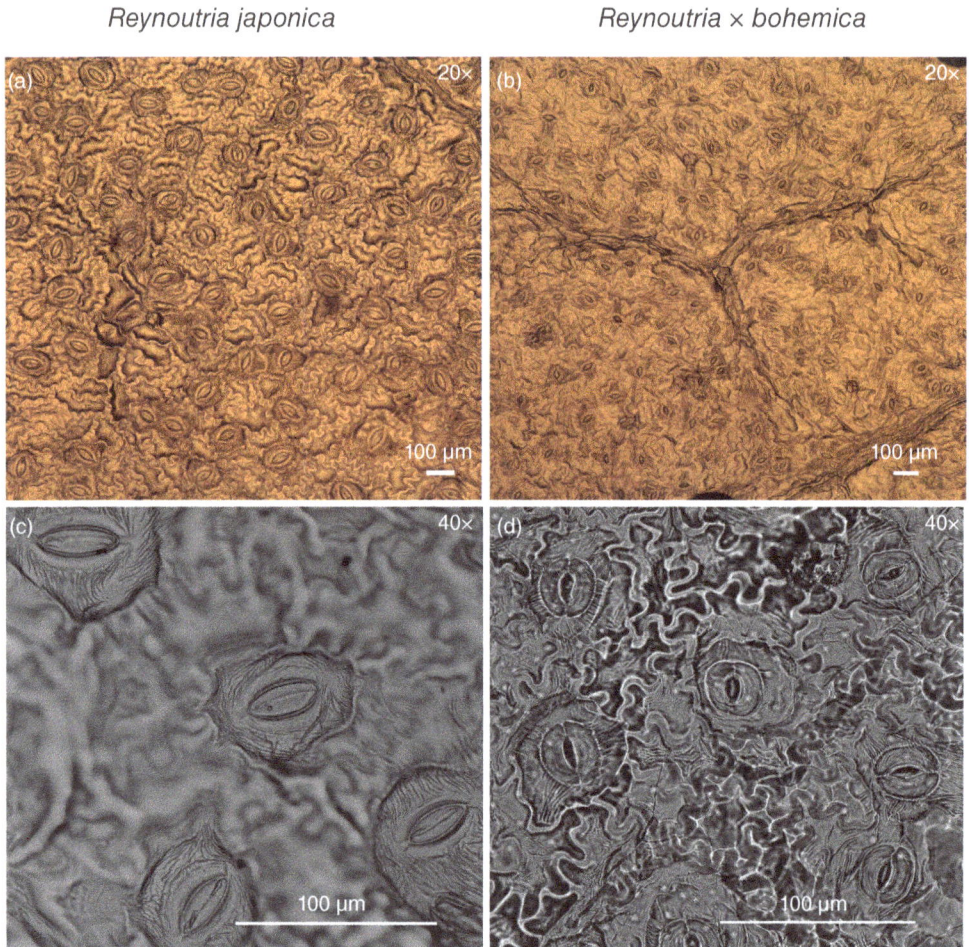

Fig. 9.5. An example of a stomatal peel made from herbarium specimens of *Reynoutria japonica* (left) and *Reynoutria × bohemica* (right). We used 20× magnification to estimate stomatal density (a, b) and 40× magnification to measure stomatal size responses (c, d) to changing conditions such as changes in carbon dioxide, temperature and precipitation over time. The pictures were taken with a Canon EOS 650D camera (and the program Canon Utilities EOS Utility 2), placed on a Carl Zeiss Axioskop 2 plus microscope. The picture size is 3456 × 3456 pixels. Image credits Christiane Karasch-Wittmann, used with permission.

Japanese knotweed populations, we found that introduced plants were larger and had denser populations in the introduced range, as expected (Irimia *et al.*, 2025). We also found support for release from natural enemies, as evidence for both herbivores and pathogens was lower in the introduced than in the native ranges (Fig. 9.6a,b; Irimia *et al.*, 2025). The patterns were associated with shifts in leaf economy (Wright *et al.*, 2004) and chemical defenses: introduced plants had higher specific leaf area but reduced leaf chlorophyll, lignin, carbon:nitrogen ratio and leaf toughness along with altered leaf tannins, flavonoids, and alkaloids compared with plants from native populations. By applying the concept of "trait syndromes" (Agrawal and Fishbein, 2006; Agrawal, 2020; Woods and Sultan, 2022), we identified three distinct multivariate knotweed syndromes primarily in the introduced ranges, and two syndromes that were mainly in the native ranges (Fig. 9.6c; Irimia *et al.*, 2025). This study provided a snapshot of the trait variation in current wild populations of knotweed, but we could not explain much of the observed

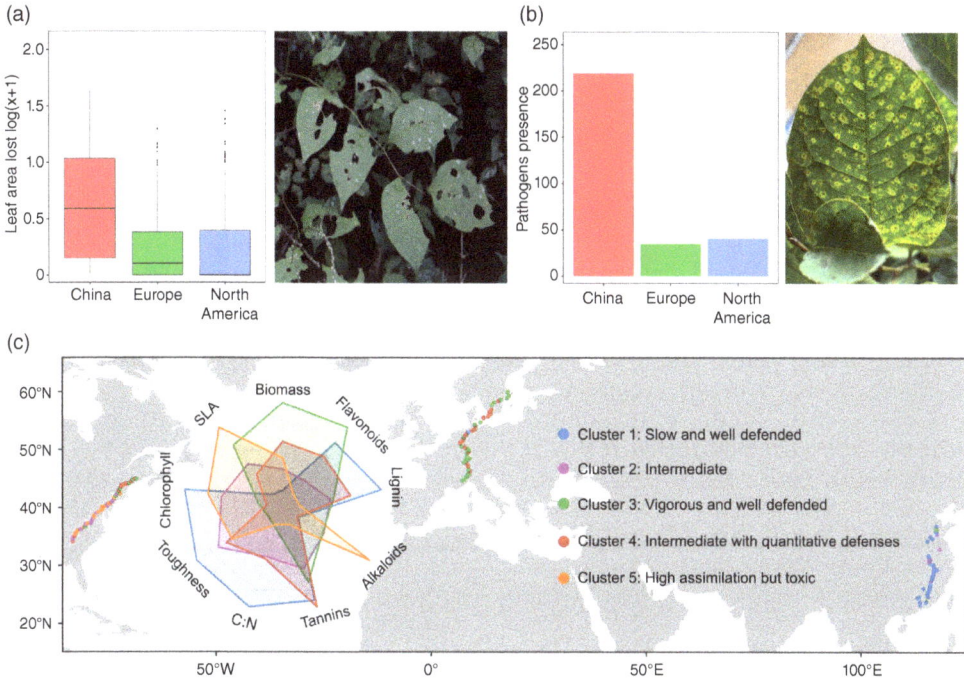

Fig. 9.6. (a) Leaf area lost to herbivores, and a photo of herbivore attack in the field and (b) number of individuals and photo showing signs of pathogen presence, in one native (China) and two introduced (Europe, North America) ranges of Japanese knotweed *s.l.* (c) Five multivariate trait syndromes identified across 150 knotweed populations surveyed across the three ranges. The radial plots show the make-up of the profiles (based on trait means) for each of the five trait syndromes. C:N, carbon:nitrogen ratio; SLA, specific leaf area. Reprinted from Irimia *et al.* (2025) as per the open-access policy of John Wiley and Sons. Figure used under license CC-BY-4.0

trait variation with the environmental conditions we evaluated (Irimia *et al.*, 2025).

We evaluated the genetic and epigenetic variation of these same plants along with individuals from several populations from the purported region of origin of the introduced individuals in the southern island of Kyushu, Japan (Zhang *et al.*, 2024) using both genotyping-by-sequencing (GBS; Elshire *et al.*, 2011) and bisulfite-converted GBS (epiGBS; van Gurp *et al.*, 2016; Gawehns, *et al.*, 2022). In previous work, we used cytology and AFLP markers to show that some populations around Long Island, New York, consisted of the single *R. japonica* genotype reported across the USA and Europe, while the majority consisted of a few *R. × bohemica* hybrid haplotypes (Richards *et al.*, 2012). Using epiGBS, we found some unexpected DNA sequence differences among individuals that we previously reported to be copies of a single haplotype

(Robertson *et al.*, 2020). Moreover, these differences appeared to be associated with habitat. In particular, beach populations were differentiated from marsh and roadside ones, similar to the patterns of trait differentiation in our reciprocal transplant study in these populations (Robertson *et al.*, 2020; Yuan *et al.*, 2024). Although another recent GBS study reported high genotypic diversity among *R. japonica* individuals (VanWallendael *et al.*, 2021), we interpreted these initial results with some caution, given the limitations of sequencing technology to detect mutations over error (Yoder and Tiley, 2021). In addition, we did not have a reference genome, and identifying polymorphisms across copies of a given region in hexaploid and octoploid species is technically challenging (Salmon and Ainouche, 2015; VanWallendael *et al.*, 2021). Such complexity requires more sequencing depth to identify all copies of the loci and *de novo* mutations

(Salmon and Ainouche, 2015; Paun *et al.*, 2019; VanWallendael *et al.*, 2021; Yoder and Tiley, 2021). This problem may be exacerbated with the short sequence fragments used in GBS approaches (Richards *et al.*, 2017; Paun *et al.*, 2019). With large, high-ploidy genomes, it is difficult to differentiate heterozygosity among multiple copies at a given locus within an individual from those that are among-individual polymorphisms (Paun *et al.*, 2019). Furthermore, the epiGBS protocol relies on bisulfite treatment of the DNA and the creation of a *de novo* reference. Genetic polymorphisms may be confounded with methylation polymorphisms during tabulation, leading to false increases in single-nucleotide polymorphism (SNP) variation (Liu *et al.*, 2012; Gao *et al.*, 2015).

In our evaluation of the collection, we attempted to address some of these important technical problems by including replicates of the same individual and evaluating SNPs without bisulfite conversion of the DNA (i.e. using the standard GBS protocol; Elshire *et al.*, 2011; Paun *et al.*, 2019). We used our technical replicates of the same samples to define a threshold below which plants were "most likely clonal replicates." We found that genetic variation among individuals within populations was almost as low as that between replicates of the same individual, and

therefore within populations, most individuals were most likely clonal replicates (Fig. 9.7).

Surprisingly, low within-population genetic variation was true not only for introduced populations but also for native populations from Japan and China (Fig. 9.7b). Genetic distances between plants from Japan and those from introduced ranges (Europe and North America) were likewise very low, supporting the reported relationship between these Japanese plants and introduced plants. In contrast, Chinese plants were clearly distinct from Japanese plants (Fig. 9.7). Further analyses showed that only a single SNP was significantly different between European and the North American populations ($<1\%$). In contrast, about 7767 SNPs (21%) were significantly different between China versus Europe and North American plants (false discovery rate <0.05 and percentage deviance explained $\geq 10\%$). We found similar results in the single-methylation polymorphism (SMP) data with no significant differences between European and North American samples, but about 3% of all SMPs were different between China versus Europe and North America (false discovery rate <0.05 and the percentage sum of squares explained $\geq 10\%$). These results suggested that in both genetic and epigenetic variation, plants from China were different from those in Japan,

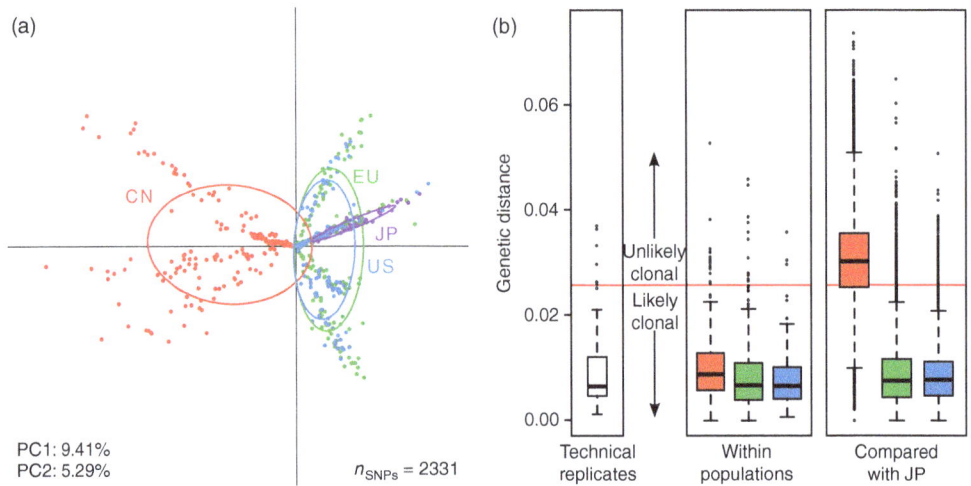

Fig. 9.7. (a) Principal component (PC) analysis of pairwise genetic distances for *Reynoutria japonica* (8×) colored by region of origin: China (CN), Europe (EU), North America (US) and Japan (JP). (b) Pairwise genetic distances between technical replicates, within populations and between individuals from CN, EU, US and JP. Unpublished data from I. van Riemsdijk *et al.* Figure used with permission.

Europe and North America, but no differences existed between plants from Japan and plants from Europe and North America (I. van Riemsdijk *et al.* unpublished data).

We evaluated whether environmental characteristics of the field sites explained variation in genomics that could indicate a response to natural selection. In Europe, *R. japonica* plants had few SNPs that were significantly associated with latitude and canopy cover (false discovery rate <0.05 and percentage deviance explained ≥10%). In China, we found more associations between SNPs and climatic variables. This may reflect adaptation over a longer period in the native range compared with the introduced range. Using the same approach but with SMPs, we found a small number of epigenetic variants associated with climate in populations in Europe (I. van Riemsdijk *et al.* unpublished data). We used multiple matrix regression with randomization to evaluate pairwise genetic, epigenetic, climate and soil differences (Wang and Bradburd, 2014; Herrera *et al.*, 2017). We found that genetic and epigenetic differences between plants were associated with soil types in the introduced range, whereas they were associated with climate variables in the native range (the overall regression explained about 10% of the variation in phenotypic distance in each range; *P* value = 0.001).

Our analysis included better and deeper coverage of the loci that we interrogated (40× compared with 10× coverage of fragments) than our previous GBS analyses of New York populations, and we mapped them to our reference *Reynoutria multiflora* genome, which provided greater confidence in the differences we identified. However, GBS still suffers from the limitation of probing only a small fraction of the genome (Paun *et al.*, 2019; van Moorsel *et al.*, 2019). It could be that *de novo* mutation provided important genomic differences that have contributed to the success of the introduced populations, but so far, we have found very few differences, and we have not yet linked these differences to corresponding differences in function. Isolating this type of mutation will require much more in-depth analysis of the genome. For this effort, we have assembled a reference genome for the diploid congener *R. multiflora* along with reference transcriptomes for *R. multiflora* and the three polyploid *Reynoutria* taxa (*R. sachalinensis* (4×), *R. × bohemica* (6×), and *R. japonica*

(the common 8×)) (A. Cavé-Radet *et al.* unpublished data). Using comparative genomics of these references along with recently published genomes for *R. multiflora* (He *et al.*, 2024; Zeng *et al.*, 2024; Zhao *et al.*, 2023) and *R. japonica* (F. Wang *et al.*, 2025; Zhang *et al.*, 2025) promises further insight into the functional changes in the genome of the invasive individuals.

9.4.3 Assessing heritability and differentiation of traits and plasticity

We have accumulated evidence from common-garden experiments at the XTBG and Fudan University that supports the possibility of divergence in important traits within the native and introduced ranges of knotweed. We found herbivore resistance in introduced populations differed from most Chinese but not Japanese populations (Fig. 9.8a, b). Our results suggested that the original plant that was introduced from Japan may have been particularly resistant to herbivores (Cao *et al.*, 2025). Similarly, we found that plants from the introduced North American and European populations differed in many other traits from those of native Chinese populations and were similar to plants from native Japanese populations (S. Wang *et al.*, 2025). We identified one clear exception: introduced North American and European plants expressed higher plasticity in ramet production than Japanese plants, suggesting the evolution of increased plasticity of clonality following introduction to the new ranges (Fig. 9.8c; S. Wang *et al.*, 2025). This pattern supports the "plasticity first" hypothesis, as initial plasticity in clonality introduced from Japan may have exposed variation among individuals in the ability to respond with clonal growth (Levis and Pfennig, 2016). Selection in the introduced range could then increase plasticity in clonality (genetic accommodation of plasticity, *sensu* Bock *et al.*, 2018).

We also discovered that *R. japonica* populations in northern China (~31–38°N) were intentionally introduced from the southern regions (~22–31°N) (Wang, 1996; Zhao, 2021; Zhao *et al.*, 2024). We confirmed this claim in interviews with indigenous residents, researchers and managers during plant collections (Zhao *et al.*, 2024). Following its introduction, *R. japonica* has successfully naturalized and expanded as a

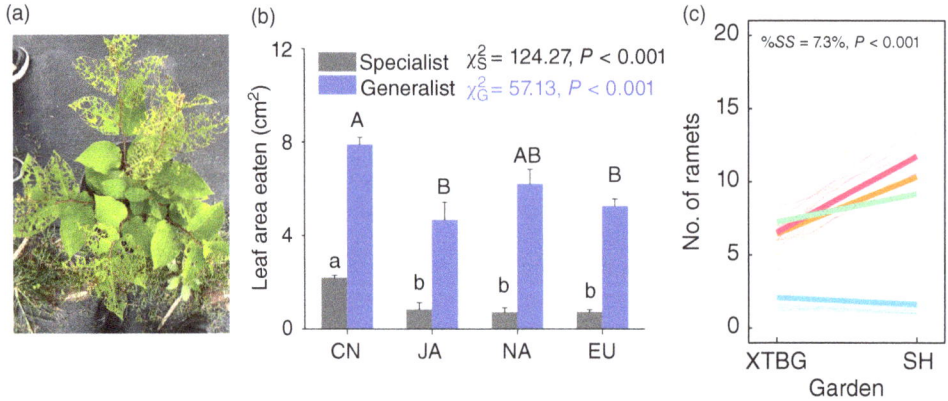

Fig. 9.8. (a) Typical damage by beetles on plants from Chinese populations in the XTBG common garden. Image credits P. Cao. (b) Leaf area eaten by a specialist (*Gallerucida bifasciata*) or a generalist (*Spodoptera litura*) on *Reynoutria japonica* leaves from different ranges according to a bioassay experiment in a common garden in Shanghai. The leaf area eaten was much higher on plants originating from China (CN) than on plants from Japan (JA), North America (NA; except for specialist, but not generalist, herbivory) and Europe (EU). We show adjusted means (± SE) from analysis of variance (different letters above error bars indicate significant group differences based on Tukey's HSD (honestly significant difference) *post-hoc* tests), χ^2 statistics and the significance level of the regression model (P. Cao *et al.*, unpublished data). (c) Reaction norms for populations from native Chinese (blue) and Japanese (green) ranges and introduced North American (orange) and European (red) ranges grown in Xishuangbanna (XTBG) and Shanghai (SH) common gardens. We show the mean response in number of ramets for each population with thin lines. The mean response for each range is indicated by thick lines. The factor "range" was split into three contrasts with one degree of freedom each (Schmid *et al.*, 2017). We provide the percentage of the sum of squares explained by the range × garden effects and the overall F-test: contrasts between CN versus JA/EU/US were highly significant ($P < 0.001$), while contrasts between EU versus US were not. Data from Wang *et al.* (2025b) as per the policy of the New Phytologist Foundation. Figures used with permission.

"native invasive" in natural habitats, resulting in significant biodiversity and agriculture losses in the introduced regions of northern China (Zhao, 2021). We observed differences between plants from northern and southern Chinese populations for several leaf traits and preference of *Spodoptera litura* and *Spodoptera exigua* caterpillars and the aphid *Aphis citricola*. In general, herbivores performed better on plants collected from the introduced range than on those from the native range. Across plants collected from the introduced range, herbivores also performed better on plants collected from high latitudes than on those from low latitudes (Fig. 9.9). These data may support the suggestion that the introduced plants have evolved post-introduction and are now more susceptible to the local herbivores in common gardens located in the native region (Zhao *et al.*, 2024). More work will be required to examine the genomic mechanisms that underlie this divergence.

Evaluating metabolites is one potentially powerful approach to understanding these types of changes in response to herbivory. To understand large-scale geographical patterns of secondary metabolites, and their environmental drivers and trade-offs, we studied 39 native populations of *R. japonica* along a large latitudinal gradient in China ($23-31°N$) (Bi *et al.*, 2024). We measured the concentrations of six polyphenols in the leaves and rhizomes of *R. japonica* and associated the variation in these metabolites with biotic and abiotic environmental factors as well as with plant traits. We found that climate was an important driver of variation in secondary metabolites, both above- and belowground. Remarkably, the patterns of climate-metabolite associations differed between leaves and rhizomes, as well as between putative low-cost versus high-cost compounds. Annual mean temperature was a strong predictor of aboveground metabolites, while annual precipitation was more frequently

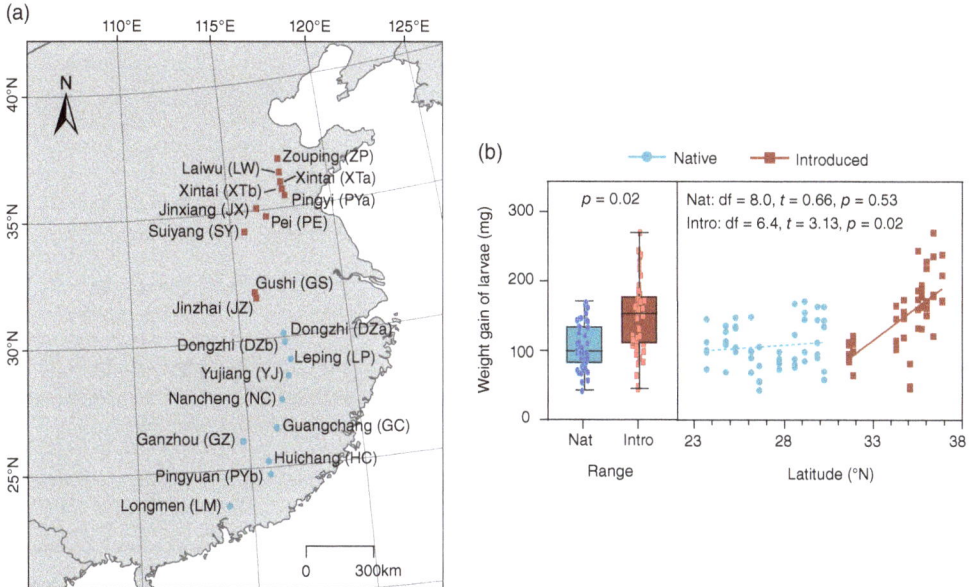

Fig. 9.9. (a) Collection sites of native (blue) and purported "native invasive" (red) *Reynoutria japonica* populations in the eastern mainland of China (Lambert map projection). (b) Variation in performance of *Spodoptera litura* feeding on leaves of *R. japonica* from it's native (Nat) and introduced (Intro) ranges (shown as boxplots, left) and across latitudes within each range (shown as scatter plots, right) in a greenhouse experiment at Fudan University. The boxplots indicate the median and interquartile ranges, with data distribution shown as scattered dots. The scatter plots show relationships between herbivore performance and plant-collection latitudes, separately for each plant range; we plotted the regression lines, with solid lines indicating significant slopes and dashed lines indicating non-significant slopes. Reprinted from Zhao *et al.* (2024) as per the policy of the Ecological Society of America. Figure used with permission.

associated with variation in belowground metabolites (Fig. 9.10; Bi *et al.*, 2024). In addition, annual temperature was positively associated with high-cost metabolites but negatively associated with low-cost metabolites. Our findings indicated that metabolite allocation strategies differed between above- and belowground tissues of *R. japonica* in the native range. As latitude increases, *R. japonica* invested relatively more into belowground metabolites. We propose that reduced high-cost metabolites in the leaves at higher latitudes may help to conserve nutrients after defoliation, while maintaining high-cost metabolites in rhizomes may be important for persistent allelopathic effects and resource conservation belowground. The divergent patterns of above- and belowground metabolite allocation thus may reflect the multiple functions of metabolites, expansion history and the plants' adaptation to different environments.

In addition to the importance for response to herbivory and pathogens, some of these secondary metabolites may be used to mediate microbial interactions. These interactions can strongly influence plant response, particularly in the face of challenging environmental conditions (van der Heide *et al.*, 2012; Kivlin *et al.*, 2013). Previous work has shown that soil conditions and root exudates strongly influence rhizosphere microbiomes (Fitzpatrick *et al.*, 2018; Hu *et al.*, 2018; Sasse *et al.*, 2018). The microbiome not only varies among plant species but is also significantly influenced by plant genotype within species (Agler *et al.*, 2016; Wagner *et al.*, 2016; Bowen *et al.*, 2017; Bergelson *et al.*, 2019; Hughes *et al.*, 2020) and by environmental context (Mandyam and Jumpponen, 2015). The microbiome could facilitate plant invasions, considering that a comparison across 94 plant invaders demonstrated that invasive plants had

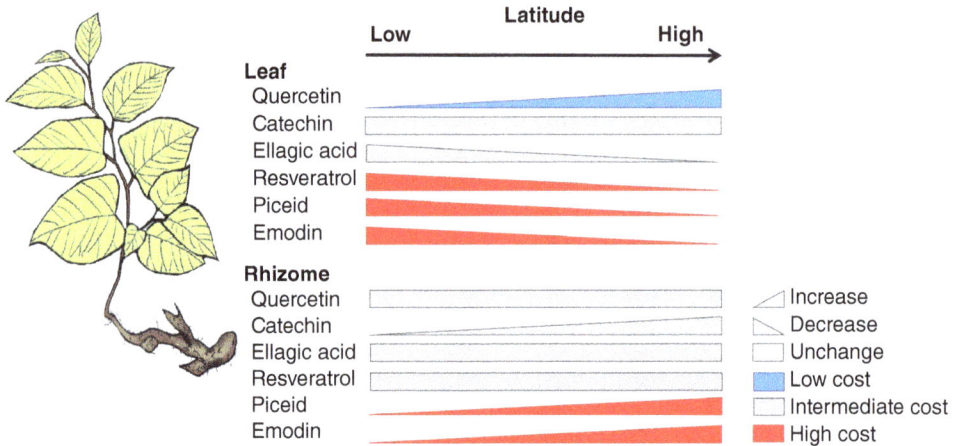

Fig. 9.10. Simplified schematic presentation of the observed latitudinal patterns of secondary metabolites in *Reynoutria japonica* leaves versus rhizomes, with different colors indicating low-, intermediate- and high-cost metabolites. From Bi *et al.* (2024) as per the policy of the British Ecological Society. Figure used with permission.

more similar microbiomes than native plants in diverse ecosystem types and geographical regions (Nunez-Mir and McCary, 2024). A study of invasive Canadian thistle, *Cirsium arvense*, reported a core group of rhizosphere-associated microorganisms compared with nearby bulk soil across a variety of climates and soils (Eberly *et al.*, 2024). Work in other invasive species (*Alternanthera philoxeroides*, *Carpobrotus edulis*, *Mikania micrantha*, *Phragmites australis*) further indicate the possibility that microbiome associations may facilitate plant invasions (Rodríguez-Caballero *et al.*, 2020; Yin *et al.*, 2020; Li *et al.*, 2022; Wang *et al.*, 2023). So far, there have been few studies on the role of the microbiome in knotweed invasions.

We compared the microbial community structure of the rhizosphere between populations of *R. japonica* in its native range in China (relatively low- and mid-latitude areas: 23–25°N and 28–31°N, respectively) and in populations that are considered to be in the recent expansion of the native range (relatively high-latitude areas: 34–36°N). We found that in the native-invasive, high-latitude areas of *R. japonica*'s range, there was a decrease in the diversity of bacterial and fungal lineages in the rhizosphere soil and a decrease in community homogenization (Q. He, J. Bi and J. Wu, unpublished data). Additionally, the abundance and relative abundance of pathogenic fungi in the rhizosphere of *R. japonica* at high latitudes were lower than in low- and mid-latitude

areas. Finally, in the high-latitude areas of *R. japonica*'s range, there was a significant enhancement in the rhizosphere soil enzymes decomposing cellulose and phenolic substances. This change in characteristics of *R. japonica* during its expansion may be adaptive in high-latitude areas, where secondary metabolism is enhanced, more indigestible substances are produced and resource acquisition efficiency is increased. Future work will continue to combine observational and experimental approaches in the native and introduced ranges with molecular and functional trait analyses.

The contribution of soil biota to the success of knotweed in introduced ranges remains unclear. Adding activated carbon to the soil greatly improved the growth of native forbs grown with knotweed propagated from the introduced range in Europe (Murrell *et al.*, 2011; Parepa *et al.*, 2012; Parepa *et al.*, 2013b). As activated carbon binds to organic compounds, including certain allelochemicals, this suggested that the negative impacts of knotweed were chemically mediated. Furthermore, different components of the soil biota in combination with activated carbon treatment showed that invasive knotweed strongly benefited from the presence of soil biota. In fact, the full natural soil biota shifted the competitive balance in favor of knotweed over six native species (Parepa *et al.*, 2013b). In this case, knotweed plants could affect beneficial mycorrhizae of the invaded communities, as has been found in other plant invaders (e.g. Stinson *et al.*, 2006; Callaway *et al.*, 2008). Garnica *et al.* (2022) isolated one root endophyte, *Serendipita herbamans*, that contributed to these responses. They found that the effects of the fungal colonization on knotweed performance depended on environmental conditions (Garnica *et al.*, 2022). The fungus increased knotweed biomass and leaf chlorophyll content in low-nutrient conditions but had the opposite effect in shaded conditions.

However, we found that *R. × bohemica*'s soil legacy in the introduced North American range may have the opposite effect: it may facilitate other species, at least during certain life stages (Lee *et al.*, 2024). We tested the impacts of *R. × bohemica*'s soil legacy using soils collected from field populations of *R. × bohemica* and from a common garden of potted *R. × bohemica* containing standardized soil that had been conditioned by *R. × bohemica* for 2 years prior to the experiment. We found seedling emergence was higher in *R. × bohemica*-conditioned soils than in unconditioned soils across five native, five introduced and five agricultural plant species. However, we did not find a clear trend for the effects of *R. × bohemica*'s soil legacy on seedling performance. In fact, across tests of both seedling emergence and seedling performance, the impacts of *R. × bohemica*-conditioned soil were highly context dependent: *R. × bohemica*'s soil legacy was highly species specific and was also mediated by the collection latitude of the *R. × bohemica* rhizomes used to condition standardized common-garden soils, suggesting site-specific effects.

Conflicting findings for the interactions between knotweeds and microbes—with important consequences for its invasion success—may be due to myriad factors, including the knotweed genotypes used in the studies, site-specific differences in soil biota, and inherent differences between testing the impact of knotweed's soil legacy versus how actively growing knotweeds alter their local soil environment. We currently have very little information about the patterns of secondary metabolites and microbial community differences across populations in introduced ranges. In addition, experiments that disentangle the biotic components of an invader's impact on soils using activated carbon (e.g. Murrell *et al.*, 2011; Parepa *et al.*, 2012; Parepa *et al.*, 2013b) versus sterilization through autoclaving (e.g.

Lee *et al.*, 2024) are known to have different unintended consequences on soil characteristics, which may lead to different interpretations. Plant–microbe interactions may also be mediated by rapid plant molecular responses to environmental conditions, but this line of inquiry is in its infancy. How plant genomic and epigenomic variation interacts with variation in the microbiome to translate into plant performance has largely been underexplored and promises to be a productive area of future research.

9.5 Predicting the Future Spread of Knotweed

Global climate change has caused warming of many ecosystems and a reduction in the constraints imposed by freezing, potentially facilitating the poleward and/or elevational spread of invasive species (Walther *et al.*, 2002; Parmesan and Yohe, 2003; Root *et al.*, 2003; Hickling *et al.*, 2006; Chen *et al.*, 2011; Christina *et al.*, 2020) and exacerbating their impact and spread (Dukes and Mooney, 1999; Weltzin *et al.*, 2003; Thuiller *et al.*, 2007). However, quantitative analyses of species' response to future climate scenarios are rare (but see Chefaoui *et al.*, 2019; Lee *et al.*, 2021; Adhikari *et al.*, 2022).

We evaluated potential range shifts in native and introduced populations of *R. japonica* under climatic change by utilizing ecological niche modeling of possible knotweed habitat in 2070 on a global scale, based on knotweed habitat in the modern climate, and predicted global climate in 2070. This modeling showed significant northward range expansion in the European and North American populations, while Asian populations were predicted to experience stable or even shrinking suitable habitats under the same climate scenario. The limited range expansion in native populations is attributable to different predicted climate-change patterns for Asia compared with Europe/North America and may be further exacerbated by the genetic distinctions between the Chinese *R. japonica* populations and those in other countries. The most recent Intergovernmental Panel on Climate Change (IPCC) report forecasted increased drought frequency in the southern region of the Yangtze River in China (Calvin *et al.*, 2023). This may pose a challenge for species like Japanese knotweed *s.l.*, which require consistent water availability (Bailey and Wisskirchen, 2004; Jovanović *et al*., 2018). Additionally, our research indicated that Chinese populations of *R. japonica* exhibited lower leaf toughness compared with those in Japan, Europe, and North America (P. Cao *et al*., unpublished data). As plants with softer leaves are generally less drought resistant (Wright *et al*., 2004), these morphological differentiations could contribute to higher susceptibility of the Chinese populations to climate change. Evaluating the genomic changes that underlie such trait responses will shed light on the future potential of these important species.

9.6 Advancing Our Understanding of Ecosystem-level Processes with Genomics

Invasion biology has characterized how specific life history traits allow species to cope with a large range of ecological conditions (van Kleunen *et al*., 2018; Gioria *et al*., 2023). Yet, despite decades of work to understand invasion, critical knowledge gaps include a lack of understanding of the biodiversity within native populations of the invasive species, the biodiversity of the native communities and the importance of diversity levels for ecosystem functioning. While our work is beginning to fill this void for Japanese knotweed, we also presently lack compelling evidence for the genomic basis of eco-evolutionary feedback processes, and how this affects the success or failure of plant invaders or their impacts on local biota.

For example, our investigations into how knotweed mediates plant community dynamics through their soil environment indicated that knotweed has a strong influence on ecosystem-level processes, but further research is necessary to understand whether these impacts can be generalizable. In a pilot mesocosm experiment, we found support for heritable differences in the decomposition of *R.* × *bohemica* from across latitudinal gradients in the introduced range (S.B. Endriss, unpublished data). Additionally, many studies have highlighted how plant invaders appear to be superior at competing for resources, an effect that could result from changes in natural enemies in the new range, such as those we

found for knotweed that "release" plants from top-down control (i.e. the enemy release hypothesis; Keane and Crawley, 2002). Introduced species may benefit tremendously if novel habitats lack biotic resistance (Mangla and Callaway, 2007; Blossey *et al*., 2021), when native plants lack mutualistic interactions with soil microbes (van der Putten *et al*., 2013), or when native plants are more strongly suppressed by pathogens and herbivores (Carpenter and Cappuccino, 2005; Kalisz *et al*., 2014; Deardorff and Gorchov, 2021; Gorchov *et al*., 2021). As we develop genomic resources to inform the study of Japanese knotweed, this global plant invasion provides an opportunity to explore many aspects of functional biodiversity because of knotweed's impacts on local biodiversity across a broad climate gradient in both native and invasive ranges. Integrating plant genomics, metabolomics and microbiome studies, across latitudinal gradients in native and introduced ranges, may enhance our understanding of the functional significance of intra- and interspecific biodiversity (Fig. 9.11). Applying such technologies allows further investigations into the role of microbial interactions, soil biota, fungal symbionts, and macroinvertebrates in invasion success and may offer insights into the complex dynamics of other invasive species.

9.7 Conclusions

Our work on the Japanese knotweed complex highlights the importance of interdisciplinary and global-scale work in developing a model system in invasion biology. We have reconstructed the Japanese knotweed invasion on a global scale using conservative chloroplast DNA markers, GBS markers and changes in traits across the introduced and native ranges to elucidate whether a particularly aggressive genotype may have become a globally successful plant invader. In addition to field collections, we have employed herbarium collections and multiple common-garden experiments to further evaluate the range of trait responses of individuals collected across introduced and native ranges. We found support for the spread of a single haplotype from Japan to most of the introduced range, but there may be some cryptic diversity that has contributed to the invasion. In particular, the increased

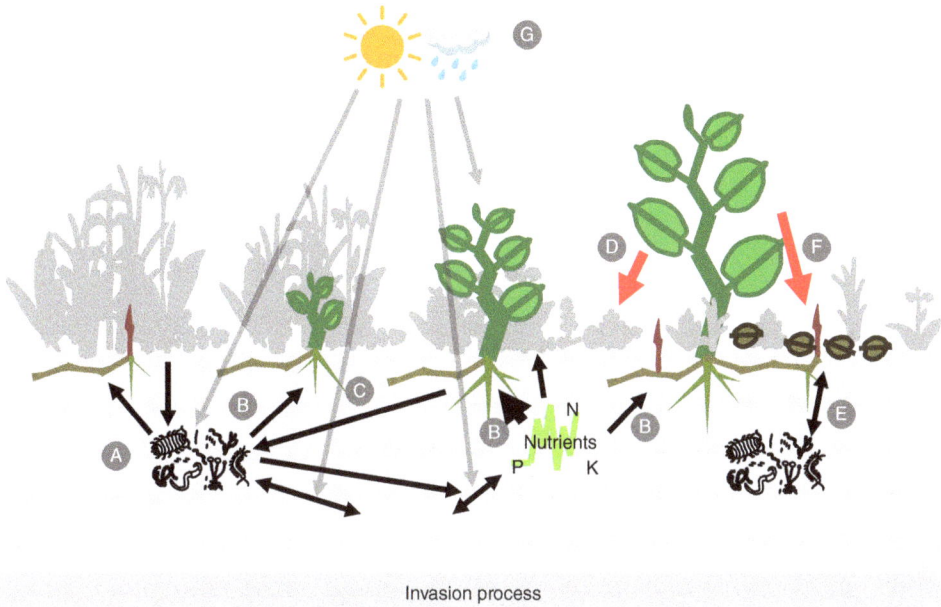

Fig. 9.11. The hypothesized invasion of Japanese knotweed responding to and driving local biodiversity and ecosystem processes progresses from left to right. Complex interactions shape soil biodiversity and plant and animal communities above and below ground (A). Soil biodiversity and nutrients impact knotweed regeneration and growth (B). Knotweed altered soil (C) and suppressed local above- (D) and belowground (E) biodiversity. We predict alterations in soil ecosystem processes such as decomposition (F). Interactions are impacted by climate (G) across latitudinal gradients and on the changing planet.

plasticity in clonality of introduced plants compared with source Japanese plants supported some changes that may have contributed to the invasion process. Much more detailed genomics work will be required to pinpoint the genomic basis of such changes. However, this knowledge of invasion history and heritable changes in trait variation is vital for understanding how genomic changes translate into functional consequences and ecosystem function. Such insight will provide powerful tools to address the response of invasive plants to rapid global change, and mitigate widespread impacts on ecosystem function and resilience.

Acknowledgments

This work was supported by the National Natural Science Foundation of China (31961133028), the German Federal Ministry of Education and Research (MOPGA Project 306055), the German Research Foundation (grant 431595342), the Chinese Academy of Sciences President's International Fellowship Initiative (2023VBB0012), the Department of Science and Technology of Yunnan Province (XDYC-QNRC-2022-0014) and the Horizon Europe research and innovation program under the Marie Skłodowska-Curie Actions (grant 101033168).

References

Adhikari, P., Kim, B.-J., Hong, S.-H. and Lee, D.-H. (2022) Climate change induced habitat expansion of nutria (*Myocastor coypus*) in South Korea. *Scientific Reports* 12: 3300. DOI: 10.1038/s41598-022-07347-5

Agler, M.T., Ruhe, J., Kroll, S., Morhenn, C., Kim, S.-T. *et al.* (2016) Microbial hub taxa link host and abiotic factors to plant microbiome variation. *PLOS Biology* 14: e1002352. DOI: 10.1371/journal.pbio.1002352

Agrawal, A.A. (2020) A scale-dependent framework for trade-offs, syndromes, and specialization in organ-ismal biology. *Ecology* 101: e02924. DOI: 10.1002/ecy.2924

Agrawal, A.A. and Fishbein, M. (2006) Plant defense syndromes. *Ecology* 87, S132–S149. DOI: 10.1890/0012-9658(2006)87[132:pds]2.0.co;2

Aguilera, A.G., Alpert, P., Dukes, J.S. and Harrington, R. (2010) Impacts of the invasive plant *Fallopia japon-ica* (Houtt.) on plant communities and ecosystem processes. *Biological Invasions* 12, 1243–1252. DOI: 10.1007/s10530-009-9543-z

Ainouche, M.L., Baumel, A., Salmon, A. and Yannic, G. (2004) Hybridization, polyploidy and speciation in *Spartina* (Poaceae). *New Phytologist* 161, 165–172. DOI: 10.1046/j.1469-8137.2003.00926.x

Ainouche, M.L., Fortune, P.M., Salmon, A., Parisod, C., Grandbastien, M.-A. *et al.* (2009) Hybridization, polyploidy and invasion: lessons from *Spartina* (Poaceae). *Biological Invasions* 11, 1159–1173. DOI: 10.1007/s10530-008-9383-2

Bailey, J.P. (2013) The Japanese knotweed invasion viewed as a vast unintentional hybridisation experiment. *Heredity* 110, 105–110. DOI: 10.1038/hdy.2012.98

Bailey, J.P. and Conolly, A.P. (2000) Prize-winners to pariahs – a history of Japanese knotweed *s.l.* (Polyg-onaceae) in the British Isles. *Watsonia* 23, 93–110.

Bailey, J.P. and Wisskirchen, R. (2004) The distribution and origins of *Fallopia* × *bohemica* (Polygonaceae) in Europe. *Nordic Journal of Botany* 24, 173–199. DOI: 10.1111/j.1756-1051.2004.tb00832.x

Bailey, J.P., Bímová, K. and Mandák, B. (2009) Asexual spread versus sexual reproduction and evolution in Japanese Knotweed *s.l.* sets the stage for the "Battle of the Clones." *Biological Invasions* 11, 1189–1203. DOI: 10.1007/s10530-008-9381-4

Baker, H.G. (1965) Characteristics and modes of origin of weeds. In: Baker, H.G. and Stebbins, G.L. (eds) *The Genetics of Colonizing Species*. Academic Press, New York, pp. 147–172.

Bergelson, J., Mittelstrass, J. and Horton, M.W. (2019) Characterizing both bacteria and fungi improves understanding of the *Arabidopsis* root microbiome. *Scientific Reports* 9: 24. DOI: 10.1038/s41598-018-37208-z

Bi, J., Bossdorf, O., Liao, Z., Richards, C.L., Parepa, M, *et al.* (2024) Divergent geographic variation in above- versus below-ground secondary metabolites of *Reynoutria japonica*. *Journal of Ecology* 112, 514–527. DOI: 10.1111/1365-2745.14248

Bímová, K., Mandák, B. and Kašparová, I. (2004) How does *Reynoutria* invasion fit the various theories of invasibility? *Journal of Vegetation Science* 15, 495–504. DOI: 10.1111/j.1654-1103.2004.tb02288.x

Blossey, B., Nuzzo, V., Dávalos, A., Mayer, M., Dunbar, R. *et al.* (2021) Residence time determines invasive-ness and performance of garlic mustard (*Alliaria petiolata*) in North America. *Ecology Letters* 24, 327–336. DOI: 10.1111/ele.13649

Bock, D.G., Caseys, C., Cousens, R.D., Hahn, M.A., Heredia, S.M. *et al.* (2015) What we still don't know about invasion genetics. *Molecular Ecology* 24, 2277–2297. DOI: 10.1111/mec.13032

Bock, D.G., Kantar, M.B., Caseys, C., Matthey-Doret, R. and Rieseberg, L.H. (2018) Evolution of invasiveness by genetic accommodation. *Nature Ecology and Evolution* 2, 991–999. DOI: 10.1038/s41559-018-0553-z

Bowen, J.L., Kearns, P.J., Byrnes, J.E.K., Wigginton, S., Allen, W.J. *et al.* (2017) Lineage overwhelms envir-onmental conditions in determining rhizosphere bacterial community structure in a cosmopolitan inva-sive plant. *Nature Communications* 8: 433. DOI: 10.1038/s41467-017-00626-0

Callaway, R.M., Cipollini, D., Barto, K., Thelen, G.C., Hallett, S.G. *et al.* (2008) Novel weapons: invasive plant suppresses fungal mutualists in America but not in its native Europe. *Ecology* 89, 1043–1055. DOI: 10.1890/07-0370.1

Calvin, K., Dasgupta, D., Krinner, G., Mukherji, A., Thorne, P.W. *et al.* (2023) *Climate Change 2023: Syn-thesis Report: Summary for Policymakers. Contribution of working groups I, II and III to the Sixth As-sessment Report of the Intergovernmental Panel on Climate Change*. IPCC, Geneva, Switzerland, pp. 35–115. Available at: www.ipcc.ch/report/ar6/syr/downloads/report/IPCC_AR6_SYR_SPM.pdf (ac-cessed 23 July 2025).

Cao, P., Liao, Z., Wang, S., Parepa, M., Zhang, L. *et al.* (2025) Cross-continental variation of herbivore resistance in a global plant invader. *Ecography* 2025: e07569. DOI: 10.1111/ecog.07569

Carpenter, D. and Cappuccino, N. (2005) Herbivory, time since introduction and the invasiveness of exotic plants. *Journal of Ecology* 93, 315–321. DOI: 10.1111/j.1365-2745.2005.00973.x

Casson, S.A. and Hetherington, A.M. (2010) Environmental regulation of stomatal development. *Current Opinion in Plant Biology* 13, 90–95. DOI: 10.1016/j.pbi.2009.08.005

Chefaoui, R.M., Serebryakova, A., Engelen, A.H., Viard, F. and Serrão, E.A. (2019) Integrating reproductive phenology in ecological niche models changed the predicted future ranges of a marine invader. *Diver-sity and Distributions* 25, 688–700. DOI: 10.1111/ddi.12910

Chen, I.-C., Hill, J.K., Ohlemüller, R., Roy, D.B. and Thomas, C.D. (2011) Rapid range shifts of species associated with high levels of climate warming. *Science* 333, 1024–1026. DOI: 10.1126/science.1206432

Chen, Z.J. and Yu, H.H. (2013) Genetic and epigenetic mechanisms for polyploidy and hybridity. In: Chen, Z.J. and Birchler, J.A. (eds) *Polyploid and Hybrid Genomics*. Wiley, Ames, Iowa, pp. 335–354. DOI: 10.1002/9781118552872.ch21

Christina, M., Limbada, F. and Atlan, A. (2020) Climatic niche shift of an invasive shrub (*Ulex europaeus*): a global scale comparison in native and introduced regions. *Journal of Plant Ecology* 13, 42–50. DOI: 10.1093/jpe/rtz041

Clark, J.W. and Donoghue, P.C.J. (2018) Whole-genome duplication and plant macroevolution. *Trends in Plant Science* 23, 933–945. DOI: 10.1016/j.tplants.2018.07.006

Deardorff, J.L. and Gorchov, D.L. (2021) Beavers cut, but do not prefer, an invasive shrub, Amur honeysuckle (*Lonicera maackii*). *Biological Invasions* 23, 193–204. DOI: 10.1007/s10530-020-02365-8

Del Tredici, P. (2017) The introduction of Japanese knotweed, *Reynoutria japonica*, into North America. *Journal of the Torrey Botanical Society* 144, 406–416. DOI: 10.3159/TORREY-D-17-00002.1

Dlugosch, K.M., Anderson, S.R., Braasch, J., Cang, F.A. and Gillette, H.D. (2015) The devil is in the details: genetic variation in introduced populations and its contributions to invasion. *Molecular Ecology* 24, 2095–2111. DOI: 10.1111/mec.13183

Dukes, J.S. and Mooney, H.A. (1999) Does global change increase the success of biological invaders? *Trends in Ecology and Evolution* 14, 135–139. DOI: 10.1016/s0169-5347(98)01554-7

Eberly, J.O., Hurd, A., Oli, D., Dyer, A.T., Seipel, T.F. and Carr, P.M. (2024) Compositional profiling of the rhizosphere microbiome of Canada thistle reveals consistent patterns across the United States Northern Great Plains. *Scientific Reports* 14: 18016. DOI: 10.1038/s41598-024-69082-3

Elshire, R.J., Glaubitz, J.C., Sun, Q., Poland, J.A., Kawamoto, K. *et al.* (2011) A robust, simple genotyping-by-sequencing (GBS) approach for high diversity species. *PLOS ONE* 6: e19379. DOI: 10.1371/journal.pone.0019379

Eschen, R., Kadzamira, M., Stutz, S., Ogunmodede, A., Djeddour, D. *et al.* (2023) An updated assessment of the direct costs of invasive non-native species to the United Kingdom. *Biological Invasions* 25, 3265–3276. DOI: 10.1007/s10530-023-03107-2

Estoup, A., Ravigné, V., Hufbauer, R., Vitalis, R., Gautier, M. and Facon, B. (2016) Is there a genetic paradox of biological invasion? *Annual Review of Ecology, Evolution, and Systematics* 47, 51–72. DOI: 10.1146/annurev-ecolsys-121415-032116

Fitzpatrick, C.R., Copeland, J., Wang, P.W., Guttman, D.S., Kotanen, P.M. and Johnson, M.T.J. (2018) Assembly and ecological function of the root microbiome across angiosperm plant species. *Proceedings of the National Academy of Sciences USA* 115, E1157–E1165. DOI: 10.1073/pnas.1717617115

Fogelman, K.J., Bilger, M.D., Holt, J.R. and Matlaga, D.P. (2018) Decomposition and benthic macroinvertebrate communities of exotic Japanese knotweed (*Fallopia japonica*) and American sycamore (*Platanus occidentalus*) detritus within the Susquehanna River. *Journal of Freshwater Ecology* 33, 299–310. DOI: 10.1080/02705060.2018.1458660

Franck, A.R. (2018) *Overview of the University of South Florida Herbarium*. Available at: www.researchgate.net/publication/310413761_Overview_of_the_University_of_South_Florida_Herbarium (accessed 23 July 2025).

Gammon, M.A. and Kesseli, R. (2010) Haplotypes of *Fallopia* introduced into the US. *Biological Invasions* 12, 421–427. DOI: 10.1007/s10530-009-9459-7

Gao, L., Geng, Y., Li, B., Chen, J. and Yang, J. (2010) Genome-wide DNA methylation alterations of *Alternanthera philoxeroides* in natural and manipulated habitats: implications for epigenetic regulation of rapid responses to environmental fluctuation and phenotypic variation. *Plant, Cell & Environment* 33, 1820–1827. DOI: 10.1111/j.1365-3040.2010.02186.x

Gao, S., Zou, D., Mao, L., Liu, H., Song, P. *et al.* (2015) BS-SNPer: SNP calling in bisulfite-seq data. *Bioinformatics* 31, 4006–4008. DOI: 10.1093/bioinformatics/btv507

Garnica, S., Liao, Z., Hamard, S., Waller, F., Parepa, M. and Bossdorf, O. (2022) Environmental stress determines the colonization and impact of an endophytic fungus on invasive knotweed. *Biological Invasions* 24, 1785–1795. DOI: 10.1007/s10530-022-02749-y

Gaskin, J.F., Schwarzländer, M., Grevstad, F.S., Haverhals, M.A., Bourchier, R.S. and Miller, T.W. (2014) Extreme differences in population structure and genetic diversity for three invasive congeners: knotweeds in western North America. *Biological Invasions* 16, 2127–2136. DOI: 10.1007/s10530-014-0652-y

Gawehns, F., Postuma, M., van Antro, M., Nunn, A., Sepers, B. *et al.* (2022) epiGBS2: improvements and evaluation of highly multiplexed, epiGBS-based reduced representation bisulfite sequencing. *Molecular Ecology Resources* 22, 2087–2104. DOI: 10.1111/1755-0998.13597

Gerber, E., Krebs, C., Murrell, C., Moretti, M., Rocklin, R. and Schaffner, U. (2008) Exotic invasive knot-weeds (*Fallopia* spp.) negatively affect native plant and invertebrate assemblages in European riparian habitats. *Biological Conservation* 141, 646–654. DOI: 10.1016/j.biocon.2007.12.009

Gioria, M., Hulme, P.E., Richardson, D.M. and Pyšek, P. (2023) Why are invasive plants successful? *Annual Review of Plant Biology* 74, 635–670. DOI: 10.1146/annurev-arplant-070522-071021

Gorchov, D.L., Blossey, B., Averill, K.M., Dávalos, A., Heberling, J.M. *et al.* (2021) Differential and interacting impacts of invasive plants and white-tailed deer in eastern U.S. forests. *Biological Invasions* 23, 2711–2727. DOI: 10.1007/s10530-021-02551-2

Grabar, H. (2019) Oh, No, Not Knotweed! *Slate.com.* 8 May. Available at: https://slate.com/technology/2019/05/japanese-knotweed-invasive-plants.html (accessed 23 July 2025).

Grimsby, J.L., Tsirelson, D., Gammon, M.A. and Kesseli, R. (2007) Genetic diversity and clonal vs. sexual reproduction in *Fallopia* spp. (Polygonaceae). *American Journal of Botany* 94, 957–964. DOI: 10.3732/ajb.94.6.957

Grotkopp, E. and Rejmánek, M. (2007) High seedling relative growth rate and specific leaf area are traits of invasive species: phylogenetically independent contrasts of woody angiosperms. *American Journal of Botany* 94, 526–532. DOI: 10.3732/ajb.94.4.526

Hamma, L. (2023) Investigating spatio-temporal responses of stomata and specific leaf area of invasive Asian knotweeds to climatic factors using herbarium collections. BSc thesis, University of Tubingen, Tubingen, Germany.

Haworth, M., Marino, G., Materassi, A., Raschi, A., Scutt, C.P. and Centritto, M. (2023) The functional significance of the stomatal size to density relationship: Interaction with atmospheric [CO_2] and role in plant physiological behaviour. *Science of the Total Environment* 863, 160908. DOI: 10.1016/j.scitotenv.2022.160908

He, Q., Miao, Y., Zheng, X., Wang, Y., Wang, Y. *et al.* (2024) The near-complete genome assembly of *Reynoutria multiflora* reveals the genetic basis of stilbene and anthraquinone biosynthesis. *Journal of Systematics and Evolution* 62, 1085–1102. DOI: 10.1111/jse.13068

Heberling, J.M. (2022) Herbaria as big data sources of plant traits. *International Journal of Plant Sciences* 183, 87–118. DOI: 10.1086/717623

Herrera, C.M., Medrano, M. and Bazaga, P. (2017) Comparative epigenetic and genetic spatial structure of the perennial herb *Helleborus foetidus*: isolation by environment, isolation by distance, and functional trait divergence. *American Journal of Botany* 104, 1195–1204. DOI: 10.3732/ajb.1700162

Hickling, R., Roy, D.B., Hill, J.K., Fox, R. and Thomas, C.D. (2006) The distributions of a wide range of taxonomic groups are expanding polewards. *Global Change Biology* 12, 450–455. DOI: 10.1111/j.1365-2486.2006.01116.x

Hodgins, K.A., Bock, D.G. and Rieseberg, L.H. (2018) Trait evolution in invasive species. *Annual Plant Reviews Online* 1, 1–37. DOI: 10.1002/9781119312994.apr0643

Hollingsworth, M.L. and Bailey, J.P. (2000) Evidence for massive clonal growth in the invasive weed *Fallopia japonica* (Japanese Knotweed). *Botanical Journal of the Linnean Society. Linnean Society of London* 133, 463–472. DOI: 10.1111/j.1095-8339.2000.tb01589.x

Hu, L., Robert, C.A.M., Cadot, S., Zhang, X., Ye, M. *et al.* (2018) Root exudate metabolites drive plant-soil feedbacks on growth and defense by shaping the rhizosphere microbiota. *Nature Communications* 9: 2738. DOI: 10.1038/s41467-018-05122-7

Hughes, A.R., Moore, A.F.P. and Gehring, C. (2020) Plant response to fungal root endophytes varies by host genotype in the foundation species *Spartina alterniflora*. *American Journal of Botany* 107, 1645–1653. DOI: 10.1002/ajb2.1573

Inamura, A., Ohashi, Y., Sato, E., Yoda, Y., Masuzawa, T. *et al.* (2000) Intraspecific sequence variation of chloroplast DNA reflecting variety and geographical distribution of *Polygonum cuspidatum* (Polygonaceae) in Japan. *Journal of Plant Research* 113, 419–426. DOI: 10.1007/PL00013950

Irimia, R.E., Zhao, W., Cao, P., Parepa, M., Liao, Z.-Y. *et al.* (2025) Cross-continental shifts of ecological strategy in a global plant invader. *Global Ecology and Biogeography* 34: e70001. DOI: 10.1111/geb.70001

Jovanović, S., Hlavati-Širka, V., Lakušić, D., Jogan, N., Nikolić, T. *et al.* (2018) *Reynoutria* niche modelling and protected area prioritization for restoration and protection from invasion: a Southeastern Europe case study. *Journal for Nature Conservation* 41, 1–15. DOI: 10.1016/j.jnc.2017.10.011

Kagawa, K. and Takimoto, G. (2018) Hybridization can promote adaptive radiation by means of transgressive segregation. *Ecology Letters* 21, 264–274. DOI: 10.1111/ele.12891

Kalisz, S., Spigler, R.B. and Horvitz, C.C. (2014) In a long-term experimental demography study, excluding ungulates reversed invader's explosive population growth rate and restored natives. *Proceedings of the National Academy of Sciences USA* 111, 4501–4506. DOI: 10.1073/pnas.1310121111

Keane, R.M. and Crawley, M.J. (2002) Exotic plant invasions and the enemy release hypothesis. *Trends in Ecology and Evolution* 17, 164–170. DOI: 10.1016/S0169-5347(02)02499-0

Kivlin, S.N., Emery, S.M. and Rudgers, J.A. (2013) Fungal symbionts alter plant responses to global change. *American Journal of Botany* 100, 1445–1457. DOI: 10.3732/ajb.1200558

Lampard, G.R., Macalister, C.A. and Bergmann, D.C. (2008) *Arabidopsis* stomatal initiation is controlled by MAPK-mediated regulation of the bHLH SPEECHLESS. *Science* 322, 1113–1116. DOI: 10.1126/science.1162263

Lang, P.L.M., Willems, F.M., Scheepens, J.F., Burbano, H.A. and Bossdorf, O. (2019) Using herbaria to study global environmental change. *New Phytologist* 221, 110–122. DOI: 10.1111/nph.15401

Lecerf, A., Patfield, D., Boiché, A., Riipinen, M.P., Chauvet, E. and Dobson, M. (2007) Stream ecosystems respond to riparian invasion by Japanese knotweed (*Fallopia japonica*). *Canadian Journal of Fisheries and Aquatic Sciences* 64, 1273–1283. DOI: 10.1139/f07-092

Lee, C.M., Lee, D.-S., Kwon, T.-S., Athar, M. and Park, Y.-S. (2021) Predicting the global distribution of *Solenopsis geminata* (Hymenoptera: Formicidae) under climate change using the MaxEnt model. *Insects* 12: 229. DOI: 10.3390/insects12030229

Lee, K., Endriss, S.B. and Blossey, B. (2024) Self-help or self-sabotage? A common invader's soil legacy does not impede, and may facilitate, potential competitors. *Ecosphere* 15: e4917. DOI: 10.1002/ecs2.4917

Levis, N.A. and Pfennig, D.W. (2016) Evaluating 'plasticity-first' evolution in nature: key criteria and empirical approaches. *Trends in Ecology and Evolution* 31, 563–574. DOI: 10.1016/j.tree.2016.03.012

Li, C., Bo, H., Song, B., Chen, X., Cao, Q. *et al.* (2022) Reshaping of the soil microbiome by the expansion of invasive plants: shifts in structure, diversity, co-occurrence, niche breadth, and assembly processes. *Plant and Soil* 477, 629–646.

Liu, Y., Siegmund, K.D., Laird, P.W. and Berman, B.P. (2012) Bis-SNP: combined DNA methylation and SNP calling for Bisulfite-seq data. *Genome Biology* 13: R61. DOI: 10.1186/gb-2012-13-7-r61

Lowe, S., Browne, M., Boudjelas, S. and de Poorter, M. (2000) *100 of the World's Worst Invasive Alien Species: A Selection from the Global Invasive Species Database*. The Invasive Species Specialist Group, Species Survival Commission, World Conservation Union, Auckland, New Zealand. First published in *Aliens* 12 (December 2000), updated and reprinted November 2004. Available at: https://portals.iucn.org/library/sites/library/files/documents/2000-126.pdf (accessed 24 July 2025).

Maerz, J.C., Blossey, B. and Nuzzo, V. (2005) Green frogs show reduced foraging success in habitats invaded by Japanese knotweed. *Biodiversity and Conservation* 14, 2901–2911. DOI: 10.1007/s10531-004-0223-0

Mandák, B., Pyšek, P., Lysak, M., Suda, J., Krahulcova, A. and Bímová, K. (2003) Variation in DNA-ploidy levels of *Reynoutria* taxa in the Czech Republic. *Annals of Botany* 92, 265–272. DOI: 10.1093/aob/mcg141

Mandák, B., Pyšek, P. and Bímová, K. (2004) History of the invasion and distribution of *Reynoutria* taxa in the Czech Republic: a hybrid spreading faster than its parents. *Preslia* 76, 15–64.

Mandák, B., Bímová, K., Pyšek, P., Štěpánek, J. and Plačková, I. (2005) Isoenzyme diversity in *Reynoutria* (Polygonaceae) taxa: escape from sterility by hybridization. *Plant Systematics and Evolution* 253, 219–230. DOI: 10.1007/s00606-005-0316-6

Mandyam, K.G. and Jumpponen, A. (2015) Mutualism–parasitism paradigm synthesized from results of root-endophyte models. *Frontiers in Microbiology* 5: 776. DOI: 10.3389/fmicb.2014.00776

Mangla, S. and Callaway, R.M. (2007) Exotic invasive plant accumulates native soil pathogens which inhibit native plants. *The Journal of Ecology* 96, 58-67. DOI: 10.1111/j.1365-2745.2007.01312.x

Matesanz, S. and Sultan, S.E. (2013) High-performance genotypes in an introduced plant: insights to future invasiveness. *Ecology* 94, 2464–2474. DOI: 10.1890/12-1359.1

Matesanz, S., Horgan-Kobelski, T. and Sultan, S.E. (2015) Evidence for rapid ecological range expansion in a newly invasive plant. *AoB Plants* 7: plv038. DOI: 10.1093/aobpla/plv038

McGaughran, A., Dhami, M.K., Parvizi, E., Vaughan, A.L., Gleeson, D.M. *et al.* (2024) Genomic tools in biological invasions: current state and future frontiers. *Genome Biology and Evolution* 16: evad230. DOI: 10.1093/gbe/evad230

Meyer, M. and Kircher, M. (2010) Illumina sequencing library preparation for highly multiplexed target capture and sequencing. *Cold Spring Harbor Protocols* 2010: pdb.prot5448. DOI: 10.1101/pdb.prot5448

Mounger, J., Ainouche, M.L., Bossdorf, O., Cavé-Radet, A., Li, B. *et al.* (2021) Epigenetics and the success of invasive plants. *Philosophical Transactions of the Royal Society B: Biological Sciences* 376: 20200117. DOI: 10.1098/rstb.2020.0117

Murrell, C., Gerber, E., Krebs, C., Parepa, M., Schaffner, U. and Bossdorf, O. (2011) Invasive knotweed affects native plants through allelopathy. *American Journal of Botany* 98, 38–43. DOI: 10.3732/ajb.1000135

Neinavaie, F., Ibrahim-Hashim, A., Kramer, A.M., Brown, J.S. and Richards, C.L. (2021) The genomic processes of biological invasions: from invasive species to cancer metastases and back again. *Frontiers in Ecology and Evolution* 9: 681100. DOI: 10.3389/fevo.2021.681100

Nicotra, A.B., Atkin, O.K., Bonser, S.P., Davidson, A.M., Finnegan, E.J. *et al.* (2010) Plant phenotypic plasticity in a changing climate. *Trends in Plant Science* 15, 684–692. DOI: 10.1016/j.tplants.2010.09.008

Nunez-Mir, G.C. and McCary, M.A. (2024) Invasive plants and their root traits are linked to the homogenization of soil microbial communities across the United States. *Proceedings of the National Academy of Sciences USA* 121: e2418632121. DOI: 10.1073/pnas.2418632121

Parepa, M., Schaffner, U. and Bossdorf, O. (2012) Sources and modes of action of invasive knotweed allelopathy: the effects of leaf litter and trained soil on the germination and growth of native plants. *NeoBiota* 13, 15–30. DOI: 10.3897/neobiota.13.3039

Parepa, M., Fischer, M. and Bossdorf, O. (2013a) Environmental variability promotes plant invasion. *Nature Communications* 4: 1604. DOI: 10.1038/ncomms2632

Parepa, M., Schaffner, U. and Bossdorf, O. (2013b) Help from under ground: soil biota facilitate knotweed invasion. *Ecosphere* 4: 31. DOI: 10.1890/ES13-00011.1

Parepa, M., Kahmen, A., Werner, R.A., Fischer, M. and Bossdorf, O. (2019) Invasive knotweed has greater nitrogen-use efficiency than native plants: evidence from a ^{15}N pulse-chasing experiment. *Oecologia* 191, 389–396. DOI: 10.1007/s00442-019-04490-1

Parmesan, C. and Yohe, G. (2003) A globally coherent fingerprint of climate change impacts across natural systems. *Nature* 421, 37–42. DOI: 10.1038/nature01286

Paun, O., Verhoeven, K.J.F. and Richards, C.L. (2019) Opportunities and limitations of reduced representation bisulfite sequencing in plant ecological epigenomics. *New Phytologist* 221, 738–742. DOI: 10.1111/nph.15388

Peng, W., Qin, R., Li, X. and Zhou, H. (2013) Botany, phytochemistry, pharmacology, and potential application of *Polygonum cuspidatum* Sieb.et Zucc.: a review. *Journal of Ethnopharmacology* 148, 729–745. DOI: 10.1016/j.jep.2013.05.007

Pyšek, P. and Hulme, P.E. (2005) Spatio-temporal dynamics of plant invasions: linking pattern to process. *Ecoscience* 12, 302–315. DOI: 10.2980/i1195-6860-12-3-302.1

Pyšek, P. and Prach, K. (1993) Plant invasions and the role of riparian habitats: a comparison of four species alien to central Europe. *Journal of Biogeography* 20, 413–420. DOI: 10.2307/2845589

Pyšek, P., Brock, J.H., Bímová, K., Mandák, B., Jarosík, V. *et al.* (2003) Vegetative regeneration in invasive *Reynoutria* (Polygonaceae) taxa: the determinant of invasibility at the genotype level. *American Journal of Botany* 90, 1487–1495. DOI: 10.3732/ajb.90.10.1487

Renčo, M., Čerevková, A. and Homolová, Z. (2021) Nematode communities indicate the negative impact of *Reynoutria japonica* invasion on soil fauna in ruderal habitats of Tatra national park in Slovakia. *Global Ecology and Conservation* 26: e01470. DOI: 10.1016/j.gecco.2021.e01470

Richards, C.L., Walls, R.L., Bailey, J.P., Parameswaran, R., George, T. and Pigliucci, M. (2008) Plasticity in salt tolerance traits allows for invasion of novel habitat by Japanese knotweed *s. l.* (*Fallopia japonica* and *F.* × *bohemica*, Polygonaceae). *American Journal of Botany* 95, 931–942. DOI: 10.3732/ajb.2007364

Richards, C.L., Schrey, A.W. and Pigliucci, M. (2012) Invasion of diverse habitats by few Japanese knotweed genotypes is correlated with epigenetic differentiation. *Ecology Letters* 15, 1016–1025. DOI: 10.1111/j.1461-0248.2012.01824.x

Richards, C.L., Alonso, C., Becker, C., Bossdorf, O., Bucher, E. *et al.* (2017) Ecological plant epigenetics: evidence from model and non-model species, and the way forward. *Ecology Letters* 20, 1576–1590. DOI: 10.1111/ele.12858

Robertson, M., Alvarez, M., van Gurp, T., Wagemaker, C.A.M., Yu, F. *et al.* (2020) Combining epiGBS markers with long read transcriptome sequencing to assess differentiation associated with habitat in *Reynoutria* (aka *Fallopia*). *bioRxiv* (preprint). DOI: 10.1101/2020.09.30.317966

Rodríguez-Caballero, G., Caravaca, F., Díaz, G., Torres, P. and Roldán, A. (2020) The invader *Carpobrotus edulis* promotes a specific rhizosphere microbiome across globally distributed coastal ecosystems. *Science of the Total Environment* 719: 137347. DOI: 10.1016/j.scitotenv.2020.137347

Root, T.L., Price, J.T., Hall, K.R., Schneider, S.H., Rosenzweig, C. and Pounds, J.A. (2003) Fingerprints of global warming on wild animals and plants. *Nature* 421, 57–60. DOI: 10.1038/nature01333

Salmon, A. and Ainouche, M.L. (2015) Next-generation sequencing and the challenge of deciphering evolution of recent and highly polyploid genomes. In: Hörandl, E and Appelhans, M. (eds) *Next Generation*

Sequencing in Plant Systematics. Regnum Vegetabile Vol. 158. Koeltz Botanical Books, Totnes, UK, pp. 93–118.

Sasse, J., Martinoia, E. and Northen, T. (2018) Feed your friends: do plant exudates shape the root microbiome? *Trends in Plant Science* 23, 25–41. DOI: 10.1016/j.tplants.2017.09.003

Schierenbeck, K.A. and Ellstrand, N.C. (2009) Hybridization and the evolution of invasiveness in plants and other organisms. *Biological Invasions* 11, 1093–1105. DOI: 10.1007/s10530-008-9388-x

Schmid, B., Baruffol, M., Wang, Z. and Niklaus, P.A. (2017) A guide to analyzing biodiversity experiments. *Journal of Plant Ecology* 10, 91–110. DOI: 10.1093/jpe/rtw107

Siemens, T.J. and Blossey, B. (2007) An evaluation of mechanisms preventing growth and survival of two native species in invasive Bohemian knotweed (*Fallopia* × *bohemica*, Polygonaceae). *American Journal of Botany* 94, 776–783. DOI: 10.3732/ajb.94.5.776

Stinson, K.A., Campbell, S.A., Powell, J.R., Wolfe, B.E., Callaway, R.M. *et al.* (2006) Invasive plant suppresses the growth of native tree seedlings by disrupting belowground mutualisms. *PLOS Biology* 4: e140. DOI: 10.1371/journal.pbio.0040140

te Beest, M., Le Roux, J.J., Richardson, D.M., Brysting, A.K., Suda, J. *et al.* (2012) The more the better? The role of polyploidy in facilitating plant invasions. *Annals of Botany* 109, 19–45. DOI: 10.1093/aob/mcr277

Thuiller, W., Richardson, D.M. and Midgley, G.F. (2007) Will climate change promote alien plant invasions? In: Nentwig, W. (ed.) *Biological Invasions. Ecological Studies, Vol. 193.* Springer, Berlin/Heidelberg, pp. 197–211. DOI: 10.1007/978-3-540-36920-2_12

Topp, W., Kappes, H. and Rogers, F. (2008) Response of ground-dwelling beetle (Coleoptera) assemblages to giant knotweed (*Reynoutria* spp.) invasion. *Biological Invasions* 10, 381–390. DOI: 10.1007/s10530-007-9137-6

van Kleunen, M., Dawson, W., Schlaepfer, D., Jeschke, J.M. and Fischer, M. (2010) Are invaders different? A conceptual framework of comparative approaches for assessing determinants of invasiveness. *Ecology Letters* 13, 947–958. DOI: 10.1111/j.1461-0248.2010.01503.x

van de Peer, Y., Mizrachi, E. and Marchal, K. (2017) The evolutionary significance of polyploidy. *Nature Reviews Genetics* 18, 411–424. DOI: 10.1038/nrg.2017.26

van de Peer, Y., Ashman, T.-L., Soltis, P.S. and Soltis, D.E. (2021) Polyploidy: an evolutionary and ecological force in stressful times. *Plant Cell* 33, 11–26. DOI: 10.1093/plcell/koaa015

van der Heide, T., Govers, L.L., de Fouw, J., Olff, H., van der Geest, M. *et al.* (2012) A three-stage symbiosis forms the foundation of seagrass ecosystems. *Science* 336, 1432–1434. DOI: 10.1126/science.1219973

van der Putten, W.H., Bardgett, R.D., Bever, J.D., Bezemer, T.M., Casper, B.B. *et al.* (2013) Plant–soil feedbacks: the past, the present and future challenges. *Journal of Ecology* 101, 265–276. DOI: 10.1111/1365-2745.12054

van Gurp, T.P., Wagemaker, N.C.A.M., Wouters, B., Vergeer, P., Ouborg, J.N.J. and Verhoeven, K.J.F. (2016) epiGBS: reference-free reduced representation bisulfite sequencing. *Nature Methods* 13, 322–324. DOI: 10.1038/nmeth.3763

van Kleunen, M., Bossdorf, O. and Dawson, W. (2018) The ecology and evolution of alien plants. *Annual Review of Ecology, Evolution, and Systematics* 49, 25–47. DOI: 10.1146/annurev-ecolsys-110617-062654

van Moorsel, S.J., Schmid, M.W., Wagemaker, N.C.A.M., van Gurp, T., Schmid, B. and Vergeer, P. (2019) Evidence for rapid evolution in a grassland biodiversity experiment. *Molecular Ecology* 28, 4097–4117. DOI: 10.1111/mec.15191

VanWallendael, A., Hamann, E. and Franks, S.J. (2018) Evidence for plasticity, but not local adaptation, in invasive Japanese knotweed (*Reynoutria japonica*) in North America. *Evolutionary Ecology* 32, 395–410. DOI: 10.1007/s10682-018-9942-7

VanWallendael, A., Alvarez, M. and Franks, S.J. (2021) Patterns of population genomic diversity in the invasive Japanese knotweed species complex. *American Journal of Botany* 108, 857–868. DOI: 10.1002/ajb2.1653

Verlaque, R., Affre, L., Diadema, K., Suehs, C.M. and Médail, F. (2011) Unexpected morphological and karyological changes in invasive *Carpobrotus* (Aizoaceae) in Provence (S-E France) compared to native South African species. *Comptes Rendus Biologies* 334, 311–319. DOI: 10.1016/j.crvi.2011.01.008

Wagner, M.R., Lundberg, D.S., Del Rio, T.G., Tringe, S.G., Dangl, J.L. and Mitchell-Olds, T. (2016) Host genotype and age shape the leaf and root microbiomes of a wild perennial plant. *Nature Communications* 7: 12151. DOI: 10.1038/ncomms12151

Walls, R.L. (2010) Hybridization and plasticity contribute to divergence among coastal and wetland populations of invasive hybrid Japanese knotweed *s.l.* (*Fallopia* spp.). *Estuaries and Coasts* 33, 902–918. DOI: 10.1007/s12237-009-9190-8

Walther, G.-R., Post, E., Convey, P., Menzel, A., Parmesan, C. *et al.* (2002) Ecological responses to recent climate change. *Nature* 416, 389–395. DOI: 10.1038/416389a

Wang, F., Li, M., Liu, Z., Li, W., He, Q. *et al.* (2025) The mixed auto-/allooctoploid genome of Japanese knotweed (*Reynoutria japonica*) provides insights into its polyploid origin and invasiveness. *Plant Journal* 121: e70005. DOI: 10.1111/tpj.70005

Wang, I.J. and Bradburd, G.S. (2014) Isolation by environment. *Molecular Ecology* 23, 5649–5662. DOI: 10.1111/mec.12938

Wang, J.W. (1996) The inexpressible plants of Polygonaceae (Part One). *Plant Journal* 2, 27–29.

Wang, S., Liao, Z.-Y., Cao, P., Schmid, M.W., Zhang, L. *et al.* (2025) General-purpose genotypes and evolution of higher plasticity in clonality underlie knotweed invasion. *New Phytologist* 246, 758-768. DOI: 10.1111/nph.20452

Wang, T., Guo, X., Yang, J., Chi, X., Zhu, Y. *et al.* (2023) The introduced lineage of *Phragmites australis* in North America differs from its co-existing native lineage in associated soil microbial structure rather than plant traits. *Plant and Soil* 493, 137–156. DOI: 10.1007/s11104-023-06216-y

Weltzin, J.F., Belote, R.T. and Sanders, N.J. (2003) Biological invaders in a greenhouse world: will elevated CO_2 fuel plant invasions? *Frontiers in Ecology and the Environment* 1, 146–153. DOI: 10.1890/1540-9295(2003)001[0146:biiagw]2.0.co;2

Wendel, J.F., Jackson, S.A., Meyers, B.C. and Wing, R.A. (2016) Evolution of plant genome architecture. *Genome Biology* 17: 37. DOI: 10.1186/s13059-016-0908-1

Willems, F.M., Scheepens, J.F. and Bossdorf, O. (2022) Forest wildflowers bloom earlier as Europe warms: lessons from herbaria and spatial modelling. *New Phytologist* 235, 52–65. DOI: 10.1111/nph.18124

Woods, E.C. and Sultan, S.E. (2022) Post-introduction evolution of a rapid life-history strategy in a newly invasive plant. *Ecology* 103: e3803. DOI: 10.1002/ecy.3803

Woodward, F.I., Lake, J.A. and Quick, W.P. (2002) Stomatal development and CO_2: ecological consequences. *New Phytologist* 153, 477–484. DOI: 10.1046/j.0028-646X.2001.00338.x

Wright, I.J., Reich, P.B., Westoby, M., Ackerly, D.D., Baruch, Z. *et al.* (2004) The worldwide leaf economics spectrum. *Nature* 428, 821–827. DOI: 10.1038/nature02403

Yin, L., Liu, B., Wang, H., Zhang, Y., Wang, S. *et al.* (2020) The rhizosphere microbiome of *Mikania micrantha* provides insight into adaptation and invasion. *Frontiers in Microbiology* 11: 1462. DOI: 10.3389/fmicb.2020.01462

Yoder, A.D. and Tiley, G.P. (2021) The challenge and promise of estimating the *de novo* mutation rate from whole-genome comparisons among closely related individuals. *Molecular Ecology* 30, 6087–6100. DOI: 10.1111/mec.16007

Yuan, W., Pigliucci, M. and Richards, C. (2024) Rapid phenotypic differentiation in the iconic Japanese knotweed *s.l.* invading novel habitats. *Scientific Reports* 14: 14640. DOI: 10.1038/s41598-024-64109-1

Zeng, S., Mo, C., Xu, B., Wang, Z., Zhang, F. *et al.* (2024) T2T genome assemblies of *Fallopia multiflora* (Heshouwu) and *F. multiflora* var. *angulata*. *Scientific Data* 11: 1103. DOI: 10.1038/s41597-024-03943-4

Zhang, J., Xu, Q., You, L., Li, B., Zhang, Z. *et al.* (2025) Chromosome-scale genome assembly and annotation of Huzhang (*Reynoutria japonica*). *Scientific Data* 12: 474. DOI: 10.1038/s41597-025-04773-8

Zhang, L., Niu, H., Wang, S., Zhu, X., Luo, C. *et al.* (2012) Gene or environment? Species-specific control of stomatal density and length. *Ecology and Evolution* 2, 1065–1070. DOI: 10.1002/ece3.233

Zhang, L., van Riemsdijk, I., Liu, M., Liao, Z., Cavé-Radet, A. *et al.* (2024) Biogeography of a global plant invader: from the evolutionary history to future distributions. *Global Change Biology* 30: e17622. DOI: 10.1111/gcb.17622

Zhang, Y.-Y., Zhang, D.-Y. and Barrett, S.C.H. (2010) Genetic uniformity characterizes the invasive spread of water hyacinth (*Eichhornia crassipes*), a clonal aquatic plant. *Molecular Ecology* 19, 1774–1786. DOI: 10.1111/j.1365-294x.2010.04609.x

Zhang, Y.-Y., Parepa, M., Fischer, M. and Bossdorf, O. (2016) Epigenetics of colonizing species? A study of Japanese knotweed in central Europe. In: Barrett, S.C.H., Colautti, R.I., Dlugosch, K.M. and Rieseberg, L.H. (eds) *Invasion Genetics: The Baker and Stebbins Legacy*. Wiley-Blackwell, Hoboken, New Jersey, pp. 328–340. DOI: 10.1002/9781119072799.ch19

Zhao, W. (2021) Biogeographic variation in growth and defense traits of *Reynoutria japonica* between invasive and native ranges. MSc thesis, Fudan University, Shanghai, China.

Zhao, Y., Yang, Z., Zhang, Z., Yin, M., Chu, S. *et al.* (2023) The first chromosome-level *Fallopia multiflora* genome assembly provides insights into stilbene biosynthesis. *Horticulture Research* 10: uhad047. DOI: 10.1093/hr/uhad047

Zhao, Y.-J., Wang, S., Liao, Z.-Y., Parepa, M., Zhang, L. *et al.* (2024) Geographic variation in leaf traits and palatability of a native plant invader during domestic expansion. *Ecology* 105: e4425. DOI: 10.1002/ecy.4425

Zhou, Z., Miwa, M., Nara, K., Wu, B., Nakaya, H. *et al.* (2003) Patch establishment and development of a clonal plant, *Polygonum cuspidatum*, on Mount Fuji. *Molecular Ecology* 12, 1361–1373. DOI: 10.1046/j.1365-294x.2003.01816.x

10 Mechanisms of Parallel Polygenic Adaptation During Habitat Invasions

Carol Eunmi Lee*

Department of Integrative Biology, University of Wisconsin-Madison, Madison, Wisconsin, USA

Abstract

Rates of biological invasions are accelerating throughout the world. However, our understanding of the evolutionary mechanisms that promote invasive success remains incomplete. Genome architecture can affect mechanisms of rapid and parallel evolution during invasions. In aquatic habitats, many invaders are crossing biogeographical boundaries from saline to freshwater habitats. The copepod *Eurytemora affinis* species complex provides a valuable model system for analyzing evolutionary mechanisms underlying these invasions, with populations that have recently invaded freshwater habitats multiple times independently from saline sources. Surprisingly, parallel selection favors the same alleles repeatedly during these invasions, even across clades with differing chromosome number, indicating that salinity adaptation proceeds in a predictable manner. Genome architecture appears to have profound impacts on the response to selection in these populations, with chromosome number likely impacting the relative importance of standing genetic variation versus positive epistasis in promoting parallel evolution. More research is needed to fully explore the role of genome architecture in impacting population responses to selection, especially during habitat invasions.

10.1 Introduction

This chapter explores the physiological and evolutionary mechanisms of invasive success, including the impacts of genome architecture on the response to selection during habitat invasions. For many invasive populations, evolutionary responses are essential for the initial and long-term survival in the novel range (Lee, 2002, 2010). The genome architecture of organisms could have profound impacts on how populations respond to natural selection and how readily they could adapt to rapid habitat change. Genome architecture refers to the arrangement and composition of genetic elements (genes, regulatory regions, non-coding DNA, chromosomes) in a genome, such as the number of genes, gene order, nucleotide composition (e.g. GC content), genome size, amount of coding versus non-coding DNA, number of transposons and repetitive DNA (e.g. microsatellites), mean number and size of introns within genes, number of chromosomes and ploidy levels. The following discussion focuses on insights gained from detailed analyses of the copepod *Eurytemora affinis* species complex, which is an established model system in invasion biology. This copepod serves as a particularly valuable model for analyzing mechanisms of invasive success because many populations have invaded freshwater habitats multiple times independently

*Corresponding author: carollee@wisc.edu

© CAB International 2025. *Invasion Genomics* (Eds D. Bock and M. Rius)
DOI: 10.1079/9781800626263.0010

from saline native habitats (Lee, 1999). This pattern of saline-to-freshwater invasions is characteristic of some of the most destructive invaders in aquatic habitats (Lee and Gelembiuk, 2008; Lee et al., 2022b). In addition, the multiple independent invasions provide replicated tests for studying the extent to which evolutionary pathways are labile or constrained. Finally, this system offers ample genomic resources, including chromosome-level genomes for multiple clades (monophyletic groups composed of a common ancestor and all of its descendants), inbred lines and reference transcriptomes (Rasch et al., 2004; Posavi et al., 2014, 2020; Eyun et al., 2017; Du et al., 2024, 2025).

Increases in globalization and expansions of trade networks are causing rates of biological invasions to accelerate throughout the world (Seebens et al., 2018; Mormul et al., 2022). In recent years, many new invasive species have emerged (Seebens et al., 2018), with some evidence that range sizes of invasive species are expanding exponentially (Mormul et al., 2022). In aquatic habitats, the shipping and dumping of large volumes of ship ballast water has been serving as a major transport vector for invaders (Carlton, 1985; Ruiz et al., 1997). Since 2006–2008, ballast water control through mid-ocean exchange, while imperfect, has been serving as an important method for lowering rates of aquatic invasions (Darling et al., 2018; Bradie et al., 2021; Bailey et al., 2022; Ricciardi and MacIsaac, 2022). This method involves exchanging freshwater or estuarine ballast with oceanic water prior to entering an inland port, such as those in the Great Lakes. While mid-ocean ballast water exchange likely prevents many species from being introduced into freshwater lakes, this strategy alone might not necessarily prevent colonization by populations that are resilient to salinity change.

Historically, populations that originate from more saline habitats, particularly brackish water habitats, have been particularly successful as invaders into freshwater habitats (de Beaufort, 1954; Miller, 1958; Mordukhai-Boltovskoi, 1979; Jażdżewski, 1980; Lee, 1999; Havel et al., 2005; May et al., 2006; Lee and Gelembiuk, 2008; Casties et al., 2016). For instance, since the opening of the St. Lawrence Seaway in 1959, a large portion of the approximately 180 invasive species in the North American Great Lakes have originated from brackish waters of the Black and Caspian Sea basin (Jażdżewski,

1980; Witt et al., 1997; Lee and Bell, 1999; Ricciardi and MacIsaac, 2000; Cristescu et al., 2001; Gelembiuk et al., 2006; May et al., 2006), much more so than expected based on transport opportunity and abundance in the native ranges (Casties et al., 2016). These invaders that are crossing salinity boundaries include some of the most prolific invaders in aquatic habitats, such as zebra mussels, quagga mussels, the fishhook waterflea (Cercopagis pengoi), and many species of amphipods (Jażdżewski, 1980; Witt et al., 1997; Ricciardi and MacIsaac, 2000; Cristescu et al., 2001; Bij de Vaate et al., 2002; Gelembiuk et al., 2006; May et al., 2006).

In aquatic habitats, salinity imposes a formidable biogeographical boundary that structures the distribution of most taxa, from bacteria to vertebrates (Hutchinson, 1957; Remane and Schlieper, 1971; Lozupone and Knight, 2007; Lee et al., 2022b). An intermediate salinity of 5 PSU (practical salinity units) tends to impose a biogeographical and physiological barrier that separates saline and freshwater communities (Remane and Schlieper, 1971; Khlebovich and Abramova, 2000). In fact, over half of all animal phyla remains strictly marine, and the majority of animal taxa have not evolved the ability to tolerate lower salinities (Little, 1983, 1990). Thus, transitions from saline to freshwater habitats represent fundamental niche expansions that have occurred only sporadically throughout evolutionary history.

Such salinity transitions are physiologically challenging, requiring the regulation of fluxes of water, ions and osmolytes between organismal extracellular fluids and the external environment (Withers, 1992; Willmer et al., 2008). As life evolved in the sea, most marine organisms are osmoconformers, lacking the need for ionic and osmotic regulation between body fluids and the environment (Withers, 1992; Willmer et al., 2008). Colonizations away from the sea into lower-salinity habitats have resulted in increases in ionic and osmotic gradients between organismal extracellular fluids and the external environment. Such habitat shifts required the evolution of body fluid regulation and mechanisms to maintain physiological homeostasis. Consequently, physiological mechanisms to perform ion uptake and prevent ionic losses became necessary (Evans and Clairborne, 2009; Lee et al., 2022a).

Given the physiological challenges of crossing salinity boundaries, how might saline populations survive the transition to freshwater habitats during rapid invasions? And might common mechanisms govern these shifts across different taxa? When the rate or extent of environmental change exceeds the physiological tolerances of organisms, the populations must migrate or evolve in response to the environmental change (Lee, 2002, 2010). Rapid evolutionary shifts would be particularly challenging to accomplish for metazoans with longer generation times, relative to shorter-lived bacteria, viruses, or protists.

The copepod *Eurytemora affinis* species complex provides a powerful model for exploring fundamental mechanisms of range expansions and niche evolution (Figs. 10.1 and 10.2). This small crustacean is a dominant member of planktonic communities in many estuaries throughout the northern hemisphere. As such, this copepod is functionally important in planktonic food webs as grazers of algae and as major food sources for larval salmon and forage fish, such as anchovy, herring, and sardine (Simenstad *et al.*, 1990; Viitasalo *et al.*, 1994, 1995, 2001;

Shaheen *et al.*, 2001; Winkler *et al.*, 2003; Kimmel *et al.*, 2006; Livdāne *et al.*, 2016).

Most notably, within the past approximately 80 years, populations of the *E. affinis* species complex have invaded freshwater habitats multiple times independently from genetically distinct saline sources (Fig. 10.2) (Lee, 1999). These freshwater invasions have been facilitated by the construction of canals and ship ballast water discharge into freshwater lakes. Multiple independent invasions, both within and among genetically distinct clades (sibling species), provide the opportunity to examine whether the same evolutionary mechanisms are involved during independent transitions into freshwater habitats. Interestingly, some clades are more invasive than others, whereas others are not invasive but remain restricted in their geographical ranges (Fig. 10.2a) (Lee, 1999; Winkler *et al.*, 2008).

During the saline-to-freshwater invasions, populations of the *E. affinis* species complex experience rapid evolutionary shifts in physiological tolerance and performance. In response to rapid changes in salinity, short-term or developmental

(a) Adult female

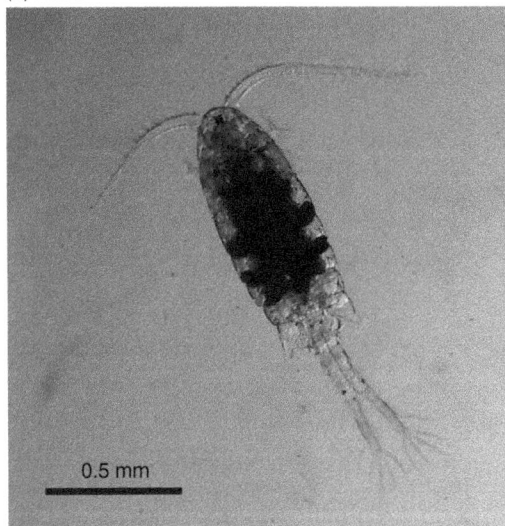

(b) Adult male

0.5 mm

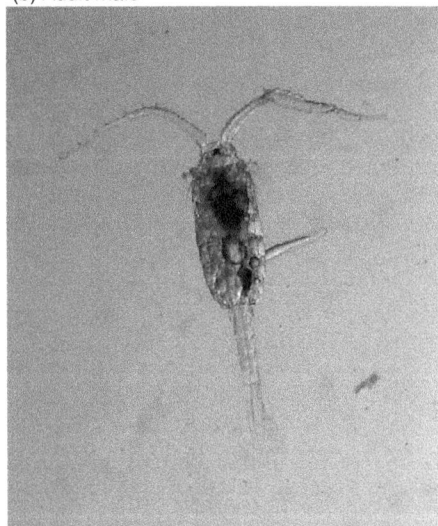

Fig. 10.1. Adult copepods of *Eurytemora carolleeae* (Atlantic clade of the *E. affinis species* complex) at 15 PSU salinity. These copepods were collected from Baie de L'Isle Verte salt marsh in the St. Lawrence estuary, Quebec, Canada, in 2022. In this habitat, salinity fluctuates seasonally between approximately 5 and 40 PSU. Generation time is about 20–30 days. (a) Adult female, approximately 1.5 mm in length. (b) Adult male holding a spermatophore. Photos by Brooke Lewis in the Lee Laboratory, used with permission.

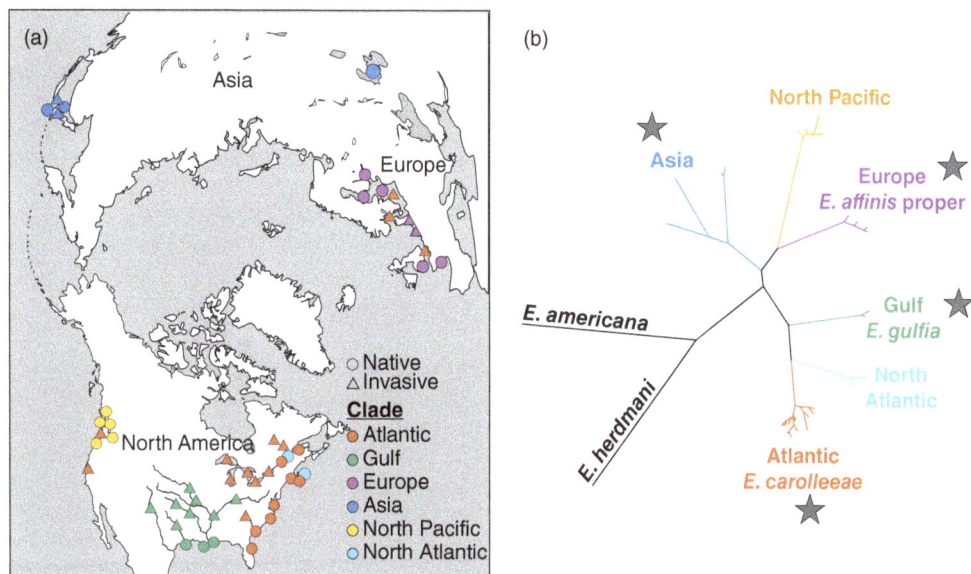

Fig. 10.2. Map and phylogeny of the *Eurytemora affinis species* complex. (a) Map showing the geographical distribution of the genetically distinct clades (sibling species, indicated by different colors) of the *E. affinis* species complex. Circles represent native distributions, whereas triangles represent the invasive populations. (b) Phylogeny of the *E. affinis* species complex based on mitochondrial *COI* and 16S rRNA sequences from Lee (2000). Stars indicate the clades that gave rise to freshwater invasions. Adapted from Du *et al.* (2025). Used with permission.

acclimation by saline populations could not account for the ability to tolerate freshwater conditions (Lee and Petersen, 2003; Lee *et al.*, 2003, 2007). Surviving the salinity transition requires an evolutionary response, with natural selection acting most strongly on the larval (naupliar) life history stages (Lee *et al.*, 2003, 2007). As a consequence of the evolutionary shifts, freshwater populations show increases in freshwater tolerance, ion uptake activity, and ion retention relative to their saline ancestors (see below for more details) (Lee *et al.*, 2003, 2007, 2011; Popp *et al.*, 2024).

Most intriguingly, populations from genetically divergent clades exhibit parallel evolutionary shifts in ionic regulation during the saline-to-freshwater invasions. Genetically distinct clades show parallel evolutionary shifts in body fluid regulation and ion-transporter enzyme activity between saline and freshwater populations (see details in next section) (Lee *et al.*, 2011, 2012). Moreover, at the genome-wide scale, striking patterns of parallel evolution are evident both within and across the clades in response to salinity change, with selection often acting on the

same loci, frequently on the same single-nucleotide polymorphisms (SNPs), in both wild populations and laboratory selection lines (Posavi *et al.*, 2020; Stern and Lee, 2020; Lee, 2021; Stern *et al.*, 2022; Diaz *et al.*, 2023).

In response to salinity change, ion-transport-related genes tend to form the dominant functional (gene ontology (GO)) categories under selection (Posavi *et al.*, 2020; Stern and Lee, 2020; Lee, 2021; Stern *et al.*, 2022; Diaz *et al.*, 2023). Other functional categories of genes under selection include those involved in energy production, gene regulation, translation, neuronal regulation and development. Notably, genes related to ion transport dominate as the largest functional category that show parallel selection (natural selection acting on the same loci independently in two or more lineages) across independent invasions in wild populations (Stern and Lee, 2020) and in replicate laboratory selection lines (Stern *et al.*, 2022). *In situ* immunolocalization of several ion transporters found that they were expressed in putative osmoregulatory organs of *E. affinis* species complex individuals, specifically in the maxillary glands, swimming legs, and digestive tract (Fig. 10.3) (Johnson *et al.*, 2014; Gerber *et al.*, 2016; Popp, 2024; Popp *et al.*, 2024). Notably, patterns of *in situ* ion-transporter expression within the osmoregulatory organs revealed evolutionary shifts in expression between saline and freshwater populations, as well as acclimatory shifts across salinities (Popp, 2024; Popp *et al.*, 2024).

The pattern of genome-wide parallel evolution across different clades is surprising given the moderate sequence divergence and variation in genome architecture among the clades (Lee, 2000; Du *et al.*, 2024, 2025). The clades show very low morphological differentiation, except for slight divergence of the Europe clade (Lee and Frost, 2002). Nevertheless, they show moderate genome-wide sequence divergences, with average nucleotide similarities ranging from of 87% to 90% among three clades (Atlantic, Gulf, and Europe) for 19–68% of the aligned genome sequences (see below) (Du *et al.*, 2025). Moreover, despite having similar gene content, the clades of *E. affinis* species complex possess large differences in chromosome number (4–15 in the Atlantic, Gulf, and Europe clades; see below) (Du

et al., 2025). Surprisingly, these genetically distinct clades are able to intermate and produce viable F_1 hybrid offspring, in at least one reciprocal cross, and F_2 offspring in some cases (Lee, 2000).

Given the stark genetic and genomic differences among the clades, what might account for the parallel evolution in different clades, with selection acting repeatedly on the same loci during salinity transitions? Why are some populations invasive, while others are not? Are the same mechanisms operating to achieve the same physiological optima, or are different mechanisms involved? And finally, does genome architecture impact how populations respond to natural selection?

The extent of parallel evolution among independent invasions within and across different clades (i.e. their predictable and reproducible nature) reveals the degree to which their evolutionary pathways are labile or constrained. Many different factors could contribute to the extent of parallel evolution among independent invasion events. These factors might include the number of loci encoding the trait under selection, whether the selected loci (i.e. loci under selection

(a) Maxillary gland

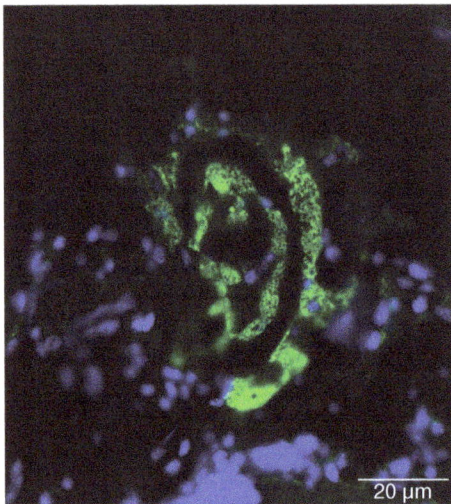

(b) Swimming legs

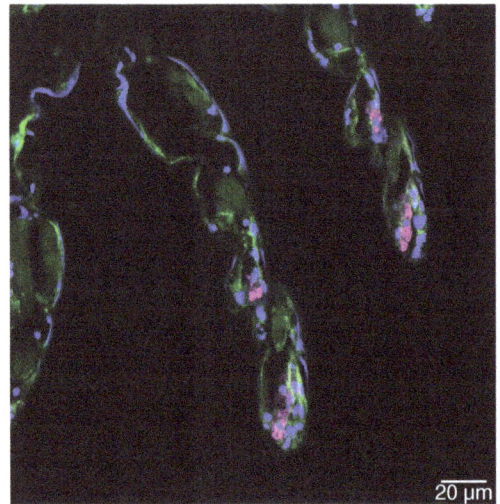

Fig. 10.3. Immunolocalization of the ion transporters Na⁺/K⁺-ATPase (NKA) and Na⁺/H⁺ antiporter, paralog 5 (NHA-5) in the osmoregulatory organs of the copepod *Eurytemora carolleeae* (Atlantic clade of the *E. affinis* species complex). The images show adult male individuals from the Lake Michigan population at 0 PSU. (a) A maxillary gland, showing staining of NKA (green). (b) Swimming legs, showing the staining of NHA-5 (pink) in the Crusalis organs and widespread staining of NKA (green). Blue indicates DAPI staining of DNA. Photographs by Teresa E. Popp in collaboration with Catherine Lorin-Nebel and Carol Lee. See Popp *et al.* (2024) and Popp (2024) for details. Scale bars, 20 µm. Used with permission.

in response to environmental change) are essential and not redundant in function, linkage among selected loci and recombination rate, starting frequencies of beneficial alleles, and interactions among the beneficial alleles, such as positive or negative epistasis (i.e. non-additive interaction effects between genes, where the effect of a mutation at a gene is affected by one or more mutations at other genes). A less explored factor is the role that genome architecture (particularly at the chromosome level) might play in promoting rapid and parallel evolution.

The important implication for invasive populations is that the factors that contribute to parallel evolution could speed up evolution and make adaptation more likely during biological invasions. For instance, the presence of high starting frequencies of beneficial alleles (potentially maintained through balancing selection, i.e. selection that maintains genetic variation within a population, including fluctuating selection, overdominance and frequency-dependent selection) would make those alleles more available for selection and would promote rapid and parallel adaptation (derived from parallel selection) during invasions. Likewise, having fewer loci encoding an adaptive trait (i.e. less polygenic) would also tend to make parallel and rapid evolution more likely. Moreover, positive synergistic epistasis among sets of alleles, in which a combination of mutations at different genes leads to higher function or fitness than expected, relative to the effects of the individual mutations alone, would facilitate selection favoring the same sets of alleles rapidly and repeatedly.

10.2 Parallel Evolutionary Responses to Salinity Change

Physiological traits and systems are typically polygenic (i.e. encoded by many genes). As the number of loci that encodes a trait increases, the extent of parallel evolution among replicate selection events is expected to go down (Barghi et al., 2019, 2020). This lack of parallel evolution for more polygenic traits is expected because, with more genes encoding a trait, the genes presumably become more redundant in function. This greater functional redundancy among genes would then allow selection to act on alternative evolutionary pathways, leading to non-parallel evolution.

How polygenic is the salinity adaptation response? The evolutionary response to salinity change is highly polygenic in E. affinis complex populations (Stern and Lee, 2020; Stern et al., 2022), involving selection at multiple loci and evolutionary shifts in multiple traits related to ionic and osmotic regulation (see below). In terms of the number of loci under selection, the salinity response involves 5000–7000 SNPs grouped into approximately 100–200 haplotype blocks (Stern and Lee, 2020; Stern et al., 2022). However, temperature adaptation might be even more polygenic, as temperature affects all physiological and metabolic reaction rates in ectotherms (Barghi et al., 2019; Clarke, 2003). For instance, we found that the temperature adaptation response of E. affinis complex populations appears to be more polygenic than salinity adaptation, and also uncorrelated with the response to salinity (Diaz et al., 2023).

In terms of the polygenic traits under selection in response to salinity change, salinity adaptation in E. affinis species complex involves several processes related to ion transport and ion retention. These include evolutionary shifts in body fluid regulation (hemolymph osmolality), ion-transporter enzyme activity (of Na^+/K^+-ATPase (NKA) and V-type H^+ ATPase (VHA)), and integument permeability (Lee et al., 2011, 2012; Popp et al., 2024). Ion-transport functions are likely localized in the copepod maxillary glands, swimming legs (Crusalis organs) and digestive tract (Johnson et al., 2014; Gerber et al., 2016; Popp, 2024; Popp et al., 2024). The maxillary glands are potentially sites of ion reabsorption from the urine, the swimming legs are likely sites of ion uptake from the environment, and the digestive tract is likely where ion uptake from food takes place. Within these organs, ion-transporter protein expression exhibits evolutionary shifts between saline and freshwater populations and acclimatory shifts across salinities (Popp, 2024; Popp et al., 2024).

Despite the polygenic nature of the evolutionary response to salinity change, populations of E. affinis complex show a remarkably high degree of parallel evolution (Fig. 10.4; Stern and Lee, 2020). Moreover, this parallelism extends to selection responses across highly divergent clades. In wild populations from two clades (Atlantic

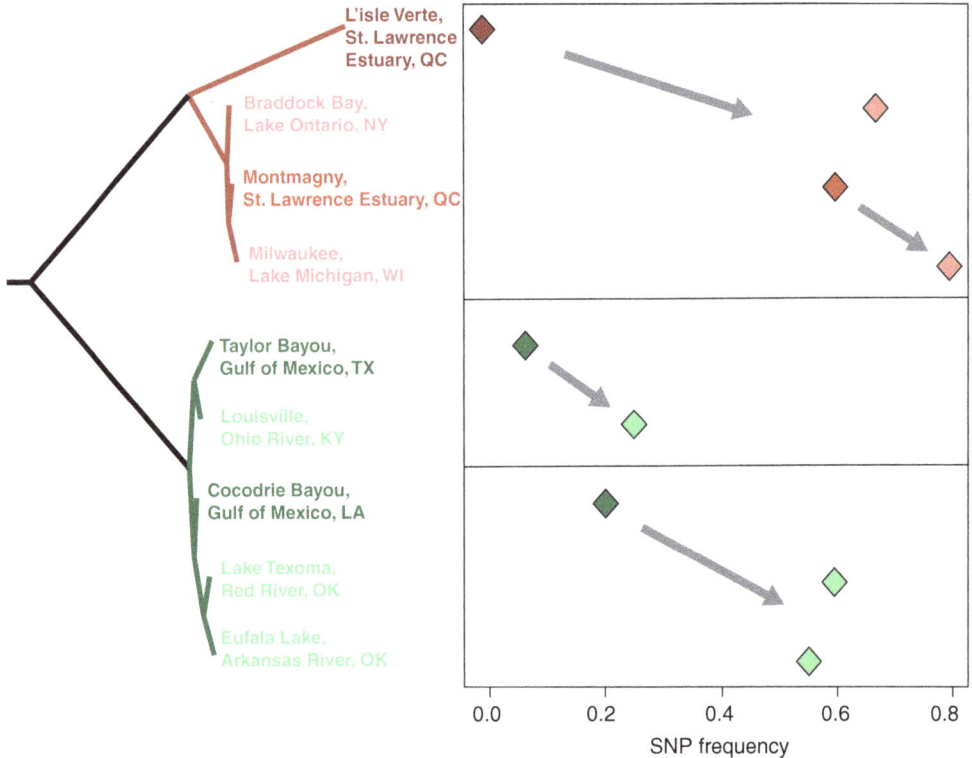

Fig. 10.4. Parallel allele frequency shifts during replicate independent saline-to-freshwater invasions. Independent invasions from saline to freshwater habitats occurred in the Atlantic clade (red) and Gulf clade (green) populations. Diamonds represent the mean single-nucleotide polymorphism (SNP) frequencies of the corresponding population on the phylogeny for SNPs under parallel selection across the independent invasions. The SNPs included here (N = 347) show both significant signatures of directional selection between saline and freshwater populations (BayeScan 3) and association with salinity change (BayPass). Gray arrows indicate the direction of change in SNP frequency from the ancestral saline populations (dark colors) to the freshwater-invading populations (light colors). Phylogeny was estimated from SNP frequency correlations using TreeMix v. 1.13. Adapted from Stern and Lee (2020). Used with permission.

and Gulf), natural selection repeatedly favors the same loci and SNPs across independent saline-to-freshwater invasions to a much greater degree than expected by chance (Stern and Lee, 2020). For instance, 42.5% of the SNPs show the same pattern of frequency shifts across three independent invasions (out of ~6981 SNPs with signals of selection across any of the invasions) in the Atlantic and Gulf clade populations (Fig. 10.4). This high degree of parallel evolution is surprising and unexpected, especially across genetically divergent clades.

Likewise, in a laboratory evolution experiment, replicate selection lines show signatures of parallel evolution during salinity decline, with selection often favoring the same alleles (SNPs) at the same sets of loci (Stern et al., 2022). Starting with a saline Baltic Sea population from Kiel, Germany (15 PSU, Europe clade), salinity was reduced to 0 PSU over ten generations (N = 10 replicate selection lines). The selection lines were genotyped by whole-genome sequencing (Pool-seq) at multiple timepoints. During salinity decline, the treatment lines exhibited significant frequency shifts in approximately 5000 SNPs grouped into 121 haplotype blocks. At the tenth generation of selection, the replicate selection lines exhibited a high degree of parallelism, with an average of 79.5% overlap in selected alleles between replicate lines (based on the Jaccard index).

For all the clades, in both the wild populations and laboratory selection lines, the SNPs under selection in response to salinity tend to occur in genomic regions heavily enriched with ion-transport-related genes (with ion transport as the largest GO category under selection) (Lee, 2021). These include paralogs (gene duplicates that have become differentiated in DNA sequence) and/or subunits of the ion-transport-related gene families Na$^+$/H$^+$ antiporter (*NHA*), Na$^+$/K$^+$ ATPase (*NKA*), carbonic anhydrase (*CA*), V-type H$^+$ ATPase (*VHA*), Na$^+$,K$^+$,2Cl$^-$ cotransporter (*NKCC*), Na$^+$/H$^+$ exchanger (*NHE*), Rh protein (*Rh*), and solute carrier 4 (*SLC4*) of the bicarbonate (HCO$_3^-$) transporter gene families, including anion exchanger (*AE*), Na$^+$, HCO$_3^-$ cotransporter (*NBC*), and Na$^+$-driven Cl$^-$/ HCO$_3^-$ exchanger (*NDCBE*) (Stern and Lee, 2020; Stern *et al.*, 2022; Diaz *et al.*, 2023). In addition, many of the same ion-transporter genes that display genetic signatures of selection also show evolutionary shifts in gene expression between saline and freshwater populations (Posavi *et al.*, 2020). These ion transporters (and carbonic anhydrase) likely form coadapted gene complexes and cooperate to perform the action of ion uptake from the environment (Fig. 10.5). Repeated selection on the same ion-transport-related loci suggest that they evolve together as a unit in response to environmental change.

In particular, the highest density of SNPs under selection in the Atlantic clade genome occurs in a genomic region near the centromere of chromosome 3 containing seven tandem paralogs (variable gene duplicates) of the Na$^+$/H$^+$ antiporter (*NHA*) gene family (Stern and Lee, 2020; Du *et al.*, 2025). Notably, the alleles at the paralogs *NHA-7* and *NHA-5* repeatedly exhibit frequency shifts and these paralogs exhibit evolution of gene expression across salinity transitions (Posavi *et al.*, 2020; Stern and Lee, 2020; Stern *et al.*,

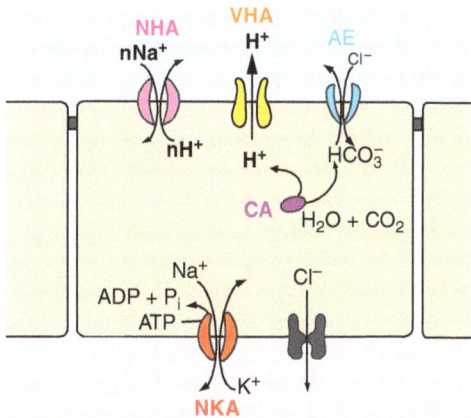

(a) **Model 1:** Na$^+$ uptake via Na$^+$/H$^+$ antiporter   (b) **Model 2:** Na$^+$ uptake via electroneutral NHE

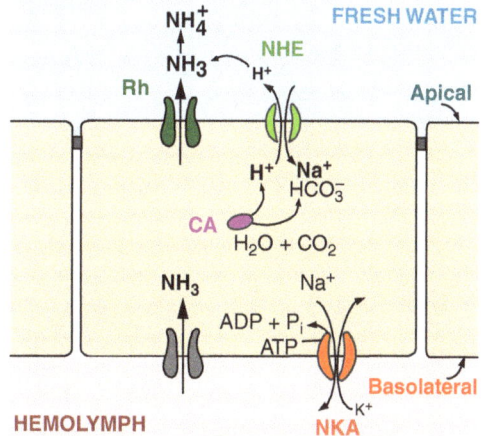

Fig. 10.5. Hypothetical models of ion uptake, particularly Na$^+$ uptake, under freshwater conditions. The models show ion uptake in ionocytes (specialized cells that perform ion transport). (a) Model 1: V-type H$^+$ ATPase (VHA) generates an electrochemical gradient by pumping protons out of the cell to facilitate uptake of Na$^+$ through an electrogenic Na$^+$ transporter (likely Na$^+$/H$^+$ antiporter in *Eurytemora affinis* species complex (NHA)). Carbonic anhydrase (CA) produces H$^+$ for VHA and HCO$_3^-$ for an anion exchanger (AE). Na$^+$ is then transported to the hemolymph through Na$^+$/K$^+$ ATPase (NKA). (b) Model 2: An ammonia transporter Rh protein exports NH$_3$ out of the cell and this NH$_3$ reacts with H$^+$ to produce NH$_4^+$. The resulting deficit of extracellular H$^+$ causes NHE to export H$^+$ in exchange for the import of Na$^+$. CA produces H$^+$ for the Na$^+$/H$^+$ exchanger (NHE). Na$^+$ is then transported to the hemolymph through NKA. These models are not comprehensive for all tissues or taxa and are not mutually exclusive. Modified from Lee *et al.* (2022a), with permission.

2022; Diaz *et al.*, 2023). In addition, these NHA paralogs show evolutionary shifts in protein expression from saline to freshwater populations within osmoregulatory organs of *E. carolleeae* (Atlantic clade) populations (Popp, 2024; Popp *et al.*, 2024). The high repeatability of selection acting on *NHA* paralogs suggests that they play critical roles in physiological functioning and essential roles in adaptation to salinity change.

Despite the polygenic nature of salinity adaptation, the critical loci under selection do not appear to be interchangeable or redundant in function. Overall, experimental results suggest the presence of functional differentiation among different paralogs of the ion-transporter gene families, with different paralogs showing divergent patterns of gene and protein expression and signatures of selection (Lee, 2021). For instance, the paralogs of NHA show differences in evolutionary and acclimatory patterns of gene expression and are expressed in different osmoregulatory organs (Posavi *et al.*, 2020; Popp, 2024; Popp *et al.*, 2024). If the loci under selection are non-redundant in function and are absolutely necessary for freshwater adaptation, then their evolutionary trajectories would more likely follow predictable and deterministic pathways, leading to parallel evolution.

Nevertheless, the high degree of parallelism across the genetically divergent clades is surprising, given their differences in genome architecture. Genome-wide sequence similarity is moderately close among the clades, with structural differences leading to low alignment between some genomes, while chromosome number varies considerably (Du *et al.*, 2025). Specifically, the Atlantic and Gulf clade genomes show an average nucleotide similarity of 90.1% for 357.2 Mb of the Atlantic clade genome (67.5%) aligned with 356.2 Mb of the Gulf clade genome (68.4%). Between the Atlantic and Europe clade genomes, average nucleotide similarity is 87.4% for 137.9 Mb of the Atlantic clade genome (26.1%) aligned with 138.3 Mb of the Europe clade genome (20.6%). Similarly, between the Gulf and Europe clade genomes, average nucleotide similarity is 87.5% for 128.1 Mb of the Gulf clade genome (24.6%) aligned with 128.5 Mb of the Europe clade genome (19.2%) (Du *et al.*, 2025). Notably, chromosome number varies among the clades, with an ancestral state of 15 chromosomes in the Europe clade (*E. affinis*

proper) and independent fusions into four chromosomes in the Atlantic clade (*E. carolleeae*) and seven chromosomes in the Gulf clade (*Eurytemora gulfia*) (Du *et al.*, 2024, 2025).

Several factors, in addition to the non-redundancy of selected loci, likely contribute to the surprisingly high degree of parallelism across replicated salinity transitions, both within and between different clades (see below). The relative importance of these different factors likely varies among clades, especially given the divergence in their genome architecture. Exploring the mechanisms underlying parallel evolution could help uncover mechanisms of rapid habitat invasions. For instance, dissecting patterns of parallel evolution can reveal the categories of trait evolution that are necessary for adaptation (e.g. ion-transporter evolution, in the case of salinity adaptation). Moreover, analyzing the genetic and genomic mechanisms underlying parallel evolution can yield fundamental insights into mechanisms that could lead to deterministic and rapid evolution (see below).

10.3   Selection on Standing Genetic Variation

Remarkably, across multiple independent saline-to-freshwater invasions in two clades, a relatively large portion of the selected alleles (SNPs) in the freshwater populations shows signatures of long-term balancing selection in the native-range populations (Stern and Lee, 2020). Specifically, of the SNPs favored by selection in parallel across independent invasions spanning the Atlantic and Gulf clades, 15–47% exhibit significant signatures of balancing selection in four ancestral saline populations in the native range. The genetic signatures of balancing selection in the native-range populations include an excess of SNPs at similar frequencies and an abundance of SNPs at intermediate frequency (Stern and Lee, 2020). This surprising result indicates that an unexpectedly large number of alleles that could facilitate freshwater adaptation are preserved in native-range saline populations of both the Atlantic and Gulf clades. This result is consistent with selection during freshwater invasions acting on standing genetic variation in the native-range saline populations (Stern and Lee, 2020).

The native saline populations of *E. affinis* species complex that give rise to freshwater-invading populations harbor conditions that would promote the action of balancing selection. Within the *E. affinis* species complex, invasive populations tend to arise from certain clades (Atlantic, Gulf, Europe, Asia) but not from other clades (North Atlantic, North Pacific) (Fig. 10.2) (Lee, 1999; Winkler *et al.*, 2008; Lee *et al.*, 2013). The native saline populations that tend to give rise to freshwater-invading populations inhabit parts of their ranges that experience fluctuating salinity, whereas the non-invasive populations tend to occur in habitats with relatively more constant conditions. For instance, in the St. Lawrence estuary, populations of the highly invasive Atlantic clade (*E. carolleeae*) and the non-invasive North Atlantic clade both occur in the central portion of the estuary, where they can vertically migrate during the tidal cycle to maintain their positions at relatively constant salinities (Winkler *et al.*, 2008). However, the invasive Atlantic clade populations have a broader habitat range extending into the marginal regions of the estuary, including coastal salt marshes that experience seasonal fluctuations in salinity. For instance, Atlantic clade populations are present in Baie de L'Isle Verte salt marsh, along the coast of the St. Lawrence estuary, where salinity can vary seasonally in the ponds between 5 and 40 PSU, with more saline conditions in the fall season (Lee *et al.*, 2003).

Several features of the fluctuating native saline habitats of the invasive clades can promote balancing selection and the maintenance of genetic variation within populations (Turelli and Barton, 2004). First, the seasonal fluctuations in salinity, due to seasonal variation in precipitation, causes the selection regime to vary across seasons. Negative genetic correlations between saltwater and freshwater survival suggest evolutionary tradeoffs between saltwater and freshwater tolerance (Lee and Petersen, 2002; Lee *et al.*, 2003, 2007). Thus, as salinity changes across seasons, selection would alternate between favoring freshwater versus saltwater-adapted genotypes. With approximately six generations per year and with a short generation time (20–30 days) relative to the period of seasonal fluctuations, different cohorts would be favored by selection in different seasons (Lee and Gelembiuk, 2008). These

different seasonal cohorts would then be stored in the sediment in a diapause egg bank (Castonguay and FitzGerald, 1990; Ban, 1992; Glippa *et al.*, 2014), preserving genetic variation in the population on which selection could act when the individuals come out of diapause.

In addition, the presence of beneficial reversal of dominance could further act to maintain genetic variation in the population by protecting the maladapted allele against negative selection (Posavi *et al.*, 2014). According to theoretical studies, beneficial reversal of dominance could greatly expand the conditions under which temporally varying selection maintains genetic variation in a population (Curtsinger *et al.*, 1994; Turelli *et al.*, 2001). Under this mechanism, dominance switches across conditions, such that freshwater tolerance is dominant under freshwater conditions, whereas saltwater tolerance is dominant under saltwater conditions. This phenomenon arises due to the more fit allele in a heterozygote compensating for the lower function of the less fit allele (Wright, 1929; Kacser and Burns, 1981). Reversal of dominance had been hypothesized to occur across conditions (Wills, 1975) but was first demonstrated empirically in the *E. affinis* species complex system (Posavi *et al.*, 2014).

The sharing of selected SNPs across native saline populations from genetically distinct clades suggests that the history of balancing selection in the *E. affinis* species complex is quite ancient, preceding the formation of the clades (Stern and Lee, 2020). Such a selection regime preserving polymorphism might date back to the ancestral range of the genus *Eurytemora* in the subarctic circumpolar region, which includes coastal Alaska (Dodson *et al.*, 2010). This geographical region encompasses a rich diversity of sympatric and parapatric species of *Eurytemora* inhabiting coastal estuaries and thousands of salt ponds, which experience salinity fluctuations on multiple time scales. A history of temporally varying salinity, along with spatial heterogeneity, has likely enabled the conditions that preserve genetic polymorphism in salinity tolerance in certain species of the genus *Eurytemora* and the *E. affinis* species complex.

The balanced polymorphisms in the ancestral saline habitats likely promote both rapid adaptation and parallel adaptation during transitions to novel salinities by *E. affinis* complex populations. The preservation of a wide range of salinity tolerance alleles in the saline habitats would provide a vast reservoir upon which natural selection could act during salinity change. In particular, the elevated frequency of low-salinity-adapted alleles in the ancestral saline populations would enable selection to act repeatedly on the same alleles during freshwater invasions. This mechanism is likely important for promoting rapid parallel adaptation in the Atlantic and Gulf clade populations. The generality of this mechanism for rapid adaptation in different clades of the *E. affinis* species complex, especially those residing in different habitat types, needs to be explored further.

10.4 The Potential Role of Positive Epistasis

Results from multiple studies indicate that the same ion-transporter alleles are under selection repeatedly in wild populations and in laboratory selection lines (Lee *et al.*, 2011; Posavi *et al.*, 2020; Stern and Lee, 2020; Lee, 2021; Stern *et al.*, 2022; Diaz *et al.*, 2023). These results are consistent with selection acting on a set of essential ion-transport-related genes that must cooperate to perform the function of ion transport between body fluids and the environment. Thus, in response to salinity change, selection might favor sets of either freshwater or saltwater alleles across a set of cooperating ion transporters (and carbonic anhydrase), such that they evolve together as a unit.

This pattern of parallel selection that repeatedly favors the same sets of ion-transporter alleles is consistent with positive synergistic epistasis driving selection on multiple specific loci encoding a polygenic trait (i.e. ion transport). Under positive epistasis, selection would tend to favor multiple beneficial alleles at functionally related loci. If a set of loci are functionally related and essential for freshwater adaptation, positive epistasis will tend to lead to parallel adaptation.

In a laboratory natural selection experiment that imposed salinity decline on a saline Baltic Sea population, the observed parallel selection acting on the same alleles across replicate selection lines was consistent with the presence of positive synergistic epistasis among the selected alleles (Stern *et al.*, 2022). Selection on a Baltic Sea population from Kiel, Germany (15 PSU) for freshwater tolerance (0 PSU) over ten generations resulted in selection favoring many of the same alleles (SNPs) among the ten replicate selection lines, much more than expected by chance (with 79.5% sharing of selected alleles between replicate lines, based on the Jaccard index). Simulations of the empirical data under different models of epistasis revealed that the high levels of parallelism between the replicate lines was matched only by the model of positive epistasis, relative to the null multiplicative model where the effects of all loci are independent. Simulations of the data indicated that the high parallelism among the replicate lines could not be explained by physical linkage among the beneficial alleles (see Supplementary Fig. 2 in Stern *et al.*, 2022). In addition, simulations indicate that selection on standing genetic variation in the replicate selection lines could contribute to parallel evolution among the selection lines but could not alone explain the high degree of parallelism among the lines (see Supplementary Fig. 3 in Stern *et al.*, 2022).

The results from this study suggest that positive synergistic epistasis might serve as an important mechanism in promoting rapid and parallel adaptation in at least some *E. affinis* complex populations. With this mechanism, selection for a polygenic trait can proceed in a repeatable and predictable manner, acting on the same complex of ion-transporter genes (and other genes related to salinity adaptation). While simulations of the data indicate that this mechanism might be contributing to freshwater adaptation, the presence of positive epistasis among alleles at different loci needs to be tested at the molecular level. For instance, to validate the role of positive epistasis, we would need to test whether the combination of freshwater-selected alleles at multiple ion-transport-related loci, such as *NHA-7*, *NKA-α1* and *CA-9*, actually increases ion-transport function under freshwater conditions significantly more than expected based on the contribution of each allele alone. In addition, it would be worthwhile testing the extent to which this mechanism operates in different clades with vastly different genome architectures (see next section).

10.5   The Potential Role of Chromosomal Evolution and Linkage

Genome architecture likely has profound impacts on how populations respond to natural selection (Lee, 2023; Du *et al.*, 2024, 2025). Within the *E. affinis* species complex, genome architecture varies considerably, with an evolutionary history marked by multiple independent chromosomal fusion events, whereby two or more separate chromosomes merge to form a single chromosome (Du *et al.*, 2024, 2025). Ancestral reconstruction of fusion events indicates that the 15 chromosomes of an ancestral Europe-like genome have fused independently into four (Atlantic clade) and seven (Gulf clade) chromosomes (Fig. 10.6). Nevertheless, gene content remains relatively constant, with the genomes of three distinct clades (Atlantic, Gulf, and Europe) all showing expansions of ion-transport-related gene families (Du *et al.*, 2024, 2025).

Patterns of chromosomal fusions suggest that the fusion events might have been favored by natural selection. Among the clades, the Atlantic clade, with only four chromosomes, is the most invasive and geographically widespread (Fig. 10.2a) (Lee, 1999; Winkler *et al.*, 2008; Sukhikh *et al.*, 2019; Słdugocki *et al.*, 2021). Notably, population genomic signatures of selection associated with contemporary salinity change in wild populations (Stern and Lee, 2020) are enriched at the chromosomal fusion sites, especially at the centromeres, of the Atlantic clade genome and to a lesser extent of the Gulf clade genome (Du *et al.*, 2025).

In the Atlantic clade genome, chromosomal fusions have brought together selected loci and repositioned them toward more central portions of the chromosomes, especially at centromeres, where recombination is low (Fig. 10.6a, red genome with four chromosomes) (Du *et al.*, 2024, 2025). For instance, the selected ion-transport-related genes were repositioned

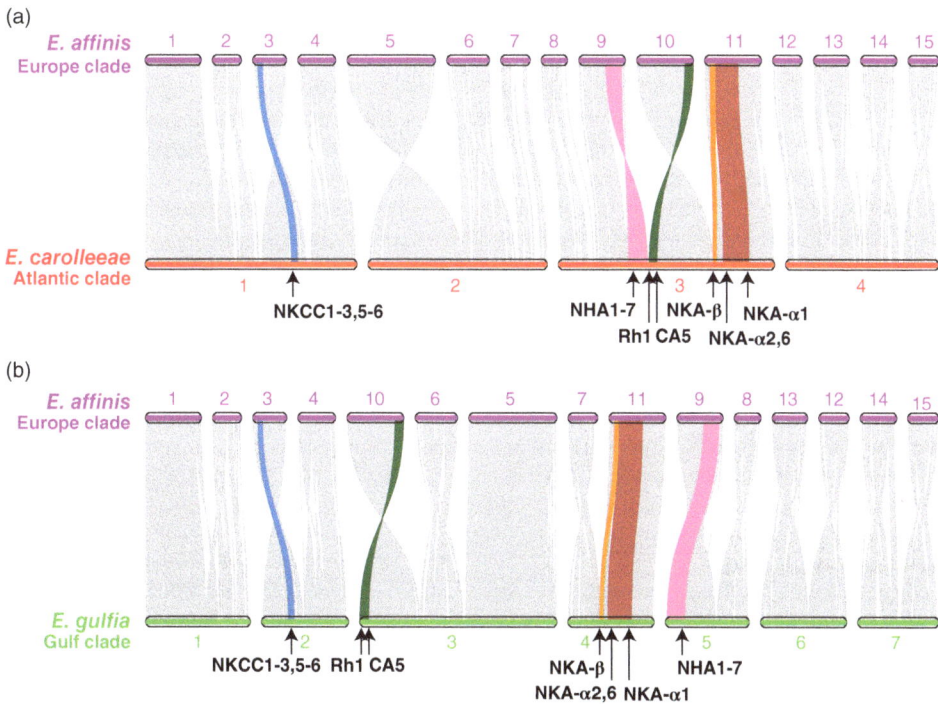

Fig. 10.6. Patterns of chromosomal fusions from 15 chromosomes (the ancestral state) of the Europe clade (*Eurytemora affinis* proper) independently into (a) the four chromosomes of the Atlantic clade (*Eurytemora carolleeae*) and (b) the seven chromosomes of the Gulf clade (*Eurytemora gulfia*). Adapted from Du *et al.* (2025). Used with permission.

much more centrally following chromosomal fusions than other ion-transporter genes not under selection. Thus, at these more central positions, beneficial alleles that are clustered together would be less likely to be separated by recombination. The potentially cooperating alleles could then be co-inherited and also more tightly coregulated in response to rapid environmental change.

These chromosomal fusion events have several noteworthy consequences for the selection responses of populations from the different clades. While we find striking patterns of parallel evolution across the clades, with often the same loci and SNPs under selection during salinity change (Stern and Lee, 2020; Stern *et al.*, 2022; Diaz *et al.*, 2023), the tempo and mode of selection likely differs among clades with distinct genome architectures. Variation in chromosome number affects the patterns of linkage among loci and the recombination landscape, altering the combination of alleles on which selection would act within populations. Fewer chromosomes for a given genome size would lead to greater linkage among loci and overall fewer recombination events. In contrast, with greater numbers of chromosomes, greater recombination rates would be more effective in bringing together beneficial alleles within genomes. In such cases, freshwater- or saltwater-adapted alleles could rapidly accumulate within genomes during selection, especially with the presence of positive epistatic interactions among the beneficial alleles.

Consequently, populations from different clades, with different chromosome numbers, would likely achieve rapid adaptation and parallel evolution through different genetic mechanisms. For instance, as the Atlantic clade genome has only four chromosomes, making the recombination rate relatively low, recombination and positive epistasis would be less likely to serve as a major mechanism in bringing beneficial alleles together within genomes. In addition, in the Atlantic clade, the selected loci are clustered in central positions of the chromosome, making recombination even less effective in bringing saline or freshwater alleles together. As the selected loci are already linked, selection would tend to act on large linkage groups.

As mentioned in the previous section, in wild populations of the Atlantic and Gulf clades, selection during salinity change is likely acting on standing genetic variation in the ancestral saline populations (Stern and Lee, 2020). The native-range saline populations of both clades exhibit genomic signatures of balancing selection, which would promote the maintenance of genetic variation upon which selection could act during invasions. Given the seasonal fluctuations in salinity in the native-range populations, selection might alternate between favoring more saltwater-versus freshwater-adapted genomes across seasons. Under such conditions, having sets of linked genes, with sets of either saline- or freshwater-adapted alleles, might be beneficial for responding to rapidly changing conditions. As such, in response to rapid habitat change, such as during biological invasions, selection could be acting on variation in pre-adapted genomes.

In sharp contrast, recombination and positive epistasis might play a much larger role in promoting rapid and parallel adaptation in populations in the genomes with greater numbers of chromosomes. Given the larger number of chromosomes, and higher recombination rate, recombination could bring together functionally related alleles across multiple loci, such as freshwater- or saltwater-adapted ion-transporter alleles. For example, for the Europe clade with 15 chromosomes, recombination and positive epistasis could be promoting parallel selection in the selection lines exposed to declining salinity (Stern *et al.*, 2022).

Thus, different genome architectures might be more advantageous for promoting rapid adaptation in different habitats with divergent selection regimes. Under temporally varying selection, where the period of fluctuation exceeds generation time, balancing selection could preserve variation in genome architecture (Lee and Gelembiuk, 2008). If the environmental fluctuations are predictable, it could be beneficial to have different sets of pre-adapted linked loci available for selection as the environment changes. Under such conditions, having fewer chromosomes, with linked loci not subjected to recombination, might be beneficial and favored by selection. However, if habitat changes are less predictable, having high recombination, and positive epistasis among functionally related loci, might be more advantageous in generating a wider array of allelic combinations available for selection. Under such conditions, having more chromosomes, with the potential for greater recombination, might be favored by selection.

10.6 Remaining Questions to Address

Many questions remain regarding the impacts of genome architecture on the probability of success of invasive populations. Given the results and predictions described above, a key question regards the conditions under which certain genome architectures are advantageous and favored by selection. For instance, how might the nature of environmental fluctuations, such as randomness or periodicity, affect genome architecture evolution? Which environmental conditions are more conducive to selection favoring higher or lower recombination, and consequently chromosomal fissions or fusions?

Another related question regards how genome architecture might shape population responses to selection during environmental change. In particular, which types of genome architectures might make certain populations more responsive to selection and more likely to become invasive? And would particular types of genome architectures be more conducive to invading certain kinds of habitats?

Additionally, how does genome architecture affect the loci under selection? Would the positioning of critical loci affect the selection response? For instance, what might be the functional consequences of positioning critical loci (e.g. ion transporters) near the telomeres versus centromeres and fusion sites? Does the clustering of ion transporters near the centromere affect ion-transport function, such as by affecting coordinated gene expression?

An additional set of questions regards the particular alleles that are favored by natural selection during invasions. During invasions from saline to freshwater habitats by E. affinis species complex populations, what particular functions are undergoing parallel selection during independent invasions? Specifically, in terms of the ion-transport alleles under selection, in what way do the SNPs under selection alter ion-transport functions during saline-to-freshwater invasions? And how do SNPs at different ion-transporter loci affect the evolution of cooperative functions among different ion transporters?

Answering these questions would address the predictions and hypotheses presented in previous sections and help uncover fundamental mechanisms of adaptation, including during rapid habitat invasions.

Acknowledgments

This project was funded by the National Science Foundation grants IOS-2412790, DEB-2055356 and OCE-1658517, and French National Research Agency ANR-19-MPGA-0004 (Macron's "Make Our Planet Great Again" award) to Carol E. Lee. Current and former Lee laboratory members Zhenyong Du, Teresa Popp and David B. Stern provided useful comments on the content of the text.

References

Bailey, S.A., Brydges, T., Casas-Monroy, O., Kydd, J., Linley, R.D. et al. (2022) First evaluation of ballast water management systems on operational ships for minimizing introductions of nonindigenous zooplankton. Marine Pollution Bulletin 182: 113947. DOI: 10.1016/j.marpolbul.2022.113947

Ban, S. (1992) Effects of photoperiod, temperature, and population density on induction of diapause egg production in Eurytemora affinis (Copepoda: Calanoida) in Lake Ohnuma, Hokkaido, Japan. Journal of Crustacean Biology 12, 361–367. DOI: 10.2307/1549029

Barghi, N., Tobler, R., Nolte, V., Jakšić, A.M., Mallard, F. et al. (2019) Genetic redundancy fuels polygenic adaptation in Drosophila. PLOS Biology 17: e3000128. DOI: 10.1371/journal.pbio.3000128

Barghi, N., Hermisson, J. and Schlötterer, C. (2020) Polygenic adaptation: a unifying framework to understand positive selection. Nature Reviews Genetics 21, 769–781. DOI: 10.1038/s41576-020-0250-z

Bij de Vaate, A., Jażdżewski, K., Ketelaars, H.A.M., Gollasch, S. and van der Veld, G. (2002) Geographical patterns in range extension of Ponto-Caspian macroinvertebrate species in Europe. Canadian Journal of Fisheries and Aquatic Sciences 59, 1159–1174. DOI: 10.1139/f02-098

Bradie, J.N., Drake, D.A.R., Ogilvie, D., Casas-Monroy, O. and Bailey, S.A. (2021) Ballast water exchange plus treatment lowers species invasion rate in freshwater ecosystems Environmental Science and Technology 55, 82–89. DOI: 10.1021/acs.est.0c05238

Carlton, J.T. (1985) Transoceanic and interoceanic dispersal of coastal marine organisms: the biology of ballast water. In: Barnes, H. and Barnes, M. (eds) *Oceanography and Marine Biology – An Annual Review*, Vol. 23. CRC Press, Boca Raton, Florida, pp. 313–371.

Casties, I., Seebens, H. and Briski, E. (2016) Importance of geographic origin for invasion success: a case study of the North and Baltic Seas versus the Great Lakes–St. Lawrence River region. *Ecology and Evolution* 6, 8318–8329. DOI: 10.1002/ece3.2528

Castonguay, M. and FitzGerald, G.J. (1990) The ecology of the calanoid copepod *Eurytemora affinis* in salt marsh tide pools. *Hydrobiologia* 202, 125–133. DOI: 10.1007/BF00006839

Clarke, A. (2003) Costs and consequences of evolutionary temperature adaptation. *Trends in Ecology and Evolution* 18, 573–581. DOI: 10.1016/j.tree.2003.08.007

Cristescu, M.E.A., Hebert, P.D.N., Witt, J.D.S., MacIsaac, H.J. and Grigorovich, I.A. (2001) An invasion history for *Cercopagis pengoi* based on mitochondrial gene sequences. *Limnology and Oceanography* 46, 224–229. DOI: 10.4319/lo.2001.46.2.0224

Curtsinger, J.W., Service, P.M. and Prout, T. (1994) Antagonistic pleiotropy, reversal of dominance, and genetic-polymorphism. *American Naturalist* 144, 210–228. DOI: 10.1086/285671

Darling, J.A., Martinson, J., Gong, Y., Okum, S., Pilgrim, E. *et al.* (2018) Ballast water exchange and invasion risk posed by intracoastal vessel traffic: an evaluation using high throughput sequencing. *Environmental Science and Technology* 52, 9926–9936. DOI: 10.1021/acs.est.8b02108

de Beaufort, L.F. (1954) *Veranderingen in de Flora en Fauna van de Zuiderzee (thans IJsselmeer) na de Afsluiting in 1932*. C. de Boer Jr, Den helder, Netherlands.

Diaz, J., Stern, D.B. and Lee, C.E. (2023) Local adaptation despite gene flow in copepod populations across salinity and temperature gradients in the Baltic and North Seas. *Authorea* (preprint). DOI: 10.22541/au.168311545.58858033/v1

Dodson, S.I., Skelly, D.A. and Lee, C.E. (2010) Out of Alaska: morphological evolution and diversity within the genus *Eurytemora* from its ancestral range (Crustacea, Copepoda). *Hydrobiologia* 653, 131–148. DOI: 10.1007/978-90-481-9908-2_11

Du, Z., Gelembiuk, G., Moss, W., Tritt, A. and Lee, C.E. (2024) The genome architecture of the copepod *Eurytemora carolleeae*, the highly invasive Atlantic clade of the *Eurytemora. affinis* species complex. *Genomics, Proteomics and Bioinformatics* 22: qzae066. DOI: 10.1093/gpbjnl/qzae066

Du, Z., Wirtz, J., Zhou, Y.J., Jenstead, A., Opgenorth, T. *et al.* (2025) Genome architecture evolution in invasive copepod species complex. *Nature Communications* (in press).

Evans, D.H. and Clairborne, J.B. (2009) Osmotic and ionic regulation in fishes. In: Evans, D.H. (ed.) *Osmotic and Ionic Regulation: Cells and Animals*, CRC Press, Boca Raton, Florida, pp. 295–366.

Eyun, S., Soh, H.Y., Posavi, M., Munro, J., Hughes, D.S.T. *et al.* (2017) Evolutionary history of chemosensory-related gene families across the Arthropoda. *Molecular Biology and Evolution* 34, 1838–1862. DOI: 10.1093/molbev/msx147

Gelembiuk, G.W., May, G.E. and Lee, C.E. (2006) Phylogeography and systematics of zebra mussels and related species. *Molecular Ecology* 15, 1033–1050. DOI: 10.1111/j.1365-294X.2006.02816.x

Gerber, L., Lee, C.E., Grousset, E., Blondeau-Bidet, E., Boucheker, N.B. *et al.* (2016) The legs have it: *in situ* expression of ion transporters V-Type H^+ ATPase and Na^+/K^+-ATPase in osmoregulating leg organs of the invading copepod *Eurytemora affinis*. *Physiological and Biochemical Zoology* 89, 233–250. DOI: 10.1086/686323

Glippa, O., Denis, L., Lesourd, S. and Souissi, S. (2014) Seasonal fluctuations of the copepod resting egg bank in the middle Seine estuary, France: impact on the nauplii recruitment. *Estuarine and Coastal Marine Science* 142, 60–67. DOI: 10.1016/j.ecss.2014.03.008

Havel, J.E., Lee, C.E. and Vander Zanden, M.J. (2005) Do reservoirs facilitate passive invasions into landscapes? *Bioscience* 55, 518–525. DOI: 10.1641/0006-3568(2005)055[0518:DRFIIL]2.0.CO;2

Hutchinson, G.E. (1957) *A Treatise on Limnology*. Wiley, New York.

Jażdżewski, K. (1980) Range extensions of some gammaridean species in European inland waters caused by human activity. *Crustaceana* 6 (Suppl.), 84–107.

Johnson, K.E., Perreau, L., Charmantier, G., Charmantier-Daures, M. and Lee, C.E. (2014) Without gills: localization of osmoregulatory function in the copepod *Eurytemora affinis*. *Physiological and Biochemical Zoology* 87, 310–324. DOI: 10.1086/674319

Kacser, H. and Burns, J.A. (1981) The molecular basis of dominance. *Genetics* 97, 639–666. DOI: 10.1093/genetics/97.3-4.639

Khlebovich, V.V. and Abramova, E.N. (2000) Some problems of crustacean taxonomy related to the phenomenon of Horohalinicum. *Hydrobiologia* 417, 109–113. DOI: 10.1023/A:1003863623267

Kimmel, D.G., Miller, W.D. and Roman, M.R. (2006) Regional scale climate forcing of mesozooplankton dynamics in Chesapeake Bay. *Estuaries and Coasts* 29, 375–387. DOI: 10.1007/BF02784987

Lee, C.E. (1999) Rapid and repeated invasions of fresh water by the saltwater copepod *Eurytemora affinis*. *Evolution* 53, 1423–1434. DOI: 10.1111/j.1558-5646.1999.tb05407.x

Lee, C.E. (2000) Global phylogeography of a cryptic copepod species complex and reproductive isolation between genetically proximate "populations". *Evolution* 54, 2014–2027. DOI: 10.1111/j.0014-3820.2000.tb01245.x

Lee, C.E. (2002) Evolutionary genetics of invasive species. *Trends in Ecology and Evolution* 17, 386–391. DOI: 10.1016/S0169-5347(02)02554-5

Lee, C.E. (2021) Ion transporter gene families as physiological targets of natural selection during salinity transitions in a copepod. *Physiology* 36, 335–349. DOI: 10.1152/physiol.00009.2021

Lee, C.E. (2023) Genome architecture underlying salinity adaptation in the invasive copepod *Eurytemora affinis* species complex: a review. *iScience* 26: 107851. DOI: 10.1016/j.isci.2023.107851

Lee, C.E. 2010. Evolution of invasive populations. *In:* Simberloff, D. and Rejmanek, M. (eds) *Encyclopedia of Biological Invasions.* University of California Press, Berkeley, California.

Lee, C.E. and Bell, M.A. (1999) Causes and consequences of recent freshwater invasions by saltwater animals. *Trends in Ecology and Evolution* 14, 284–288. DOI: 10.1016/S0169-5347(99)01596-7

Lee, C.E. and Frost, B.W. (2002) Morphological stasis in the *Eurytemora affinis* species complex (Copepoda: Temoridae). *Hydrobiologia* 480, 111–128. DOI: 10.1023/A:1021293203512

Lee, C.E. and Gelembiuk, G.W. (2008) Evolutionary origins of invasive populations. *Evolutionary Applications* 1, 427–448. DOI: 10.1111/j.1752-4571.2008.00039.x

Lee, C.E. and Petersen, C.H. (2002) Genotype-by-environment interaction for salinity tolerance in the freshwater invading copepod *Eurytemora affinis*. *Physiological and Biochemical Zoology* 75, 335–344. DOI: 10.1086/343138

Lee, C.E. and Petersen, C.H. (2003) Effects of developmental acclimation on adult salinity tolerance in the freshwater-invading copepod *Eurytemora affinis*. *Physiological and Biochemical Zoology* 76, 296–301. DOI: 10.1086/375433

Lee, C.E., Remfert, J.L. and Gelembiuk, G.W. (2003) Evolution of physiological tolerance and performance during freshwater invasions. *Integrative and Comparative Biology* 43, 439–449. DOI: 10.1093/icb/43.3.439

Lee, C.E., Remfert, J.L. and Chang, Y.-M. (2007) Response to selection and evolvability of invasive populations. *Genetica* 129, 179–192. DOI: 10.1007/s10709-006-9013-9

Lee, C.E., Kiergaard, M., Gelembiuk, G.W., Eads, B.D. and Posavi, M. (2011) Pumping ions: rapid parallel evolution of ionic regulation following habitat invasions. *Evolution* 65, 2229–2244. DOI: 10.1111/j.1558-5646.2011.01308.x

Lee, C.E., Posavi, M. and Charmantier, G. (2012) Rapid evolution of body fluid regulation following independent invasions into freshwater habitats. *Journal of Evolutionary Biology* 25, 625–633. DOI: 10.1111/j.1420-9101.2012.02459.x

Lee, C.E., Moss, W.E., Olson, N., Chau, K.F., Chang, Y.-M. and Johnson, K.E. (2013) Feasting in fresh water: impacts of food concentration on freshwater tolerance and the evolution of food x salinity response during the expansion from saline into freshwater habitats. *Evolutionary Applications* 6, 673–689. DOI: 10.1111/eva.12054

Lee, C.E., Charmantier, G. and Lorin-Nebel, C. (2022a) Mechanisms of Na^+ uptake from freshwater habitats in animals. *Frontiers in Physiology* 13: 1006113. DOI: 10.3389/fphys.2022.1006113

Lee, C.E., Downey, K., Colby, R.S., Freire, C.A., Nichols, S. *et al.* (2022b) Recognizing salinity threats in the climate crisis. *Integrative and Comparative Biology* 62, 441–460. DOI: 10.1093/icb/icac069

Little, C. (1983) *The Colonisation of Land: Origins and Adaptations of Terrestrial Animals.* Cambridge University Press Cambridge, UK.

Little, C. (1990) *The Terrestrial Invasion: An Ecophysiological Approach to the Origins of Land Animals.* Cambridge University Press Cambridge, UK.

Livdāne, L., Putnis, I., Rubene, G., Elferts, D. and Ikauniece, A. (2016) Baltic herring prey selectively on older copepodites of *Eurytemora affinis* and *Limnocalanus macrurus* in the Gulf of Riga. *Oceanologia* 58, 46–53. DOI: 10.1016/j.oceano.2015.09.001

Lozupone, C.A. and Knight, R. (2007) Global patterns in bacterial diversity. *Proceedings of the National Academy of Sciences USA* 104: 11436–11440. DOI: 10.1073/pnas.0611525104

May, G.E., Gelembiuk, G.W., Panov, V.E., Orlova, M.I. and Lee, C.E. (2006) Molecular ecology of zebra mussel invasions. *Molecular Ecology* 15, 1021–1031. DOI: 10.1111/j.1365-294X.2006.02814.x

Miller, R.C. (1958) The relict fauna of Lake Merced, San Francisco. *Journal of Marine Research* 17, 375–382.

Mordukhai-Boltovskoi, P.D. (1979) Composition and distribution of Caspian fauna in the light of modern data. *Internationale Revue der Gesamten Hydrobiologie und Hydrographie* 64, 1–38. DOI: 10.1002/iroh.19790640102

Mormul, R.P., Vieira, D.S., Bailly, D., Fidanza, K., da Silva, V.F.B. *et al.* (2022) Invasive alien species records are exponentially rising across the Earth. *Biological Invasions* 24, 3249–3261. DOI: 10.1007/s10530-022-02843-1

Popp, T.E. (2024) *Evolution of Ion Transporter Expression During Salinity Decline.* PhD thesis, University of Wisconsin-Madison, Madison, Wisconsin.

Popp, T.E., Hermet, S., Fredette-Roman, J., McKeel, E., Zozaya, W. *et al.* (2024) Evolution of ion transporter Na+/K+-ATPase expression in the osmoregulatory maxillary glands of an invasive copepod. *iScience* 27: 110278. DOI: 10.1016/j.isci.2024.110278

Posavi, M., Larget, B., Gelembiuk, G.W. and Lee, C.E. (2014) Testing for beneficial reversal of dominance during salinity shifts in the invasive copepod *Eurytemora affinis*, and implications for the maintenance of genetic variation. *Evolution* 68, 3166–3183. DOI: 10.1111/evo.12502

Posavi, M., Gulisija, D., Munro, J.B., Silva, J.C. and Lee, C.E. (2020) Rapid evolution of genome-wide gene expression and plasticity during saline to freshwater invasions by the copepod *Eurytemora affinis* species complex. *Molecular Ecology* 29, 835–4856. DOI: 10.1111/mec.15681

Rasch, E.M., Lee, C.E. and Wyngaard, G.A. (2004) DNA–feulgen cytophotometric determination of genome size for the freshwater-invading copepod *Eurytemora affinis*. *Genome* 47, 559–564. DOI: 10.1139/g04-014

Remane, A. and Schlieper, C. (1971) *Biology of Brackish Water*. Wiley, New York.

Ricciardi, A. and MacIsaac, H.J. (2000) Recent mass invasion of the North American Great Lakes by Ponto-Caspian species. *Trends in Ecology and Evolution* 15, 62–65. DOI: 10.1016/S0169-5347(99)01745-0

Ricciardi, A. and MacIsaac, H.J. (2022) Vector control reduces the rate of species invasion in the world's largest freshwater ecosystem. *Conservation Letters* 15: e12866. DOI: 10.1111/conl.12866

Ruiz, G.M., Carlton, J.T., Grosholz, E.D. and Hines, A.H. (1997) Global invasions of marine and estuarine habitats by non-indigenous species: mechanisms, extent, and consequences. *American Zoologist* 37, 621–632. DOI: 10.1093/icb/37.6.621

Seebens, H., Blackburn, T.M., Dyer, E.E. and Essl, F. (2018) Global rise in emerging alien species results from increased accessibility of new source pools. *Proceedings of the National Academy of Sciences USA* 115, E2264–E2273. DOI: 10.1073/pnas.1719429115

Shaheen, P.A., Stehlik, L.L., Meise, C.J., Stoner, A.W., Manderson, J.P. and Adams, D.L. (2001) Feeding behavior of newly settled winter flounder (*Pseudopleuronectes americanus*) on calanoid copepods. *Journal of Experimental Marine Biology and Ecology* 257, 37–51. DOI: 10.1016/S0022-0981(00)00335-X

Simenstad, C.A., Small, L.F. and McIntire, C.D. (1990) Consumption processes and food web structure in the Columbia River estuary. *Progress In Oceanography* 25, 271–297. DOI: 10.1016/0079-6611(90)90010-Y

Sługocki, Ł., Rymaszewska, A. and Kirczuk, L. (2021) To fit or to belong: characterization of the non-native invader *Eurytemora carolleeae* (Copepoda: Calanoida) in the Oder River system (Central Europe). *Aquatic Invasions* 16, 443–460. DOI: 10.3391/ai.2021.16.3.04

Stern, D.B. and Lee, C.E. (2020) Evolutionary origins of genomic adaptations in an invasive copepod. *Nature Ecology and Evolution* 4, 1084–1094. DOI: 10.1038/s41559-020-1201-y

Stern, D.B., Anderson, N.W., Diaz, J.A. and Lee, C.E. (2022) Genome-wide signatures of synergistic epistasis during parallel adaptation in a Baltic Sea copepod. *Nature Communications* 13: 4024. DOI: 10.1038/s41467-022-31622-8

Sukhikh, N., Souissi, A., Souissi, S., Holl, A.-C., Schizas, N.V. and Alekseev, V. (2019) Life in sympatry: coexistence of native *Eurytemora affinis* and invasive *Eurytemora carolleeae* in the Gulf of Finland (Baltic Sea). *Oceanologia* 61, 227—238. DOI: 10.1016/j.oceano.2018.11.002

Turelli, M. and Barton, N.H. (2004) Polygenic variation maintained by balancing selection: pleiotropy, sex-dependent allelic effects and G × E interactions. *Genetics* 166, 1053–1079. DOI: 10.1534/genetics.166.2.1053

Turelli, M., Schemske, D.W. and Bierzychudek, P. (2001) Stable two-allele polymorphisms maintained by fluctuating fitnesses and seed banks: protecting the blues in *Linanthus parryae*. *Evolution* 55, 1283–1298. DOI: 10.1111/j.0014-3820.2001.tb00651.x

Viitasalo, M., Katajisto, T. and Vuorinen, I. (1994) Seasonal dynamics of *Acartia bifilosa* and *Eurytemora affinis* (Copepoda: Calanoida) in relation to abiotic factors in the northern Baltic Sea. *Hydrobiologia* 292, 415–422. DOI: 10.1007/BF00229967

Viitasalo, M., Vuorinen, I. and Saesmaa, S. (1995) Mesozooplankton dynamics in the northern Baltic Sea: implications of variations in hydrography and climate. *Journal of Plankton Research* 17, 1857–1878. DOI: 10.1093/plankt/17.10.1857

Viitasalo, M., Flinkman, J. and Viherluoto, M. (2001) Zooplanktivory in the Baltic Sea: a comparison of prey selectivity by *Clupea harengus* and *Mysis mixta*, with reference to prey escape reactions. *Marine Ecology Progress Series* 216, 191–200. DOI: 10.3354/meps216191

Willmer, P., Stone, G. and Johnston, I. (2008) *Environmental Physiology of Animals,* Oxford, Malden.

Wills, C. (1975) Marginal overdominance in *Drosophila*. *Genetics* 81, 177–189. DOI: 10.1093/genetics/81.1.177

Winkler, G., Dodson, J.J., Bertrand, N., Thivierge, D. and Vincent, W.F. (2003) Trophic coupling across the St. Lawrence River estuarine transition zone. *Marine Ecology Progress Series* 251, 59–73. DOI: 10.3354/meps251059

Winkler, G., Dodson, J.J. and Lee, C.E. (2008) Heterogeneity within the native range: population genetic analyses of sympatric invasive and noninvasive clades of the freshwater invading copepod *Eurytemora affinis*. *Molecular Ecology* 17, 415–430. DOI: 10.1111/j.1365-294X.2007.03480.x

Withers, P.C. (1992) *Comparative Animal Physiology*. Saunders College Publishing. New York.

Witt, J.D.S., Hebert, P.D.N. and Morton, W.B. (1997) *Echinogammarus ischnus*: another crustacean invader in the Laurentian Great Lakes basin. *Canadian Journal of Fisheries and Aquatic Sciences* 54, 264–268. DOI: 10.1139/f96-292

Wright, S. (1929) Fisher's theory of dominance. *American Naturalist* 63, 274–279. DOI: 10.1086/280260

11 Genomics-informed Modelling: Advancing Our Understanding of Non-indigenous Species' Colonization and Spread

Marc Rius[1,2]* and Marta Pascual[3,4]

[1]*Department of Marine Ecology, Centre for Advanced Studies of Blanes (CEAB), Spanish Research Council (CSIC), Blanes, Spain;* [2]*Department of Zoology, Centre for Ecological Genomics and Wildlife Conservation, University of Johannesburg, Auckland Park Johannesburg, South Africa;* [3]*Departament de Genètica, Microbiologia i Estadística, Facultat de Biologia, Universitat de Barcelona, Barcelona, Spain;* [4]*Institut de Recerca de la Biodiversitat (IRBio), Universitat de Barcelona, Barcelona, Spain*

Abstract

Recent methodological advancements are revolutionizing key research fields such as population genomics and predictive modelling. Research progress in these fields is greatly enhancing the scope and breadth of non-indigenous species (NIS) studies. Population genomics allow detection of both genetic connectivity and adaptation signals that may aid NIS spread. At the same time, predictive modelling, including approaches such as ecological niche modelling (ENM), are improving our predictions of future NIS distributions. Despite clear complementarity between population genomics and predictive modelling, the combined use of these approaches is not yet common-place. Here, we review how the combined use of ENM and genomics (e.g. genomics-informed modelling studies) are both improving our understanding of NIS' colonization and establishment, and helping our predictions of NIS spread and origin. Future research using genomics and ENM in combination is expected to significantly advance our knowledge of the invasion process and help develop more effective NIS management actions.

11.1 Introduction

The human-mediated introduction of non-indigenous species (NIS) away from their native range creates many biodiversity and societal challenges. NIS directly compete with long-established native species for limited resources (Corbin and d'Antonio, 2004), alter entire community dynamics (Rius *et al.*, 2017; Levine, 2000) and incur huge economic losses, especially when NIS become pests and/or pathogens (e.g. Pimentel *et al.*, 2005; Asplen *et al.*, 2015; Nuñez *et al.*, 2020). In turn, many NIS introductions represent a rich source of knowledge for both ecologists and evolutionary biologists. For example, NIS introductions provide invaluable

*Corresponding author: mrius@ceab.csic.es

information on key ecological and evolutionary mechanisms, such as species colonization, parallel evolution, rapid adaptation and community assembly (Prevosti *et al.*, 1988; Stachowicz and Tilman, 2005; Trumbo *et al.*, 2016). In addition, newly generated eco-evolutionary knowledge of NIS helps improve biodiversity managers' practices aiming at mitigating the impacts of NIS on native biota and local economies (Sherpa and Després, 2021).

Recent technological developments have greatly improved the depth and quality of high-throughput sequencing (HTS), at the same time that HTS costs have drastically plunged (Rius *et al.*, 2015). This has profoundly impacted several scientific disciplines, including invasion science (Viard *et al.*, 2016; Ricciardi *et al.*, 2017). Genomics has increased our ability to identify population differentiation at different spatial scales, as well as to detect loci under selection that contribute to local adaptation (Carreras *et al.*, 2020; Torrado *et al.*, 2020) and/or enhanced survival across different ontogenetic stages (Torrado *et al.*, 2022). Moreover, HTS has provided the basis for identifying adaptive standing variation that facilitates NIS colonization across a wide array of environmental conditions (Chen *et al.*, 2024b; Galià-Camps *et al.*, 2025). Finally, HTS has enabled the analysis of both ancient and recent evolutionary dynamics of NIS (e.g. deep lineage divergence, population connectivity, signatures of rapid adaptation) (Welles and Dlugosch, 2018; Casso *et al.*, 2019; Galià-Camps *et al.*, 2024).

Researchers have long used a variety of modelling approaches to study NIS (Zimmermann *et al.*, 2010). Among these approaches, ecological niche modelling (ENM) has emerged as a widely applicable analytical tool to understand and predict both present and future NIS introductions. ENM uses sophisticated statistical and computer modelling algorithms (Guisan and Zimmermann, 2000; Elith *et al.*, 2006) and different data sources, which often include species records and environmental databases. ENM helps foresee geographical areas where NIS may find suitable habitats (Alvarado-Serrano and Knowles, 2014; Melo-Merino *et al.*, 2020), and also understand current NIS distributions when areas of the introduced range remain unsampled. The growing availability of global online data sets of species records has dramatically improved the scope of ENM, moving from studies with a restricted

geographical scope to predictions at a global scale (Melo-Merino *et al.*, 2020).

NIS may either invade ecological niches that match the native range's environmental space (niche stability) or shift the environmental space occupied in the new areas (Liu *et al.*, 2020; Pack *et al.*, 2022). Consequently, the environmental space in the introduced range can be reduced with niche unfilling (i.e. a proportion of the native niche that remains unoccupied in the introduced niche) or increased with niche expansion (i.e. the introduced niche includes novel environmental dimensions) (Frederico *et al.*, 2019). The size of the niche shifts in the introduced range may reduce transferability of the models, and the number of species records used may have a strong effect on the ENM results (Liu *et al.*, 2022). In addition, the successful establishment and spread of NIS often rely on genetic composition that provides a colonizing advantage to NIS (Verhoeven *et al.*, 2011; Rius and Darling, 2014). Consequently, genetic data is often required for obtaining an accurate interpretation of realized and predicted niches of NIS (Sillero *et al.*, 2020; Hudson *et al.*, 2021).

In this chapter, we review current knowledge emerging from studies combining ENM and genomic data. We first explore the importance of obtaining accurate coverage of species records/genotypes across the species range using both ENM and population genomics. We then investigate how this combined approach has advanced our understanding of the colonization and spread of NIS. Finally, we provide an overview of the role of niche shifts on invasion success.

11.2 Data Sets on Species Records and Population Genomics

A starting point of any study on NIS is to identify and delineate the section of the range that is native and the one that is introduced. The native range is where the study species has evolved naturally. The introduced range is where the NIS has been artificially transported (Fig. 11.1a) and where it may subsequently become naturalized or invasive (Richardson *et al.*, 2000). This often happens through the aid of high levels of propagule pressure (i.e. the number of individuals being artificially introduced), which facilitates the establishment of many native lineages in the

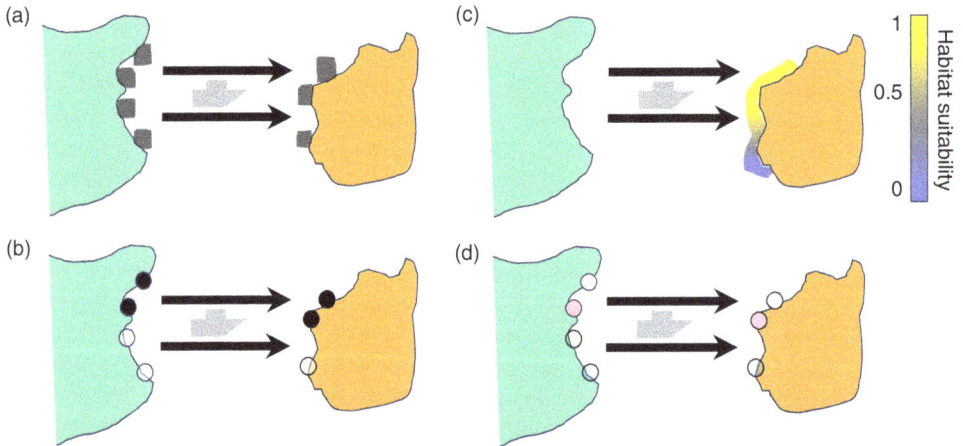

Fig. 11.1. The importance of a comprehensive coverage of species records across the species ranges for ecological niche modelling (ENM) and population genomic analyses. Native genotypes of a hypothetical coastal species are artificially transported (arrows and ship cartoon) and become established in the introduced range (orange island). (a) Sites where species are distributed across both the native and introduced ranges (i.e. unknown species records, before field sampling). (b) The observed distribution (as a result of recent field sampling effort) is indicated with black circles. Geographical areas that remain unobserved and thus absent in the species records data set are shown with unfilled circles. (c) Habitat suitability as revealed by ENM based on the observed distribution of the native range in (b). Note that part of the introduced range appears as unsuitable habitat. (d) The actual distribution of genotypes across the native and introduced ranges. The genotypes detected in the introduced range come from only two known sampling sites of the native range (i.e. the filled circles in (b)). There are two undetected genotypes in the native range (unfilled circles at (b)) with one matching an unobserved genotype in the introduced range (i.e. the unobserved site in (b)).

introduced range (e.g. Rey *et al.*, 2012; Rius and Shenkar, 2012; Osborne *et al.*, 2013). Another possibility is that a few individuals with high adaptive potential reach the introduced range and are able to expand rapidly (Pascual *et al.*, 2007). Although identifying the native and introduced ranges may seem trivial, it is often a tedious task. For example, scarcity of species records or other informative data (e.g. population genomic data) may prevent accurate designation of range types with enough resolution. The only option then will be to generate new data from scratch, which may be challenging in species with a complex invasion history and/or when it is difficult to obtain comprehensive geographical sampling coverage (Casso *et al.*, 2019; Hudson *et al.*, 2020). In addition, if the focal NIS was introduced centuries ago, population genomic studies may fail to detect both the entirety of the introduced lineages and/or the exact geographical location of the native range (Hudson *et al.*, 2022; Galià-Camps *et al.*, 2024).

In order to effectively run ENM, studies should be able to first differentiate NIS range types. A comprehensive representation of the species range is key for incorporating all informative environmental data and the species records required to provide accurate predictions of habitat suitability. Lack of accuracy and availability of species records (Fig. 11.1a) pose serious limitations for ENM studies (Hughes *et al.*, 2021). Unobserved species records in the native range (Fig. 11.1b) may lead to inaccurate ENM predictions of suitable habitats (Fig. 11.1c). Another consequence of unobserved species' records is data sets with missing genotypes (Fig. 11.1d), which could impact the analysis and interpretation of population genomics and ENM studies.

11.3   The Combined Use of Ecological Niche Modelling and Population Genetics

Pioneering studies aiming to better understand and predict species distributions (e.g. Rey *et al.*,

2012; Razgour *et al.*, 2014) paved the way for combining ENM with population genetic data. Subsequently, many studies were successful at using both approaches (Table 11.1). These studies focused on a wide variety of taxonomic groups, and increasingly include NIS. Genomics have been used to understand the distribution of NIS lineages across NIS ranges and to identify the origin of introduced lineages (e.g. Hudson *et al.*, 2021; Chen *et al.*, 2024a; Putra *et al.*, 2024). Moreover, genomics have improved our capacity to identify candidate loci that contribute to local adaptation (Carreras *et al.*, 2020). The presence of adaptive standing genetic variation in founder populations can increase their ability to colonize a wide variety of ecological niches and can thus promote dispersal to large new geographical regions (Prevosti *et al.*, 1988; Sillero *et al.*, 2020). Genomic information is therefore invaluable, as it allows researchers to select the species records that are relevant for ENM. Not considering genomic data may lead to inaccurate interpretation of ENM results, as previously suspected (Gotelli and Stanton-Geddes, 2015), but empirical evidence on this possibility has emerged more recently (Sillero *et al.*, 2020; Hudson *et al.*, 2021). Studies comparing models that consider different data sets of NIS records showed that the actual introduced range can only be accurately predicted if the model considers the environmental and biotic conditions found in the part of the native range where the introduced genotypes originated (Sillero *et al.*, 2020; Hudson *et al.*, 2021). Thus, as colonizers often originate from a reduced part of the native range, NIS may not be able to occupy all available habitats in the new geographical area within which the introduced range is located (Fig. 11.2). However, if both native and introduced ranges show environmental matching, NIS are expected to perform well and flourish across the newly colonized area. This could be partly explained by low levels of biotic resistance in the introduced range (e.g. Freestone *et al.*, 2013) and/or biotic acceptance (Stohlgren *et al.*, 2006), although considering these mechanisms in ENM remains a challenge.

Several scenarios can emerge when using ENM to understand how NIS have been introduced into new geographical areas. A first scenario is when NIS introductions come from one or two lineages that have diverged in the native range due to, for example, a physical barrier across an environmental gradient (Fig. 11.2). If only a subset of the native genotypes is artificially introduced and successfully establishes in the novel range, they may colonize habitats where conditions are similar to the part of the native range where they originated. In this case, ENM including all available species records (i.e. from the entire native and introduced ranges) may be misleading, as models may infer an incorrect introduced range size. For example, Hudson *et al.* (2021) generated population genomic data to reconstruct the invasion route of the Australian tunicate *Pyura praeputialis* (Chordata, Tunicata), and the selection of species records for the ENM turned out to be essential. The results first showed that only one native lineage (originating from eastern Australia) was introduced (similar to Fig. 11.2a, b). ENM was then used to predict the introduced range along the south-western American coastline considering three different data sets: (i) all available species records; (ii) the species records of eastern Australia only; and (iii) the species records of a south-eastern Australian lineage only. The results of the ENM showed that the current introduced range in Chile could only be identified as suitable habitat when the selected species records included the section of the native lineage where the origin of the introduced range was located (i.e. the above-mentioned data set (ii)). In turn, when data sets (i) and (iii) (i.e. not considering genomics-informed modelling) were used, predictions of the introduced range were inaccurate. In another example, Sillero *et al.* (2020) predicted the spread of the insect NIS *Drosophila subobscura* to potentially suitable habitats based on information on chromosomal arrangement polymorphisms known to be adaptive. The study considered different chromosomal inversions associated with warm and cold climates (Prevosti *et al.*, 1988) and used presence records from: (i) only the native range; (ii) all available species records; and (iii) a restricted origin of the colonizers considering the frequency of warm- and cold-adapted inversions. Depending on the model used, the predictions of the introduced range revealed: (i) large niche stability with not much unfilling (i.e. unoccupied predicted areas) when native areas had a high frequency of warm-adapted inversions; or (ii) range expansions (occupying unsuitable areas) when native areas had a high frequency of cold-adapted inversions. Similarly, Chen *et al.*

Table 11.1. Studies (in chronological order) using both predictive modelling and population genetics. Abbreviations are: ENM, ecological niche modelling; NIS: non-indigenous species; SNP, single-nucleotide polymorphism.

Study species	NIS status	Genetic marker(s)	Use of ENM and population genetics	Genetics informed ENM (or vice versa)	Reference
Wasmannia auropunctata	Yes	Microsatellite loci	Combined	Yes	Rey et al. (2012)
Plecotus austriacus	No	Microsatellite loci	Combined	Yes	Razgour et al. (2014)
Asplenium fontanum	No	Allozyme and chloroplast DNA sequence markers	Separate	No	Bystriakova et al. (2014)
Populus balsamifera	No	SNPs	Combined	Yes	Fitzpatrick and Keller (2015)
Aedes albopictus	Yes	Cytb gene	Separate	No	Pech-May et al. (2016)
Populus fremontii	No	Microsatellite loci	Combined	Yes	Ikeda et al. (2017)
Gracilaria vermiculophylla	Yes	Microsatellite loci	Combined	Yes	Sotka et al. (2018)
Myotis escalerai	No	SNPs	Combined	Yes	Razgour et al. (2019)
Drosophila subobscura	Yes	Chromosomal arrangement	Combined	Yes	Sillero et al. (2020)
Emydoidea blandingii	No	Microsatellite loci	Combined	Yes	Byer et al. (2020)
Populus angustifolia	No	Microsatellite loci	Combined	Yes	Bothwell et al. (2021)
Pyura praeputialis	Yes	SNPs	Combined	Yes	Hudson et al. (2021)
Populus angustifolia	No	Microsatellite loci	Combined	Yes	Bayliss et al. (2022)
Zosterops sp.	Yes	SNPs	Separate	No	DeRaad et al. (2024)
Ambrosia artemisiifolia	Yes	SNPs	Combined	Yes	Putra et al. (2024)
Molgula manhattensis	Yes	SNPs	Combined	Yes	Chen et al. (2024a)
Botryllus schlosseri	Yes	SNPs	Combined	Yes	Chen et al. (2024b)

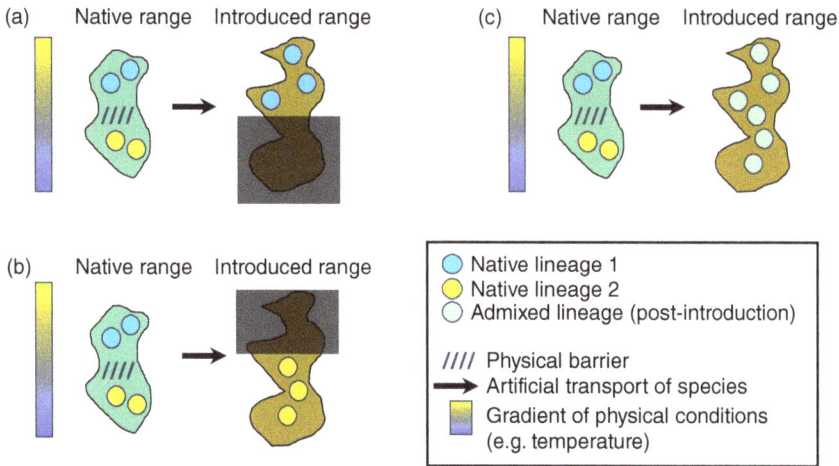

Fig. 11.2. Different scenarios of non-indigenous species' introductions, considering a latitudinal environmental gradient equally affecting the different types of species ranges (i.e. native and introduced). Note that this environmental gradient shows temperature changes in line with many regions in the southern hemisphere (see colour-gradient bars on the left, indicating warmer regions in the north and colder in the south). (a, b) Scenarios where only one native lineage is introduced and establishes successfully only in a section of the new geographical area where environmental matching is high (i.e. area occupied by coloured circles). Note that shaded areas indicate where no NIS genotypes have been introduced and thus are not part of the introduced range. In these scenarios, selecting species records of certain lineages from the native range could mislead the outcome of ecological niche modelling (ENM). (c) Scenario where both native lineages are initially introduced and subsequently outperformed by an admixed lineage that performed better across a wide range of environmental conditions in the introduced range. In this case, ENM should consider the influence of genetic admixture.

(2024a) obtained both habitat suitability data from ENM and an adaptive genomic variation data set to predict future spread of the tunicate NIS *Molgula manhattensis*. As in previous studies (e.g. Hudson *et al.*, 2021, 2022), two latitudinal genetic clusters were detected and the combined use of ENM and population genomic data indicated that certain clusters presented a higher invasion risk.

Another possible scenario that NIS studies using ENM could consider is when genetic admixture leads to a clear fitness superiority compared with parental genotypes in the introduced range. This can be facilitated by introductions from a particularly successful invasive population (i.e. bridgehead population; Lombaert *et al.*, 2010) or when two divergent native lineages are introduced into the same new geographical area. In this case, admixed genotypes could not have emerged in the native range, due to the presence of a physical barrier (Fig. 11.2c). Several studies have analysed the fitness consequences of genetic admixture and showed that admixed lineages can have a fitness advantage over

parental lineages (Turgeon *et al.*, 2011; Verhoeven *et al.*, 2011; Rius and Darling, 2014). As a result, NIS may be able to considerably expand their niche breadth and occupy extensive geographical areas (Fig. 11.2c). Fitness advantage of hybrids could lead to greater colonization success across the entire gradient of environmental conditions of the introduced range (Fig. 11.2c) compared with the parental genotypes. This scenario could eventually lead to the complete exclusion of parental lineages from the introduced range (Fig. 11.2c). Thus, ENM inferences using all available species records across species ranges may be incorrect. It is important to note that genetic admixture is known to affect different parts of the genome, including specific parental genomic combinations (Rius and Darling, 2014). In addition, chromosomal inversions may inhibit recombination (Pegueroles *et al.*, 2010), which may affect both survival and adaptation of admixed individuals.

Genomics-informed modelling can be used to explore other scenarios that have not been considered here. This could include complex demographic scenarios, typically found in studies reconstructing invasion routes of widespread NIS (Estoup and Guillemaud, 2010; Rius *et al.*, 2012; Galià-Camps *et al.*, 2024). In addition, other analytical tools such as reverse projections can be used to predict the invasion potential of different populations under a wide array of scenarios. An example of this comes from studies on terrestrial insects and marine invertebrates (Sillero *et al.*, 2020; Chen *et al.*, 2024a). These studies inferred the origin of colonizers through reverse projections using ENM and population genomic data to understand the genetic composition of the native range. In addition, the genetic content of NIS was used to detect several native environmental matching regions (Sillero *et al.*, 2020). As sympatric genetic lineages have been shown to have different levels of invasiveness (Casso *et al.*, 2019), identifying such differential potential is key for correctly interpreting ENM. Moreover, information on the abundance of adaptive markers found in the introduced range is essential to discard geographical areas that are not relevant, and to include regions where NIS may have been introduced but have not yet been detected. Considering that the origin of NIS and their genetic content can determine invasiveness across different environments, understanding the mechanisms that underpin biological invasions is key for predicting NIS spread (McGaughran *et al.*, 2024).

The studies reviewed in Table 11.1 support the idea that ENM incorporating genomic information (e.g. population structure and connectivity) results not only in more accurate predictions of current and future range expansions but also in new insights into the mechanisms that govern the invasion process. Genomics-informed modelling thus represents a rich source of knowledge to predict potential areas of introduction of NIS worldwide and to infer their origin, which is fundamental to mitigate the effects of future NIS introductions.

11.4 Range Size and Environmental Matching Between Native and Introduced Ranges

Several studies have shown clear size differences between introduced and native ranges (e.g. Lombaert *et al.*, 2010; Rius and Shenkar, 2012;

Casso *et al.*, 2019; Hudson *et al.*, 2021, 2022). Native ranges are often restricted to a specific geographical section where climatic conditions are favourable (e.g. a section of the Mediterranean Sea), while the introduced range can involve many regions around the world with analogous climatic conditions (e.g. other world regions with Mediterranean climatic conditions). This pattern is often magnified when species occupy manmade structures such as marine infrastructures and human settlements (e.g. Airoldi *et al.*, 2015; Santana Marques *et al.*, 2020). Whenever regions with environmental matching fail to become part of the introduced range, niche unfilling occurs. This often happens when only a subset of all native genetic lineages is represented in the introduced range due to a founder effect (i.e. an initial bottleneck reduces adaptive genetic variation necessary for widespread colonization), or when the dispersal into a new area is restricted and/or suitable environments are hard to reach (Strubbe *et al.*, 2013; Poursanidis *et al.*, 2020). Thus, the size of the native and introduced ranges can be affected by several factors and genomics-informed modelling should consider different range sizes (Fig. 11.3).

When the environmental space of the introduced range matches the native one (Fig. 11.3), it is expected that NIS, if bearing pre-adapted genomic combinations to the environmental space, will spread across all potential suitable habitats as long as enough time and genetic diversity are available. Hudson *et al.* (2021) showed high levels of population genomic diversity in a very small introduced range. This suggested that if there are changes in future conditions (e.g. oceanography or shipping activity), a NIS range expansion to nearby regions may occur. In line with this, pre-adaptation to local conditions may determine their colonization success and even enhance NIS spread to similar environmental areas of the introduced range (Sillero *et al.*, 2020; Hudson *et al.*, 2021).

Studies have demonstrated niche shifts during species invasions (Liu *et al.*, 2022; Pack *et al.*, 2022). These shifts can modify the suitability of habitats across the species ranges (Fig. 11.3) and thus affect environmental matching. Future studies should explore how to incorporate these shifts into genomics-informed ENM with reverse projections from the introduced range (Sillero *et al.*, 2020). In addition, adaptive genomic variation is a key factor that is increasingly being

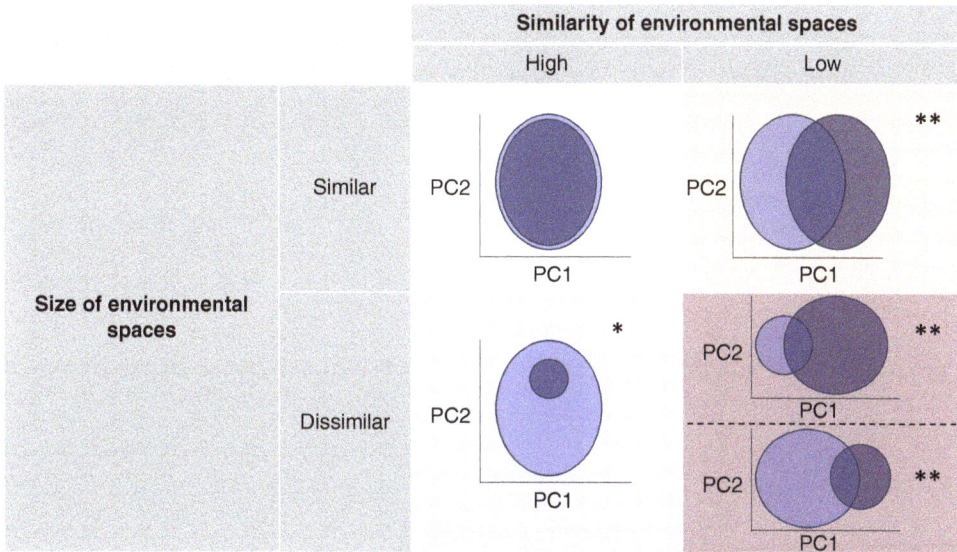

Fig. 11.3. Effects of size and similarity in environmental space of native and introduced ranges. Different environmental spaces can completely or partially overlap with the different range types. An asterisk indicates that only a subset of the native genotypes has been introduced. Two asterisks indicate the presence of a niche shift (i.e. the native range's environmental space partly differs from that of the introduced range). PC1 and PC2 indicate axes of hypothetical principal component analyses.

used to refine the overall prediction of invasion risk (Chen *et al.*, 2024a,b). ENM in combination with data on candidate loci can unravel signatures of local adaptation. This approach has been applied to forecast NIS introductions but also to understand the effects of contemporary climate change on species distributions (e.g. Marcer *et al.*, 2016; Ikeda *et al.*, 2017; Razgour *et al.*, 2019). Because both NIS introductions and changes in climate may induce range shifts (Rius *et al.*, 2014; Chown *et al.*, 2015), incorporating the role of pre-adaptation to the combined use of ENM and population genomics may be key. In addition, genomic and epigenomic variation may influence climate adaptation and invasion risk such as in a recent study of the widespread colonial ascidian *Botryllus schlosseri* (Chen *et al.*, 2024b). In this study, the authors showed that genomic and epigenomic climate maladaptations occurred in several distant regions, suggesting complementary roles in environmental response. By integrating both data sets into a genomic–epigenomic index, they found that certain populations exhibited higher adaptive potential and posed a greater risk of future invasions. This study highlighted how combining genomic and epigenomic data can improve

predictions of species' responses to climate change and enhance assessments of invasion risk.

Over the past few decades there has been an increase in research devoted to understanding the intrinsic role of both environmental matching and range size on range shifts (Sexton *et al.*, 2009; Angert *et al.*, 2020). The general principles that have emerged from this research apply to studies of NIS. but some extra particularities need to be taken into consideration. Studies of NIS often show the presence of genetic bottlenecks, admixture, gene flow, niche shifts and/or adaptation during or after introduction to novel geographical areas (Verhoeven *et al.*, 2011; Colautti and Barrett, 2013; Besnard *et al.*, 2014; Rius and Darling, 2014; Ørsted and Ørsted, 2019; Pack *et al.*, 2022). Incorporating all these mechanisms for the study of NIS across range types is thus fundamental, and future genomics-informed ENM should follow this direction.

11.5   Conclusions and Future Directions

Studies that use ENM and population genomics in isolation or in combination are growing in importance in the literature (Table 11.1), and those combining both ENM and genomic analyses are becoming essential to predict future species distributions and to infer the role of adaptation during colonization. Such studies are showing how population genomic data help refine ENM predictions and how, in turn, ENM aids researchers to conduct more efficient sample acquisition for population genomics studies. In addition, these studies are providing significant advancements for our understanding of species invasions. Together with strong genetic bottlenecks, admixture among founders, pre-adaptation, and release of pathogens and predators influence the geographical expansion of NIS (Asplen *et al.*, 2015; Ørsted and Ørsted, 2019). Thus, understanding aspects of population genomics of NIS in order to interpret ENM results is key. The studies reviewed here (Table 11.1) support the idea that a combined approach advances knowledge into key aspects of the invasion process. In addition, this approach should be more used frequently to accurately predict the distribution of NIS and inform effective management strategies. For example, such studies help unravel both the origin and the genetic composition of colonizers that are key determinants of the geographical spread and invasion success of NIS (Sillero *et al.*, 2020; Hudson *et al.*, 2021; Putra *et al.*, 2024). This is key information for biodiversity managers, especially when they need to prioritize actions to prevent NIS spread or to mitigate the negative effects of NIS.

This chapter has also described some challenges that should be explored in future research endeavours. For example, studies should generate new information on how genomics-informed ENM may take into consideration unknown environmental tolerance of admixed lineages. In addition, studies should focus on deepening our understanding of niche shifts and how these can be incorporated into ENM and population genomic analyses. Additional aspects relate to the adaptive capacity of NIS to mediate range expansions and should be incorporated into predictive models, including genomic structure and

variation, plasticity of gene expression and epigenetics (e.g. Chen *et al.*, 2024b). Another fruitful area of investigation is the use of ENM-informed adaptive genomics and adaptive genomics-informed ENMs (e.g. Jeon *et al.*, 2024). New research avenues are further opened up by health science research. For example, researchers have used integrated spatio-temporal and phylogenetic approaches to study viruses responsible for serious health issues in humans (e.g. Sun *et al.*, 2024).

Taken together, we have demonstrated that studies are increasingly using population genomics and ENM in combination (Table 11.1). This approach deepens our understanding of future shifts in habitat suitability for NIS and helps more accurately identify the origin of colonizers. In addition, genomics-informed ENM can inform biodiversity managers, particularly if certain genetic lineages have higher invasion potential than others (Casso *et al.*, 2019; Chen *et al.*, 2024b; Galià-Camps *et al.*, 2024; Putra *et al.*, 2024). Furthermore, this approach can aid management actions by making scientifically sound predictions of NIS origin and future range expansions (e.g. Sillero *et al.*, 2020; Hudson *et al.*, 2021). Surveillance efforts at range edges and/or in high-risk areas may also be required to inform and update current and future projections.

Acknowledgements

This work was supported by the research grants PID2020-118550RB (MICIU/AEI/https://doi.org/10.13039/501100011033) and PID2023-14630 7OB-C22 (MICIU/AEI/10.13039/501100011033 and FEDER, UE) of the Spanish Ministry of Science, Innovation, and Universities. MR and MP are members of the research groups SGR2021-00405 and SGR2021-01271 respectively, both funded by the Generalitat de Catalunya. Special thanks to D. Bock, Y. Chen and A. Zhan for their comments and suggestions on an earlier draft of the manuscript.

References

Airoldi, L., Turon, X., Perkol-Finkel, S. and Rius, M. (2015) Corridors for aliens but not for natives: effects of marine urban sprawl at a regional scale. *Diversity Distributions* 21, 755–768. DOI: 10.1111/ddi.12301

Alvarado-Serrano, D.F. and Knowles, L.L. (2014) Ecological niche models in phylogeographic studies: applications, advances and precautions. *Molecular Ecology Resources* 14, 233–248. DOI: 10.1111/1755-0998.12184

Angert, A.L., Bontrager, M.G. and Ågren, J. (2020) What do we really know about adaptation at range edges? *Annual Review Ecology Evolution Systematics* 51, 341–361. DOI: 10.1146/annurev-ecolsys-012120-091002

Asplen, M.K., Anfora, G., Biondi, A., Choi, D.-S., Chu, D. *et al.* (2015) Invasion biology of spotted wing Drosophila (*Drosophila suzukii*): a global perspective and future priorities. *Journal of Pest Science* 88, 469–494. DOI: 10.1007/s10340-015-0681-z

Bayliss, S.L.J., Papeş, M., Schweitzer, J.A. and Bailey, J.K. (2022) Aggregate population-level models informed by genetics predict more suitable habitat than traditional species-level model across the range of a widespread riparian tree. *PLOS ONE*, 17: e0274892. DOI: 10.1371/journal.pone.0274892

Besnard, G., Dupuy, J., Larter, M., Cuneo, P., Cooke, D. and Chikhi, L. (2014) History of the invasive African olive tree in Australia and Hawaii: evidence for sequential bottlenecks and hybridization with the Mediterranean olive. *Evolutionary Applications* 7, 195–211. DOI: 10.1111/eva.12110

Bothwell, H.M., Evans, L.M., Hersch-Green, E.I., Woolbright, S.A., Allan, G.J. and Whitham, T.G. (2021) Genetic data improves niche model discrimination and alters the direction and magnitude of climate change forecasts. *Ecological Applications* 31: e02254. DOI: 10.1002/eap.2254

Byer, N.W., Reid, B.N. and Peery, M.Z. (2020) Genetically-informed population models improve climate change vulnerability assessments. *Landscape Ecology* 35, 1215–1228. DOI: 10.1007/s10980-020-01011-x

Bystriakova, N., Ansell, S.W., Russell, S.J., Grundmann, M., Vogel, J.C. and Schneider, H. (2014) Present, past and future of the European rock fern *Asplenium fontanum*: combining distribution modelling and population genetics to study the effect of climate change on geographic range and genetic diversity. *Annals of Botany* 113, 453–465. DOI: 10.1093/aob/mct274

Carreras, C., García-Cisneros, A., Wangensteen, O.S., Ordóñez, V., Palacín, C. *et al.* (2020) East is East and West is West: population genomics and hierarchical analyses reveal genetic structure and adaptation footprints in the keystone species *Paracentrotus lividus* (Echinoidea). *Diversity and Distributions* 26, 382–398. DOI:

Casso, M., Turon, X. and Pascual, M. (2019) Single zooids, multiple loci: independent colonisations revealed by population genomics of a global invader. *Biological Invasions* 21, 3575–3592. DOI: 10.1111/ddi.13016

Chen, Y., Gao, Y., Huang, X., Li, S., Zhang, Z. and Zhan, A. (2024a) Incorporating adaptive genomic variation into predictive models for invasion risk assessment. *Environmental Science and Ecotechnology* 18: 100299. DOI: 10.1016/j.ese.2023.100299

Chen, Y., Gao, Y., Zhang, Z. and Zhan, A. (2024b) Multi-omics inform invasion risks under global climate change. *Global Change Biology* 30: e17588. DOI: 10.1111/gcb.17588

Chown, S.L., Hodgins, K.A., Griffin, P.C., Oakeshott, J.G., Byrne, M. and Hoffmann, A.A. (2015) Biological invasions, climate change and genomics. *Evolutionary Applications* 8, 23–46. DOI: 10.1111/eva.12234

Colautti, R.I. and Barrett, S.C.H. (2013) Rapid adaptation to climate facilitates range expansion of an invasive plant. *Science* 342, 364–366. DOI: doi: 10.1126/science.1242121

Corbin, J.D. and d'Antonio, C.M. (2004) Competition between native perennial and exotic annual grasses: implications for an historical invasion. *Ecology* 85, 1273–1283. DOI: 10.1890/02-0744

DeRaad, D.A., Cobos, M.E., Hofmeister, N.R., DeCicco, L.H., Venkatraman, M.X. *et al.* (2024) On the brink of explosion? Identifying the source and potential spread of introduced *Zosterops* white-eyes in North America. *Biological Invasions* 26, 1615–1639. DOI: 10.1007/s10530-024-03268-8

Elith, J., Graham, C.H., Anderson, R.P., Dudík, M., Ferrier, S. *et al.* (2006) Novel methods improve prediction of species' distributions from occurrence data. *Ecography* 29: 129–151. DOI: 10.1111/j.2006.0906-7590.04596.x

Estoup, A. and Guillemaud, T. (2010) Reconstructing routes of invasion using genetic data: why, how and so what? *Molecular Ecology* 19, 4113–4130. DOI: 10.1111/j.1365-294X.2010.04773.x

Fitzpatrick, M.C. and Keller, S.R. (2015) Ecological genomics meets community-level modelling of biodiversity: mapping the genomic landscape of current and future environmental adaptation. *Ecology Letters* 18, 1–16. DOI: 10.1111/ele.12376

Frederico, R.G., Torquato, G.V., Salvador, G.N., Andrade, A. and Rosa, G.R. (2019) Freshwater ecosystem vulnerability: is native climatic niche good enough to predict invasion events? *Aquatic Conservation: Marine and Freshwater Ecosystems* 29, 1890–1896. DOI: 10.1002/aqc.3223

Freestone, A.L., Ruiz, G.M. and Torchin, M.E. (2013) Stronger biotic resistance in tropics relative to temperate zone: effects of predation on marine invasion dynamics. *Ecology* 94, 1370–1377. DOI: 10.1890/12-1382.1

Galià-Camps, C., Enguídanos, A., Turon, X., Pascual, M. and Carreras, C. (2024) The past, the recent, and the ongoing evolutionary processes of the worldwide invasive ascidian *Styela plicata*. *Molecular Ecology* 33: e17502. DOI: 10.1111/mec.17502

Galià-Camps, C., Schell, T., Pegueroles, C., Baranski, D., Hamadou, A.B. *et al.* (2025) A worldwide genomic perspective on the invasive species *Styela plicata*: de novo genome assembly, inversion detection, and adaptation. Scientific Reports (in press).

Gotelli, N.J. and Stanton-Geddes, J. (2015) Climate change, genetic markers and species distribution modelling. *Journal of Biogeography* 42, 1577–1585. DOI: 10.1111/jbi.12562

Guisan, A. and Zimmermann, N.E. (2000) Predictive habitat distribution models in ecology. *Ecological Modelling* 135, 147–186. DOI: 10.1016/S0304-3800(00)00354-9

Hudson, J., Johannesson, K., McQuaid, C.D. and Rius, M. (2020) Secondary contacts and genetic admixture shape colonization by an amphiatlantic epibenthic invertebrate. *Evolutionary Applications* 13, 600–612. DOI: 10.1111/eva.12893

Hudson, J., Castilla, J.C., Teske, P.R., Beheregaray, L.B., Haigh, I.D. *et al.* (2021) Genomics-informed models reveal extensive stretches of coastline under threat by an ecologically dominant invasive species. *Proceedings of the National Academy of Sciences USA* 118: e2022169118. DOI: 10.1073/pnas.2022169118.

Hudson, J., Bourne, S.D., Seebens, H., Chapman, M.A. and Rius, M. (2022) The reconstruction of invasion histories with genomic data in light of differing levels of anthropogenic transport. *Philosophical Transactions of the Royal Society B: Biological Sciences* 377: 20210023. DOI: 10.1098/rstb.2021.0023

Hughes, A.C., Orr, M.C., Ma, K., Costello, M.J., Waller, J. *et al.* (2021) Sampling biases shape our view of the natural world. *Ecography* 44, 1259–1269. DOI: 10.1111/ecog.05926

Ikeda, D.H., Max, T.L., Allan, G.J., Lau, M.K., Shuster, S.M. and Whitham, T.G. (2017) Genetically informed ecological niche models improve climate change predictions. *Global Change Biology* 23, 164–176. DOI: 10.1111/gcb.13470

Jeon, J.Y., Shin, Y., Mularo, A.J., Feng, X. and DeWoody, J.A. (2024) The integration of whole-genome resequencing and ecological niche modelling to conserve profiles of local adaptation. *Diversity and Distributions* 30: e13847. DOI: 10.1111/ddi.13847

Levine, J.M. (2000) Species diversity and biological invasions: relating local process to community pattern. *Science* 288, 852–854. DOI: 10.1126/science.288.5467.852

Liu, C., Wolter, C., Xian, W. and Jeschke, J.M. (2020) Most invasive species largely conserve their climatic niche. *Proceedings of the National Academy of Sciences USA* 117, 23643–23651. DOI: 10.1073/pnas.2004289117

Liu, C., Wolter, C., Courchamp, F., Roura-Pascual, N. and Jeschke, J.M. (2022) Biological invasions reveal how niche change affects the transferability of species distribution models. *Ecology* 103: e3719. DOI: 10.1002/ecy.3719

Lombaert, E., Guillemaud, T., Cornuet, J.-M., Malausa, T., Facon, B. and Estoup, A. (2010) Bridgehead effect in the worldwide invasion of the biocontrol harlequin ladybird. *PLOS ONE*, 5: e9743. DOI: 10.1371/journal.pone.0009743

Marcer, A., Méndez-Vigo, B., Alonso-Blanco, C. and Picó, F.X. (2016) Tackling intraspecific genetic structure in distribution models better reflects species geographical range. *Ecology and Evolution* 6, 2084–2097. DOI: 10.1002/ece3.2010.

McGaughran, A., Dhami, M.K., Parvizi, E., Vaughan, A.L., Gleeson, D.M. *et al.* (2024) Genomic tools in biological invasions: current state and future frontiers. *Genome Biology and Evolution* 16: evad230. DOI: 10.1093/gbe/evad230

Melo-Merino, S.M., Reyes-Bonilla, H. and Lira-Noriega, A. (2020) Ecological niche models and species distribution models in marine environments: a literature review and spatial analysis of evidence. *Ecological Modelling* 415: 108837. DOI: 10.1016/j.ecolmodel.2019.108837

Nuñez, M.A., Pauchard, A. and Ricciardi, A. (2020) Invasion science and the global spread of SARS-CoV-2. *Trends in Ecology and Evolution* 35, 642–645. DOI: 10.1016/j.tree.2020.05.004

Ørsted, I.V. and Ørsted, M. (2019) Species distribution models of the spotted wing *Drosophila* (*Drosophila suzukii*, Diptera: Drosophilidae) in its native and invasive range reveal an ecological niche shift. *Journal of Applied Ecology* 56, 423–435. DOI: 10.1111/1365-2664.13285

Osborne, M.J., Diver, T.A. and Turner, T.F. (2013) Introduced populations as genetic reservoirs for imperiled species: a case study of the Arkansas River Shiner (*Notropis girardi*). *Conservation Genetics* 14, 637–647. DOI: 10.1007/s10592-013-0457-z

Pack, K.E., Mieszkowska, N. and Rius, M. (2022) Rapid niche shifts as drivers for the spread of a non-indigenous species under novel environmental conditions. *Diversity and Distributions* 28, 596–610. DOI: 10.1111/ddi.13471

Pascual, M., Chapuis, M.P., Mestres, F., Balanyà, J., Huey, R.B. *et al.* (2007) Introduction history of *Drosophila subobscura* in the New World: a microsatellite-based survey using ABC methods. *Molecular Ecology* 16, 3069–3083. DOI: 10.1111/j.1365-294X.2007.03336.x

Pech-May, A., Moo-Llanes, D.A., Puerto-Avila, M.B., Casas, M., Danis-Lozano, R. *et al.* (2016) Population genetics and ecological niche of invasive *Aedes albopictus* in Mexico. *Acta Tropica* 157, 30–41. DOI: 10.1016/j.actatropica.2016.01.021

Pegueroles, C., Ordóñez, V., Mestres, F. and Pascual, M. (2010) Recombination and selection in the maintenance of the adaptive value of inversions. *Journal of Evolutionary Biology* 23, 2709–2717. DOI: 10.1111/j.1420-9101.2010.02136.x

Pimentel, D., Zuniga, R. and Morrison, D. (2005) Update on the environmental and economic costs associated with alien-invasive species in the United States. *Ecological Economics* 52, 273–288. DOI: 10.1016/j.ecolecon.2004.10.002

Poursanidis, D., Kalogirou, S., Azzurro, E., Parravicini, V., Bariche, M. and Zu Dohna, H. (2020) Habitat suitability, niche unfilling and the potential spread of *Pterois miles* in the Mediterranean Sea. *Marine Pollution Bulletin* 154: 111054. DOI: 10.1016/j.marpolbul.2020.111054

Prevosti, A., Ribo, G., Serra, L., Aguade, M., Balaña, J. *et al.* (1988) Colonization of America by *Drosophila subobscura*: experiment in natural populations that supports the adaptive role of chromosomal-inversion polymorphism. *Proceedings of the National Academy of Sciences USA* 85, 5597–5600. DOI: 10.1073/pnas.85.15.5597

Putra, A.R., Hodgins, K.A. and Fournier-Level, A. (2024) Assessing the invasive potential of different source populations of ragweed (*Ambrosia artemisiifolia* L.) through genomically informed species distribution modelling. *Evolutionary Applications* 17: e13632. DOI: 10.1111/eva.13632

Razgour, O., Rebelo, H., Puechmaille, S.J., Juste, J., Ibáñez, C. *et al.* (2014) Scale-dependent effects of landscape variables on gene flow and population structure in bats. *Diversity and Distributions* 20, 1173–1185. DOI: 10.1111/ddi.12200

Razgour, O., Forester, B., Taggart, J.B., Bekaert, M., Juste, J. *et al.* (2019) Considering adaptive genetic variation in climate change vulnerability assessment reduces species range loss projections. *Proceedings of the National Academy of Sciences USA* 116, 10418–10423. DOI: 10.1073/pnas.1820663116

Rey, O., Estoup, A., Vonshak, M., Loiseau, A., Blanchet, S. *et al.* (2012) Where do adaptive shifts occur during invasion? A multidisciplinary approach to unravelling cold adaptation in a tropical ant species invading the Mediterranean area. *Ecology Letters* 15, 1266–1275. DOI: 10.1111/j.1461-0248.2012.01849.x.

Ricciardi, A., Blackburn, T.M., Carlton, J.T., Dick, J.T.A., Hulme, P.E. *et al.* (2017) Invasion science: a horizon scan of emerging challenges and opportunities. *Trends in Ecology and Evolution* 32, 464–474. DOI: 10.1016/j.tree.2017.03.007

Richardson, D.M., Py, P., Rejmánek, M., Barbour, M.G., Panetta, F.D. and West, C.J. (2000) Naturalization and invasion of alien plants: concepts and definitions. *Diversity and Distributions* 6, 93–107. DOI: 10.1046/j.1472-4642.2000.00083.x

Rius, M. and Darling, J.A. (2014) How important is intraspecific genetic admixture to the success of colonising populations? *Trends in Ecology and Evolution* 29, 233–242. DOI: 10.1016/j.tree.2014.02.003

Rius, M. and Shenkar, N. (2012) Ascidian introductions through the Suez Canal: the case study of an Indo-Pacific species. *Marine Pollution Bulletin* 64, 2060–2068. DOI: 10.1016/j.marpolbul.2012.06.029

Rius, M., Turon, X., Ordóñez, V. and Pascual, M. (2012) Tracking invasion histories in the sea: facing complex scenarios using multilocus data. *PLOS ONE*, 7: e35815. DOI: 10.1371/journal.pone.0035815

Rius, M., Clusella-Trullas, S., McQuaid, C.D., Navarro, R.A., Griffiths, C.L. *et al.* (2014) Range expansions across ecoregions: interactions of climate change, physiology and genetic diversity: range shifts across ecoregions. *Global Ecology and Biogeography* 23, 76–88. DOI: 10.1111/geb.12105

Rius, M., Bourne, S., Hornsby, H.G. and Chapman, M.A. (2015) Applications of next-generation sequencing to the study of biological invasions. *Current Zoology* 61, 488–504. DOI: 10.1093/czoolo/61.3.488

Rius, M., Teske, P.R., Manríquez, P.H., Suárez-Jiménez, R., McQuaid, C.D. and Castilla, J.C. (2017) Ecological dominance along rocky shores, with a focus on intertidal ascidians. In: Hawkins, S.J., Evans, A.J., Dale, A.C., Firth, L.B., Hughes, D.J. and Smith, I.P. (eds) *Oceanography and Marine Biology – An Annual Review, Vol.* 55. CRC Press, Boca Raton, Florida, pp. 55–85.

Santana Marques, P., Resende Manna, L., Clara Frauendorf, T., Zandonà, E., Mazzoni, R. and El-Sabaawi, R. (2020) Urbanization can increase the invasive potential of alien species. *Journal of Animal Ecology* 89, 2345–2355. DOI: 10.1111/1365-2656

Sexton, J.P., McIntyre, P.J., Angert, A.L. and Rice, K.J. (2009) Evolution and ecology of species range limits. *Annual Review Ecology Evolution Systematics* 40, 415–436. DOI: 10.1146/annurev.ecolsys.110308.120317

Sherpa, S. and Després, L. (2021) The evolutionary dynamics of biological invasions: a multi-approach perspective. *Evolutionary Applications* 14, 1463–1484. DOI: 10.1111/eva.13215

Sillero, N., Huey, R.B., Gilchrist, G., Rissler, L. and Pascual, M. (2020) Distribution modelling of an introduced species: do adaptive genetic markers affect potential range? *Proceedings of the Royal Society B: Biological Sciences* 287: 20201791. DOI: 10.1098/rspb.2020.1791

Sotka, E.E., Baumgardner, A.W., Bippus, P.M., Destombe, C., Duermit, E.A. *et al.* (2018) Combining niche shift and population genetic analyses predicts rapid phenotypic evolution during invasion. *Evolutionary Applications* 11, 781–793. DOI: 10.1111/eva.12592

Stachowicz, J.J. and Tilman, D. (2005) Species invasions and the relationships between species diversity, community saturation, and ecosystem functioning. In: Sax, D.F., Stachowicz, J.J. and Gaines, S.D. (eds) *Species Invasions: Insights into Ecology, Evolution, and Biogeography*. Sinauer Associates, Sunderland, Massachusetts, pp. 41–64.

Stohlgren, T.J., Jarnevich, C., Chong, G.W. and Evangelista, P.H. (2006) Scale and plant invasions: a theory of biotic acceptance. *Preslia* 78, 405–426.

Strubbe, D., Broennimann, O., Chiron, F. and Matthysen, E. (2013) Niche conservatism in non-native birds in Europe: niche unfilling rather than niche expansion. *Global Ecology and Biogeography* 22, 962–970. DOI: 10.1111/geb.12050

Sun, Y.-Q., Zhang, Y.-Y., Liu, M.-C., Chen, J.-J., Li, T.-T. *et al.* (2024) Mapping the distribution of Nipah virus infections: a geospatial modelling analysis. *Lancet Planetary Health* 8: e463–e475. DOI: 10.1016/S2542-5196(24)00119-0

Torrado, H., Carreras, C., Raventos, N., Macpherson, E. and Pascual, M. (2020) Individual-based population genomics reveal different drivers of adaptation in sympatric fish. *Scientific Reports* 10: 12683. DOI: 10.1038/s41598-020-69160-2

Torrado, H., Pegueroles, C., Raventos, N., Carreras, C., Macpherson, E. and Pascual, M. (2022) Genomic basis for early-life mortality in sharpsnout seabream. *Scientific Reports* 12: 17265. DOI: 10.1038/s41598-022-21597-3

Trumbo, D.R., Epstein, B., Hohenlohe, P.A., Alford, R.A., Schwarzkopf, L. and Storfer, A. (2016) Mixed population genomics support for the central marginal hypothesis across the invasive range of the cane toad (*Rhinella marina*) in Australia. *Molecular Ecology* 25, 4161–4176. DOI: 10.1111/mec.13754

Turgeon, J., Tayeh, A., Facon, B., Lombaert, E., de Clercq, P. *et al.* (2011) Experimental evidence for the phenotypic impact of admixture between wild and biocontrol Asian ladybird (*Harmonia axyridis*) involved in the European invasion. *Journal of Evolutionary Biology* 24, 1044–1052. DOI: 10.1111/j.1420-9101.2011.02234.x

Verhoeven, K.J.F., Macel, M., Wolfe, L.M. and Biere, A. (2011) Population admixture, biological invasions and the balance between local adaptation and inbreeding depression. *Proceedings of the Royal Society B: Biological Sciences* 278, 2–8. DOI: 10.1098/rspb.2010.1272

Viard, F., David, P. and Darling, J.A. (2016) Marine invasions enter the genomic era: three lessons from the past, and the way forward. *Current Zoology* 62, 629–642. DOI: 10.1093/cz/zow053

Welles, S.R. and Dlugosch, K.M. (2018) Population genomics of colonization and invasion. In: *Population Genomics*. Springer, Cham, Switzerland, pp. 655–683.

Zimmermann, N.E., Edwards, T.C., Graham, C.H., Pearman, P.B. and Svenning, J. (2010) New trends in species distribution modelling. *Ecography* 33, 985–989. DOI: 10.1111/j.1600-0587.2010.06953.x

12 Using Genomics to Inform the Management of Biological Invasions: from Sequencing Genomes to Sequencing Biomes

Dan G. Bock[1]* and Marc Rius[2,3]

[1]*School of Environment and Science, Griffith University, Nathan, Queensland, Australia;* [2]*Department of Marine Ecology, Centre for Advanced Studies of Blanes (CEAB), Spanish Research Council (CSIC), Blanes, Spain;* [3]*Department of Zoology, Centre for Ecological Genomics and Wildlife Conservation, University of Johannesburg, Auckland Park Johannesburg, South Africa*

Abstract

Over the past decade, researchers have increasingly relied on genomics to address fundamental questions in invasion science. This has fostered progress on understanding key ecological and evolutionary mechanisms that shape biological invasions. In addition, genomics has been used to inform the management of invasive species. This chapter explores these applied uses. After reviewing how molecular markers were initially implemented to inform control actions on biological invasions, it discusses how recent advances in genome sequencing are expanding the scope of managers' interventions. These applications include increasing the efficiency of chemical, biological or genetic control, and helping to prioritize management for the subset of introduced populations likely to perform well in future climates. Moving beyond the sequencing of focal species, the chapter covers how biodiversity genomics, which includes environmental DNA metabarcoding and related methods that allow entire biological communities to be surveyed, can improve both the early detection of invaders and the speed and accuracy with which mitigation measures are implemented. The chapter concludes with a discussion of current barriers to the widespread implementation of genomics for invasion management.

12.1 Introduction

For most of the early history of invasion biology research, studies focused almost exclusively on exploring basic science. By relying on invasive species as natural unplanned experiments, these studies have been contributing important breakthroughs in our understanding of fundamental ecological and evolutionary processes (Sax *et al.*, 2007). During the 1980s and 1990s, evidence of the global environmental and economic impacts of invasive species started to accumulate (Carlton and Geller, 1993; Lodge, 1993; McKinney and Lockwood, 1999). In response to these challenges and faced with a rapidly accelerating rate of new invasions (Seebens *et al.*, 2017), research progressively broadened in focus to also include management strategies (Simberloff *et al.*, 2013).

---

*Corresponding author: dan.bock@griffith.edu.au

DOI: 10.1079/9781800626263.0012

Some of the first efforts to control invasive species focused on prevention, which remains the most cost-effective management measure (Pyšek and Richardson, 2010; Simberloff *et al.*, 2013). Notable approaches include those that are based on risk assessment, such as the ones used as part of the strict biosecurity measures implemented in Australia and New Zealand (Meyerson and Reaser, 2002). The Australian Weed Risk Assessment System (Pheloung *et al.*, 1999), for example, is known to considerably reduce the rate of new biological invasions (Auld *et al.*, 2003), resulting in substantial economic benefits (Keller *et al.*, 2007). Other cost-effective interventions include those that aim to constrict introduction pathways, such as the requirement to perform mid-ocean exchange of ballast water or to clean ship hulls (Zhan *et al.*, 2015; Ricciardi and MacIsaac, 2022). Another way to control ongoing invasions is to use mitigation or even eradication measures (see review by Mack *et al.*, 2000, for examples), and studies show that these interventions – albeit much more costly and often less effective – should still be preferred over no action at all (Simberloff *et al.*, 2013).

Genetic and genomic data have been essential for guiding these management efforts (Le Roux and Wieczorek, 2009; Pyšek and Richardson, 2010; Comtet *et al.*, 2015; Stuart *et al.*, 2023). This chapter explores DNA-based approaches that are used to inform management of bioinvasions by first reviewing early efforts based on molecular markers and then covering the current use of high-throughput sequencing in this context. The chapter concludes by discussing how genomic technologies can enhance traditional invasion control measures, while also providing new options to contain the spread and impacts of invasive species.

12.2 Molecular Markers and the Management of Bioinvasions

Genetic data have been used to understand biological invasions since the early 1900s (reviewed by Bock *et al.*, 2015). At this time, analyses of chromosome numbers allowed inferences regarding the origin of invasive species or their hybrid progeny (e.g. Huskins, 1931). Subsequent studies of invasion genetics were based on the large polytene chromosomes in the salivary glands of Diptera. These unique systems enabled range-wide quantifications of chromosomal inversion frequencies using microscope analyses (e.g. Carson, 1965; Dobzhansky, 1965; reviewed by Wellenreuther and Bernatchez, 2018). In the 1960s, electrophoretic techniques were developed that allowed genetic variation to be interrogated at much finer scales, and from natural populations of any species (Harris, 1966; Lewontin and Hubby, 1966; reviewed by Allendorf, 2017). In this context, researchers in invasion biology started to investigate how invasion success could be shaped by factors such as hybridization (e.g. Brown and Burdon, 1983; reviewed by Rius and Darling, 2014) or founder effects (e.g. Warwick, 1990; Schierenbeck *et al.*, 1995; reviewed by Dlugosch and Parker, 2008). The management of biological invasions was also an important consideration from the very beginning of invasion genetics (e.g. Burdon *et al.*, 1980), and as a result, research on this topic quickly accumulated (reviewed by Le Roux and Wieczorek, 2009; Le Roux, 2021). This section reviews key inferences that were made during this time based on molecular marker data, highlighting examples from a range of invasive species, along with the management implications of these studies.

12.2.1 Species identification and the management of biological invasions

Molecular markers can fast-track taxonomic identification, even though these methods have seen a slower uptake for groups of organisms such as plants, for which classifying individuals to species level based on molecular data alone is often more challenging (Hollingsworth *et al.*, 2011). Nevertheless, molecular methods for species identification have been used with great success in many animal groups (Hebert *et al.*, 2003), for which they can complement and enhance morphology-based species diagnosis (e.g. Rius and Teske, 2013; Sheth and Thaker, 2017). They also perform substantially better than traditional taxonomy when the available biological material is damaged or fragmented, or when samples are only available for propagules that are hard to diagnose morphologically. Briski *et al.* (2011) illustrated this point using diapausing eggs of aquatic invertebrates recovered from the ballast sediment of commercial ships. In this case, the use of two mitochondrial genes (cytochrome *c* oxidase subunit I and 16S rDNA) resulted in the identification of nearly double the number of species identified using morphology-based approaches. The same study also estimated that molecular identification was over four times faster and 30% less expensive than morphological identification. These continue to be important considerations given the need for ongoing, fast and accurate species detection as part of biosecurity efforts at national and supranational levels (Armstrong and Ball, 2005; Pyšek and Richardson, 2010).

Occasionally, molecular markers have led to the identification of invasive species by chance, illustrating the value of long-term genetic monitoring. The discovery of an invasive dreissenid mussel in the Great Lakes provides a useful example. In August 1991, routine genetic surveys of invasive zebra mussels in the Erie Canal (New York) identified one individual with a distinct genotypic profile (May and Marsden, 1992). Follow-up sampling produced many more individuals with the same profile, with established populations later confirmed in Lake Ontario, Niagara River and St Lawrence River (May and Marsden, 1992). These individuals were identified as the quagga mussel, *Dreissena bugensis*. Native from Eastern Europe, *D. bugensis* is an ecosystem engineer that quickly attains large densities once introduced to new locales. Eradication is often infeasible, especially when introduced populations are well established (Karatayev and Burlakova, 2022). For this reason, ongoing management has focused on monitoring and preventing the spread of this species to new water bodies (Karatayev and Burlakova, 2022).

12.2.2 Intra-specific lineage identification and the management of biological invasions

Molecular markers also allow more fine-grained information to be accessed, including on intra-specific genetic ancestry, with important implications for the management of bioinvasions. For example, genetic ancestry can help explain why efforts for biocontrol may fail, by revealing poor pairing between an invasive species and its biocontrol agent. This occurred during initial attempts to eradicate invasive erect prickly pear, *Opuntia stricta*, in South Africa, which unknowingly relied on an incorrect lineage of the cactoblastis moth (*Cactoblastis cactorum*) biocontrol agent (Le Roux, 2021). In addition, genetic ancestry analyses can highlight whether a recently introduced population is part of a lineage known to be capable of rapid invasive spread. One of the best-known examples is the invasion of the marine green alga *Caulerpa taxifolia* in southern California. Two populations of this species were discovered in 2000 in coastal embayments between Los Angeles and San Diego (Kaiser, 2000). Genetic analyses were soon able to confirm that Californian *C. taxifolia* belonged to the same lineage that had escaped from the aquarium trade since at least the mid-1980s in the Mediterranean, spreading rapidly across thousands of hectares (Jousson *et al.*, 1998, 2000). This finding prompted rapid action by representatives of five federal and state agencies in the USA, which began eradication efforts within 6 months of initial detection (Simberloff *et al.*, 2013). This programme, which succeeded after 2 years, illustrates the utility of molecular information but also the value of rapid management responses to new incursions (Simberloff *et al.*, 2013).

12.2.3 Invasion routes, post-establishment spread and the management of biological invasions

Inferences of genetic diversity and gene flow based on molecular markers have been routinely used to reconstruct invasion routes and to reveal the nature of post-establishment spread (Estoup and Guillemaud, 2010; Cristescu, 2015). These analyses often focus on introduced species that are already widely distributed (e.g. Rius *et al.*, 2012; Bertelsmeier and Keller, 2018), and therefore eradication is usually impractical or too costly to achieve. Nevertheless, molecular data can provide management-relevant information, including on the characteristics of introduction pathways, the contribution of clonal propagation to spread (e.g. Darling and Folino-Rorem, 2009), the vectors that are accelerating secondary spread and the presence of genetic admixture (Hudson *et al.*, 2022). This information is critical for managers and allows informative decisions on quarantine methods that can be implemented. In invasive tunicates, for example, molecular data indicated that recreational boating and the exchange of contaminated aquaculture gear are important vectors of regional spread (Lacoursiere-Roussel *et al.*, 2012; reviewed by Zhan *et al.*, 2015). This information has helped the design of management policies, including those that restrict the transfer of aquaculture material or mussels between farming zones (Zhan *et al.*, 2015). Taken together, reconstructions of invasion routes using molecular markers have been key for guiding specific management actions on the vectors facilitating species introductions.

12.3 Tracking the Origin and Demography of Invasive Populations Using Genomics

In contrast to traditional molecular markers, which – at most – provide information for tens of sparsely distributed loci along the genome, high-throughput sequencing allows the genome to be surveyed end to end (Bock *et al.*, 2023). High-throughput sequencing has seen a dramatic decrease in associated research costs, which has propitiated a vast expansion of the scale at which biodiversity and invasion science studies can be conducted (Hebert *et al.*, 2018, 2025; Rius *et al.*, 2015). This sequencing technology enables inferences on the ancestry and demography of populations to be obtained for independent loci (e.g. Hudson *et al.*, 2022), thereby accounting for the confounding effect of recombination, which breaks down ancestry tracts and reshuffles genetic variation in each generation (reviewed by North *et al.*, 2021).

Particularly when paired with temporal sampling over the course of the invasion (Kim *et al.*, 2023), these methods can provide answers to important questions on the ancestry of invasive populations (e.g. What were the lineages that seeded the invasion? How are invasive populations connected by gene flow? Is the invasion the result of single or multiple introductions? Are these introductions independent – i.e. different

origins and/or introductions events – of each other?) and their demography (e.g. Did the invasion coincide with important reductions in effective population size? Are such reductions in population size more pronounced at range edges?). Addressing these questions can provide key information to assist stakeholders tasked with managing bioinvasions. Such information can include the identification of genetically distinct units that can be targeted for eradication, an outcome that has been achieved by sequencing molecular markers (e.g. Robertson and Gemmell, 2004; Rollins *et al.*, 2009, 2011) and by using high-throughput sequencing (e.g. Sjodin *et al.*, 2020). For example, Osmond and Coop (2024) described a method for reconstructing the origin and demography of populations that relies on locus-specific reconstructions of genealogy across the genome. Simulations, as well as empirical data consisting of over 1000 *Arabidopsis thaliana* genomes, have highlighted the immense amount of information that can be mined from recombining genomes. This includes inferring the location of historical ancestors for contemporary populations, as well as detecting changes in dispersal rates over time, looking back hundreds of generations.

12.4 Identifying Loci under Selection Using Genomics

One of the most exciting applications of genomic data in the context of bioinvasions is the identification of loci that have been targets of natural selection over the course of the invasion. This can be achieved using a combination of top-down (also known as forward) and bottom-up (also known as reverse) genetics (reviewed by Bock *et al.*, 2015). Forward genetics starts with knowledge of adaptive traits and attempts to identify the causal genetic polymorphisms using approaches such as quantitative trait locus (QTL) mapping or genome-wide association studies (GWAS). By contrast, reverse genetics relies on genome-wide screens of genetic variants to detect the signature of natural selection and therefore does not need a priori information on the phenotypic changes that may have contributed to the success of an invasion. With a list of candidate adaptive genes in hand, inferences can be made on the traits that selection has probably been acting on. In the context of bioinvasion management, top-down and bottom-up genetic analyses can provide critical information on the mechanisms that allow invasive populations to escape efforts for chemical or biological control, as well as the speed with which evolution can result in the emergence of resistant populations (e.g. Alves *et al.*, 2019; Kreiner *et al.*, 2022). Moreover, knowledge of adaptive loci can assist with the design of new methods for invasion control and can also improve the predictive accuracy of invasive species distribution models. This section reviews these applications.

12.4.1 Loci under selection and the genetic basis of resistance to chemical or biological control

Studies of pest populations exposed to chemical control agents such as insecticides (Bass *et al.*, 2015), herbicides (Powles and Yu, 2010; Délye *et al.*, 2013), fungicides (Lucas *et al.*, 2015, or rodenticides (Berny, 2011) have provided some of the most in-depth information on the genetics of adaptation. This is because, in these systems, knowledge of the proteins targeted by chemical control agents can speed up the identification of causal genes (e.g. Ffrench-Constant *et al.*, 2004). Indeed, pesticide resistance research has contributed influential discoveries that have shaped our understanding of adaptation (Kreiner *et al.*, 2018; Baucom, 2019). These contributions have been recognized since the Evolutionary Synthesis in the 1930s–1950s (Dobzhansky, 1937). In his seminal volume *Genetics and the Origin of Species*, for instance, Theodosius Dobzhansky referred to insect pest resistance as the 'best proof of the effectiveness of natural selection yet obtained' (Dobzhansky, 1937; see also Baucom *et al.*, 2021). Studies that followed since have tackled questions at the forefront of evolutionary biology (Kreiner *et al.*, 2018; Baucom, 2019), including the repeatability of adaptation (Baucom, 2016), the costs of adaptation (Vila-Aiub *et al.*, 2009), the role of fluctuating selection (Lagator *et al.*, 2014) and the importance of hybridization (Song *et al.*, 2011).

The main focus of pesticide resistance evolution research, however, is applied science (Ravet *et al.*, 2018; Baucom, 2019). Specifically,

studies in this field aim to diagnose the scale of resistance and understand the underlying mechanisms, so that strategies can be deployed to prevent or delay costly bioinvasions, such as those impacting agroecosystems (Ravet *et al.*, 2018; Baucom, 2019). The urgency of this research is unquestionable, given the massive loss in yield caused by agricultural pests each year (e.g. Pimentel *et al.*, 2000; Gianessi and Reigner, 2007) and the need to provide food security for a rapidly growing human population (McCouch *et al.*, 2013).

Genetic studies of pesticide resistance began with insect species in the 1940s (Dickson, 1941), with resistance to insecticides starting to be viewed as a growing environmental and economic problem by 1957 (Newman, 1957). In the years leading up to the 1980s, reports of resistance started to accumulate for weedy plants (Harper, 1956; Hilton, 1957; Baker, 1974), fungal pathogens (Delp, 1980) and rodent species (reviewed by Berny, 2011), motivating accelerated genetic research. These early studies were performed using experimental crosses to evaluate the heritability of resistance (e.g. Holliday and Putwain, 1980) or to identify the genomic location of resistance loci (e.g. Dickson, 1941). Availability of genetic and genomic tools in recent years has accelerated our understanding of the role of genes and mutations that underpin the evolution of resistance (e.g. Kreiner *et al.*, 2022; North *et al.*, 2024). In turn, these developments are revolutionizing the management of pest species in two important directions (Ravet *et al.*, 2018). First, they are enabling molecular-based surveys of resistance for pest populations, which are considerably faster than traditional resistance screens. This is important because it allows the rapid deployment of effective control methods. For instance, herbicides can be selected with knowledge of the resistance-associated genetic variants that are segregating in a target weedy population (Ravet *et al.*, 2018). Second, genomics and knowledge of resistance genes can allow novel technologies to be deployed for pest control, such as gene-drive technologies (see also section 12.4.2 below), which have so far been successfully used as proof-of-principle to reverse the evolution of pesticide resistance under laboratory settings (e.g. Kaduskar *et al.*, 2022).

Genomic technologies have also illuminated the source of resistance evolution for invasive populations spreading in natural environments. A well-known example is adaptation to biocontrol in European rabbit populations in Australia. During the 1950s, the myxoma virus was released as a means of controlling invasive rabbits, which had substantial negative effects on agriculture, forestry and the natural environment (Fenner and Ratcliffe, 1965). After the initial release, the virus spread rapidly, leading to the collapse of host populations across the continent (Fenner and Ratcliffe, 1965). Subsequent years, however, saw the emergence of attenuated virus strains that benefited from increased transmission, as well as selection for increased resistance in rabbit populations (Marshall and Fenner, 1958; Kerr, 2012). Genomic technologies have been essential

for understanding the dynamics of host–parasite coevolution in this system and have helped characterize the polygenic architecture of resistance in the host species, while also revealing that similar genetic loci have been involved in resistance evolution for isolated rabbit populations in France and the UK (Alves *et al.*, 2019).

12.4.2 Genetic biocontrol: new options for managing bioinvasions

Genetic biocontrol refers to the implementation of a set of technologies with or without genetic engineering that alter the genetic composition of invasive populations and ultimately trigger the population collapse (Fig. 12.1; see Teem *et al.*, 2020, for an in-depth review). Despite

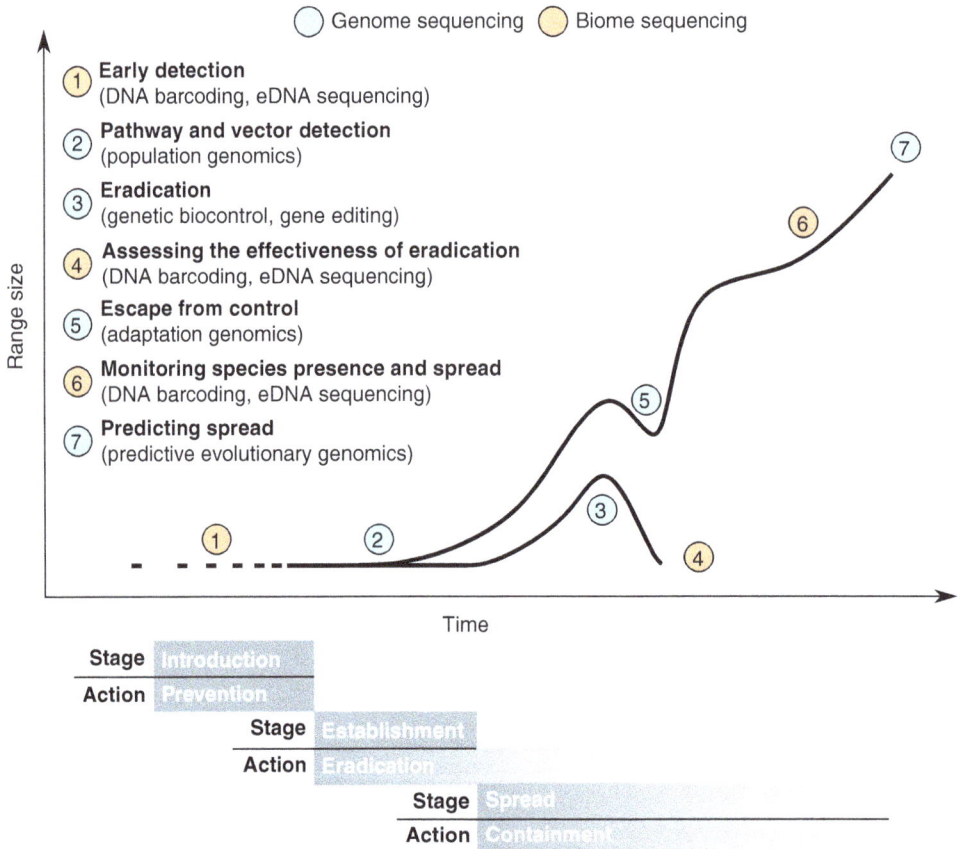

Fig. 12.1. Range size of a hypothetical invasive species over time, illustrating how genome sequencing and biome sequencing can assist efforts to prevent or mitigate invasive spread. The lower section highlights the different stages of the invasion and corresponding management actions applicable at each stage, with the grey colour gradient indicating decreasing effectiveness of management with time. eDNA, environmental DNA. Modified from Bell *et al.* (2024). Figure used with permission.

some limitations (see discussion below on risk of transgene escape), these methods allow invasive populations to be controlled without the release of chemical agents in the environment. This is important because chronic exposure to pesticides has been shown to impact human health (Frank, 2024) and to directly contribute to landscape-scale declines of wild insect populations, which are vitally important for pollination, ecosystem health and key economic activities (Goulson, 2019; Guzman et al., 2024).

The first attempts of genetic biocontrol date from the 1930s and 1950s and involved the sustained release of large numbers of sterile insects into wild insect pest populations (Alphey and Bonsall, 2018). Known as the sterile insect technique, this approach relied on interspecific crosses or on ionizing radiation to cause sterility without otherwise severely affecting the performance of released individuals (Teem et al., 2020). Ideally, a single sex is released, such that sterile individuals will mate exclusively with resident wild individuals (Rendón et al., 2004). This approach has been used with great success to manage invasions of major agricultural insect pests, such as the screw-worm fly in the USA (Knipling, 1960).

Other approaches for genetic biocontrol do not change any aspects of the genome but instead rely on natural segregating deleterious variants. A well-known example is the Trojan female technique (Gemmell et al., 2013). This approach relies on the evolutionary effect known as 'mother's curse', which involves the accumulation of deleterious mutations in the mitochondrial genomes of males (Gemmell et al., 2004). Because mitochondria are predominantly maternally inherited, mitochondrial DNA mutations that impact only male fitness, such as those involving male fertility and male-specific patterns of ageing (Gemmell et al., 2004; Camus et al., 2012), are not exposed to natural selection and can therefore accumulate in populations. From the perspective of bioinvasion management, simulation studies have shown that releases of female carriers of such mutations (i.e. 'Trojan females') can represent an effective means of population control (Gemmell et al., 2013). A persisting hurdle in the application of the Trojan female technique for invasion management, however, is the identification of mitochondrial genome mutations that have asymmetric fitness effects in males and females (Gemmell

et al., 2013). High-throughput sequencing, which enables mitogenome-wide screens for mutations that impact fitness (Vasta et al., 2009), is therefore critical for ensuring the utility of this method for invasion control.

Recent research on genetic biocontrol has increasingly relied on clustered regularly interspaced short palindromic repeats (CRISPR)-based gene-editing systems (Teem et al., 2020; Harvey-Samuel et al., 2017). These systems can introduce targeted modifications into the genomes of invasive species to control their populations (Fig. 12.1; approach #3). Depending on the design, gene edits have the potential to induce sterility, skew sex ratios by disrupting male or female development, impair essential metabolic pathways to cause mortality, or alter traits that contribute to invasion success or disease transmission (Harvey-Samuel et al., 2017). Advances in high-throughput sequencing now allow researchers to rapidly map the genomes of target species, enabling precise gene-editing interventions. Depending on the mode of propagation, biocontrol strategies based on gene editing can be classified as self-limiting or self-sustaining (Harvey-Samuel et al., 2017). Self-limiting strategies typically do not make use of biased inheritance and therefore remain effective for only a limited number of generations. They can also be reversed if no further releases of gene-edited individuals occur (Harvey-Samuel et al., 2017; Noble et al., 2019). In contrast, self-sustaining strategies involve gene-drive systems that can spread genetic modifications throughout the population without ongoing intervention due to biased inheritance mechanisms.

Gene-editing systems designed to suppress invasive species have so far been tested primarily through in silico modelling and confined laboratory experiments (e.g. Manser et al., 2019; Gierus et al., 2022; Kaduskar et al., 2022). Field trials, ongoing for genetically engineered mosquitoes since 2009 (Enserink, 2010), require extensive planning, public consultation and regulatory approvals. This is because the potential for unintended consequences could be high, especially for self-sustaining gene-drive systems, which can spread rapidly beyond the target population (Esvelt and Gemmell, 2017). In extreme cases, such escape events could result in the global extinction of the targeted species – including of populations within the native range – or affect non-target species if hybridization enables transgene escape (Teem *et al.*, 2020). Moreover, because gene-editing approaches can lead to the rapid collapse of invasive populations, thorough ecological studies are essential to anticipate and mitigate undesired outcomes, such as population explosions of previously rare species (Caut *et al.*, 2009).

12.4.3 Predictive evolutionary genomics: forecasting the future evolutionary success of invasive species

While some of the approaches highlighted above look at evolution that occurred at some point in the past (e.g. the emergence of an allele that confers pesticide resistance or that enhances dispersal ability), it is also possible to leverage genomic data to predict how populations will perform in the future. Recent years have seen increased use of these methods, often with the goal of enhancing the climate resilience of crops, or of understanding the extent to which natural populations will be impacted by future climate change (e.g. Fitzpatrick and Keller, 2015; Bay *et al.*, 2017; Capblancq *et al.*, 2020; Waldvogel *et al.*, 2020; Layton *et al.*, 2024). The most widely used approach in this category is known as genomic offset (GO; Fitzpatrick and Keller, 2015). Briefly, GO methods estimate the genetic change required to maintain local adaptation in the future. Therefore, given genome-wide marker data and estimates for a current and a future set of environmental conditions, the extent of genomic maladaptation can be quantified (Capblancq *et al.*, 2020).

GOs offer information that was previously inaccessible, but it is important to note that this approach, like many others, has its own limitations (Capblancq *et al.*, 2020). The most notable limitation is that GOs simplify the relationships between genotype and environment. For example, they assume that alleles associated with current environmental conditions provide a selective advantage, without measuring adaptive phenotypes or fitness (Capblancq *et al.*, 2020; but see Exposito-Alonso *et al.*, 2019, for an alternative approach that does model fitness). In addition, sources of adaptive variants that are not represented in sampled populations, such as new mutations or gene flow, are currently not considered in GOs (Capblancq *et al.*, 2020).

Despite these limitations, GOs and related approaches in predictive evolutionary genomics have tremendous relevance for managing bioinvasions. Indeed, these methods are starting to be validated in the context of predicting invasion risk (e.g. Camus *et al.*, 2024; Chen *et al.*, 2024). Therefore, they could help focus resources that are deployed for the management of ongoing invasions by identifying populations that have the greatest evolutionary potential under a set of future environments (Fig. 12.1; approach #7). Conversely, they may help reveal populations that – barring new introductions or post-introduction gene flow – are likely to represent evolutionary sinks (i.e. populations with limited adaptation potential that are unlikely to persist under future environments). Thus, the implementation of these methods could represent a genomic alternative to recent evolution-informed management actions that were proposed based on trait data (Shine and Baeckens, 2023).

12.5 Biodiversity Genomics: Sequencing Biomes to Manage Invasions

Invasion scientists and biodiversity managers can now leverage genomic technologies at unprecedented scales, extending far beyond single species analyses to include entire biological communities. Advancements in biodiversity science, building on foundational methods that analyse short genomic segments (Hebert *et al.*, 2003), now enable high-accuracy identification and detection of invasive species. DNA barcoding – the identification of species based on a short stretch of DNA – has been a cornerstone of biodiversity research for decades (Hebert *et al.*, 2003; Sheth and Thaker, 2017). This approach has enabled significant research progress, from describing new species to detecting cryptic introduced species (e.g. Hebert *et al.*, 2004; Armstrong and Ball, 2005; Rius and Teske, 2013). In parallel, the number of studies focusing on the detection of environmental DNA (eDNA; DNA shed by organisms into their environment) from a wide array of samples (e.g. soil, water) has surged in recent years (Bohmann *et al.*, 2014). These methods allow invasive species to be identified without the need for direct sample collection or visual species confirmation in the field (Bell *et al.*, 2024).

Growing research interest in eDNA has driven the optimization of several molecular techniques, including traditional, quantitative and digital droplet polymerase chain reaction (PCR) assays targeting both DNA and RNA (Nathan *et al.*, 2014). These methods enable the detection of one or a limited number of species (Collins *et al.*, 2018), including invasive species (Wood *et al.*, 2019). In contrast, a high-throughput sequencing technique known as eDNA metabarcoding has emerged as an effective approach for detecting entire communities of species from a single sample (Comtet *et al.*, 2015; Deiner *et al.*, 2017; Holman *et al.*, 2021; Keck *et al.*, 2022). Beyond species detection, eDNA metabarcoding offers insights into biodiversity across taxonomic levels – from microbes to large metazoans – as well as community dynamics over time and space.

eDNA metabarcoding has several applications for managing invasive species (Fig. 12.2) (Viard and Comtet, 2015). One key use is the detection of invasive species before they are observed through traditional surveys (Fig. 12.1, approach #1; Bell *et al.*, 2024). For example, eDNA metabarcoding has identified newly introduced species even in well-studied regions where frequent and detailed biodiversity studies have historically been undertaken (Holman *et al.*, 2019). As noted above, early detection of invasive species is particularly important, as it is during this stage that management measures are most likely to be effective (Reaser *et al.*, 2020).

Similarly, targeted eDNA metabarcoding can help with the early detection of specific invasive taxa (e.g. Wu *et al.*, 2023), enabling assessments of the effectiveness of management practices, such as eradication. This approach thus complements traditional or genomics-enabled invasive species removal (Fig. 12.1, approach #4; Bell *et al.*, 2024).

eDNA metabarcoding is particularly useful for detecting community changes over time due to the presence of introduced species (Fig. 12.2) (Pukk *et al.*, 2021). While this research area is still under development, it has the potential to provide valuable information, including on ecological mechanisms driving community shifts during invasions. In addition, comparing current and past community states provides key information for managers, who can find the most effective targeted strategies to mitigate the effects of invasive species on native species.

The use of eDNA to detect of eDNA to invasive species does not come without methodological challenges (Larson *et al.*, 2020; Darling and Mahon, 2011). For example, the decay of eDNA and the effects of eDNA dispersal (Collins *et al.*, 2018; Holman *et al.*, 2022) need to be taken into consideration when eDNA results are interpreted. In addition, the availability of reliable reference DNA databases and accessible bioinformatic pipelines (Darling *et al.*, 2017) can be a major challenge for implementing eDNA-based surveillance methods for the rapid detection of invasive species (Hatzenbuhler *et al.*, 2017). Finally, the standardization of laboratory methods remains a challenge (Bowers *et al.*,

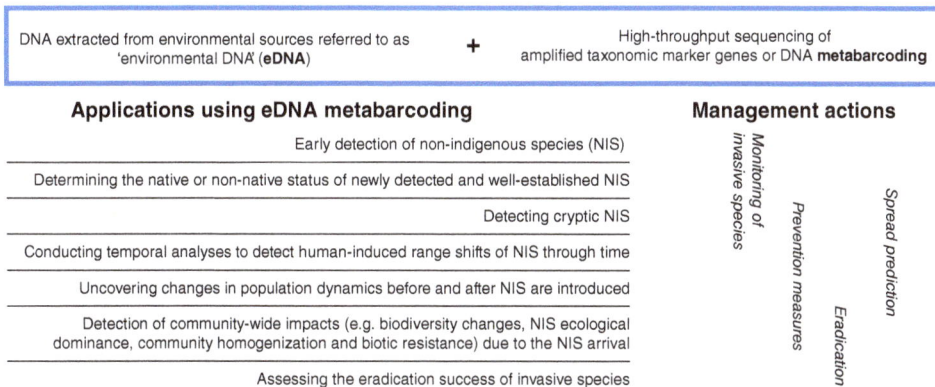

Fig. 12.2. Applications and management actions associated with environmental DNA metabarcoding.

2021). Despite all these difficulties, ongoing research is expected to improve methodologies and minimize roadblocks in the near future.

12.6 Conclusions

Genomic technologies have quickly become essential for modern invasion biology research, helping to generate novel insights into the ecology, genetics and evolution of invasive species (Bock *et al.*, 2015; North *et al.*, 2021; McGaughran *et al.*, 2024; Hodgins *et al.*, 2025). In addition, genomic tools have significantly improved the information available for managing biological invasions (Figs 12.1 and 12.2). These practical applications have seen tremendous development over the past two decades and now include the sequencing of entire genomes and biomes. The sequencing of genomes provides a wealth of unprecedented information that can help us locate the source(s) of bioinvasions (Osmond and Coop, 2024), identify regions of the genome that allow invasive species to evade chemical or biological control (Alves *et al.*, 2019; Kreiner *et al.*, 2022) and pinpoint genetic variants that can be targeted to induce the rapid collapse of invasive populations (Manser *et al.*, 2019; Kaduskar *et al.*, 2022). The sequencing of biomes can help us detect new incursions of invasive species (Larson *et al.*, 2020), assess the effectiveness of eradication efforts (Bell *et al.*, 2024), track the movement of the leading edge of ongoing invasions (Lodge, 2024) and monitor changes in species composition that follow the introduction of invasive species and other sources of anthropogenic disturbance (Pukk *et al.*, 2021).

In spite of the many advantages of using genomic tools to assist the management of invasions, most of these methods still require optimization and testing. For example, genomic forecasts of the performance of invasive populations under future climates will benefit from complementary experiments that can evaluate the effects of candidate adaptive variants on fitness, and from more sophisticated analytics that can account for *de novo* mutations and gene flow among populations (Capblancq *et al.*, 2020). Likewise, the deployment of gene-drive technologies for the management of invasions will require gene-drive systems that are only active in specific populations (Esvelt *et al.*, 2014), or that at least will not spread indefinitely in the wild (Noble *et al.*, 2019), as well as rigorous pre-deployment validation under controlled settings (Esvelt and Gemmell, 2017). In the arena of eDNA-based biomonitoring, there is an urgent need to expand and validate DNA barcode reference libraries (e.g. Blagoev *et al.*, 2016), and to verify the sensitivity and potential sources of experimental error (Darling and Mahon, 2011). Lastly, even when all technology-related hurdles are overcome, the successful implementation of any genomics-informed method for the control of invasive species will depend on effective collaboration between researchers and biodiversity managers, as well as appropriate consultation with policy makers and the public at large. Once all these conditions are met, genomics will be poised to reach its full potential in informing bioinvasion management.

References

Allendorf, F.W. (2017) Genetics and the conservation of natural populations: allozymes to genomes. *Molecular Ecology* 26, 420–430. DOI: 10.1111/mec.13948

Alphey, N. and Bonsall, M.B. (2018) Genetics-based methods for agricultural insect pest management. *Agricultural and Forest Entomology* 20, 131–140. DOI: 10.1111/afe.12241

Alves, J.M., Carneiro, M., Cheng, J.Y., Lemos de Matos, A., Rahman, M.M. *et al.* (2019) Parallel adaptation of rabbit populations to myxoma virus. *Science* 363, 1319–1326. DOI: 10.1126/science.aau7285

Armstrong, K.F. and Ball, S.L. (2005) DNA barcodes for biosecurity: invasive species identification. *Philosophical Transactions of the Royal Society B: Biological Sciences* 360, 1813–1823. DOI: 10.1098/rstb.2005.1713

Auld, B., Morita, H., Nishida, T., Ito, M. and Michael, P. (2003) Shared exotica: plant invasions of Japan and south eastern Australia. *Cunninghamia* 8, 147–152.

Baker, H.G. (1974) The evolution of weeds. *Annual Review of Ecology and Systematics* 5, 1–24.

Bass, C., Denholm, I., Williamson, M.S. and Nauen, R. (2015) The global status of insect resistance to neonicotinoid insecticides. *Pesticide Biochemistry and Physiology* 121, 78–87. DOI: 10.1016/j.pestbp.2015.04.004

Baucom, R.S. (2016) The remarkable repeated evolution of herbicide resistance. *American Journal of Botany* 103, 181–183. DOI: 10.3732/ajb.1500510

Baucom, R.S. (2019) Evolutionary and ecological insights from herbicide-resistant weeds: what have we learned about plant adaptation, and what is left to uncover? *New Phytologist* 223, 68–82. DOI: 10.1111/nph.15723

Baucom, R.S., Iriart, V., Kreiner, J.M. and Yakimowski, S. (2021) Resistance evolution, from genetic mechanism to ecological context. *Molecular Ecology* 30, 5299–5302. DOI: 10.1111/mec.16224

Bay, R.A., Rose, N., Barrett, R., Bernatchez, L., Ghalambor, C.K. *et al.* (2017) Predicting responses to contemporary environmental change using evolutionary response architectures. *American Naturalist* 189, 463–473. DOI: 10.1086/691233

Bell, K.L., Campos, M., Hoffmann, B.D., Encinas-Viso, F., Hunter, G.C. and Webber, B.L. (2024) Environmental DNA methods for biosecurity and invasion biology in terrestrial ecosystems: progress, pitfalls, and prospects. *Science of the Total Environment* 926: 171810. DOI: 10.1016/j.scitotenv.2024.171810

Berny, P. (2011) Challenges of anticoagulant rodenticides: resistance and ecotoxicology. In: Stoytcheva, M. (ed.) *Pesticides in the Modern World – Pests Control and Pesticides Exposure and Toxicity Assessment*. IntechOpen, London, pp. 441–468. Available at: www.intechopen.com/chapters/20791 (accessed 24 July 2025).

Bertelsmeier, C. and Keller, L. (2018) Bridgehead effects and role of adaptive evolution in invasive populations. *Trends in Ecology and Evolution* 14, 527–534. DOI: 10.1016/j.tree.2018.04.014

Blagoev, G.A., Dewaard, J.R., Ratnasingham, S., deWaard, S.L., Lu, L. *et al.* (2016) Untangling taxonomy: a DNA barcode reference library for Canadian spiders. *Molecular Ecology Resources* 16, 325–341. DOI: 10.1111/1755-0998.12444

Bock, D.G., Caseys, C., Cousens, R.D., Hahn, M.A., Heredia, S.M. *et al.* (2015) What we still don't know about invasion genetics. *Molecular Ecology* 24, 2277–2297. DOI: 10.1111/mec.13032

Bock, D.G., Liu, J., Novikova, P. and Rieseberg, L.H. (2023) Long-read sequencing in ecology and evolution: understanding how complex genetic and epigenetic variants shape biodiversity. *Molecular Ecology* 32, 1229–1235. DOI: 10.1111/mec.16884

Bohmann, K., Evans, A., Gilbert, M.T., Carvalho, G.R., Creer, S. *et al.* (2014) Environmental DNA for wildlife biology and biodiversity monitoring. *Trends in Ecology and Evolution* 29, 358–367. DOI: 10.1016/j.tree.2014.04.003

Bowers, H.A., Pochon, X., von Ammon, U., Gemmell, N., Stanton, J.A. *et al.* (2021) Towards the optimization of eDNA/eRNA sampling technologies for marine biosecurity surveillance. *Water* 13: 1113. DOI: 10.3390/w13081113

Briski, E., Cristescu, M.E., Bailey, S.A. and MacIsaac, H.J. (2011) Use of DNA barcoding to detect invertebrate invasive species from diapausing eggs. *Biological Invasions* 13, 1325–1340. DOI: 10.1007/s10530-010-9892-7

Brown, A.H.D. and Burdon, J.J. (1983) Multilocus diversity in an outbreeding weed, *Echium piantagineum* L. *Australian Journal of Biological Sciences* 36, 503–510. DOI: 10.1071/BI9830503

Burdon, J.J., Marshall, D.R. and Groves, R.H. (1980) Isozyme variation in *Chondrilla juncea* L. in Australia. *Australian Journal of Botany* 28, 193–198. DOI: 10.1071/BT9800193

Camus, L., Gautier, M. and Boitard, S. (2024) Predicting species invasiveness with genomic data: is genomic offset related to establishment probability? *Evolutionary Applications* 17: e13709. DOI: 10.1111/eva.13709

Camus, M.F., Clancy, D.J. and Dowling, D.K. (2012) Mitochondria, maternal inheritance, and male aging. *Current Biology* 22, 1717–1721. DOI: 10.1016/j.cub.2012.07.018

Capblancq, T., Fitzpatrick, M.C., Bay, R.A., Exposito-Alonso, M. and Keller, S.R. (2020) Genomic prediction of (mal)adaptation across current and future climatic landscapes. *Annual Review of Ecology, Evolution, and Systematics* 51, 245–569. DOI: 10.1146/annurev-ecolsys-020720-042553

Carlton, J.T. and Geller, J.B. (1993) Ecological roulette: the global transport of nonindigenous marine organisms. *Science* 261, 78–82. DOI: 10.1126/science.261.5117.78

Carson, H.L. (1965) Chromosomal morphism in geographically widespread species of *Drosophila*. In: Baker, H.G. and Stebbins, G.L. (eds) *The Genetics of Colonizing Species*. Academic Press, New York, pp. 503–531.

Caut, S., Angulo, E. and Courchamp, F. (2009) Avoiding surprise effects on Surprise Island: alien species control in a multitrophic level perspective. *Biological Invasions* 11, 1689–1703. DOI: 10.1007/s10530-008-9397-9

Chen, Y., Gao, Y., Zhang, Z. and Zhan, A. (2024) Multi-omics inform invasion risks under global climate change. *Global Change Biology* 30: e17588. DOI: 10.1111/gcb.17588

Collins, R.A., Wangensteen, O.S., O'Gorman, E.J., Mariani, S., Sims, D.W. and Genner, M.J. (2018) Persistence of environmental DNA in marine systems. *Communications Biology* 1: 185. DOI: 10.1038/s42003-018-0192-6

Comtet, T., Sandionigi, A., Viard, F. and Casiraghi, M., (2015) DNA (meta) barcoding of biological invasions: a powerful tool to elucidate invasion processes and help managing aliens. *Biological Invasions* 17, 905–922. DOI: 10.1007/s10530-015-0854-y

Cristescu, M.E. (2015) Genetic reconstructions of invasion history. *Molecular Ecology* 24, 2212–2225. DOI: 10.1111/mec.13117

Darling, J.A. and Folino-Rorem, N.C. (2009) Genetic analysis across different spatial scales reveals multiple dispersal mechanisms for the invasive hydrozoan *Cordylophora* in the Great Lakes. *Molecular Ecology* 18, 4827–4840. DOI: 10.1111/j.1365-294X.2009.04405.x

Darling, J.A. and Mahon, A.R. (2011) From molecules to management: adopting DNA-based methods for monitoring biological invasions in aquatic environments. *Environmental Research* 111, 978–988. DOI: 10.1016/j.envres.2011.02.001

Darling, J.A., Galil, B.S., Carvalho, G.R., Rius, M., Viard, F. and Piraino, S. (2017) Recommendations for developing and applying genetic tools to assess and manage biological invasions in marine ecosystems. *Marine Policy* 85, 54–64. DOI: 10.1016/j.marpol.2017.08.014

Deiner, K., Bik, H.M., Mächler, E., Seymour, M., Lacoursière-Roussel, A. *et al.* (2017) Environmental DNA metabarcoding: transforming how we survey animal and plant communities. *Molecular Ecology* 26, 5872–5895. DOI: 10.1111/mec.14350

Delp, C.J. (1980) Coping with resistance to plant disease. *Plant Disease* 64, 652–657. DOI: 10.1094/PD-64-652

Délye, C., Jasieniuk, M. and Le Corre, V. (2013) Deciphering the evolution of herbicide resistance in weeds. *Trends in Genetics* 29, 649–658. DOI: 10.1016/j.tig.2013.06.001

Dickson, R.C. (1941) Inheritance of resistance to hydrocyanic acid fumigation in the California red scale. *Hilgardia* 13, 515–522. DOI:10.3733/hilg.v13n09p513

Dlugosch, K.M. and Parker, I.M. (2008) Founding events in species invasions: genetic variation, adaptive evolution, and the role of multiple introductions. *Molecular Ecology* 17, 431–449. DOI: 10.1111/j.1365-294X.2007.03538.x

Dobzhansky, T. (1937) *Genetics and the Origin of Species*. Columbia University Press, New York.

Dobzhansky, T. (1965) "Wild" and "domestic" species of Drosophila. In: Baker, H.G. and Stebbins, G.L. (eds) *The Genetics of Colonizing Species*. Academic Press, New York, pp. 533–546.

Enserink, M. (2010) GM mosquito trial alarms opponents, strains ties in Gates-funded project. *Science* 330, 1030–1031. DOI:10.1126/science.330.6007.1030

Estoup, A. and Guillemaud, T. (2010) Reconstructing routes of invasion using genetic data: why, how and so what? *Molecular Ecology* 19, 4113–4130. DOI: 10.1111/j.1365-294X.2010.04773.x

Esvelt, K.M. and Gemmell, N.J. (2017) Conservation demands safe gene drive. *PLOS Biology* 15: e2003850. DOI: doi.org/10.1371/journal.pbio.2003850

Esvelt, K.M., Smidler, A.L., Catteruccia, F. and Church, G.M. (2014) Concerning RNA-guided gene drives for the alteration of wild populations. *elife* 3: e03401. DOI: 10.7554/eLife.03401

Exposito-Alonso, M., Burbano, H.A., Bossdorf, O., Nielsen, R. and Weigel, D. (2019) Natural selection on the *Arabidopsis thaliana* genome in present and future climates. *Nature* 573, 126–129. DOI: 10.1038/s41586-019-1520-9

Fenner, F. and Ratcliffe, F.N. (1965) *Myxomatosis*. Cambridge University Press, London.

Ffrench-Constant, R.H., Daborn, P.J. and Le Goff, G. (2004) The genetics and genomics of insecticide resistance. *Trends in Genetics* 20, 163–170. DOI: 10.1016/j.tig.2004.01.003

Fitzpatrick, M.C. and Keller, S.R. (2015) Ecological genomics meets community-level modelling of biodiversity: mapping the genomic landscape of current and future environmental adaptation. *Ecology Letters* 18, 1–16. DOI: 10.1111/ele.12376

Frank, E.G. (2024) The economic impacts of ecosystem disruptions: costs from substituting biological pest control. *Science* 385: eadg0344. DOI: 10.1126/science.adg0344

Gemmell, N.J., Jalilzadeh, A., Didham, R.K., Soboleva, T. and Tompkins, D.M. (2013) The Trojan female technique: a novel, effective and humane approach for pest population control. *Proceedings of the Royal Society B: Biological Sciences* 280: 20132549. DOI: 10.1098/rspb.2013.2549

Gemmell, N.J., Metcalf, V.J. and Allendorf, F.W. (2004) Mother's curse: the effect of mtDNA on individual fitness and population viability. *Trends in Ecology and Evolution* 19, 238–244. DOI: 10.1016/j. tree.2004.02.002

Gianessi, L.P. and Reigner, N.P. (2007) The value of herbicides in U.S. crop production. *Weed Technology* 21, 559–566. DOI: 10.1614/WT-06-130.1

Gierus, L., Birand, A., Bunting, M.D., Godahewa, G.I., Piltz, S.G. *et al.* (2022) Leveraging a natural murine meiotic drive to suppress invasive populations. *Proceedings of the National Academy of Sciences USA* 119: e2213308119. DOI: 10.1073/pnas.2213308119

Goulson, D. (2019) The insect apocalypse, and why it matters. *Current Biology* 29, R967–R971. DOI: 10.1016/j.cub.2019.06.069

Guzman, L.M., Elle, E., Morandin, L.A., Cobb, N.S., Cheshire, P.R. *et al.* (2024) Impact of pesticide use on wild bee distributions across the United States. *Nature Sustainability* 7, 1324–1334. DOI: 10.1038/s41893-024-01413-8

Harper, J.L. (1956) The evolution of weeds in relation to resistance to herbicides. In: *Proceedings of the Third British Weed Control Conference*. British Weed Control Council, London, pp. 179–188.

Harris, H. (1966) C. Genetics of Man Enzyme polymorphisms in man. *Proceedings of the Royal Society B: Biological Sciences* 164, 298–310.

Harvey-Samuel, T., Ant, T. and Alphey, L. (2017) Towards the genetic control of invasive species. *Biological Invasions* 19, 1683–1703. DOI: 10.1007/s10530-017-1384-6

Hatzenbuhler, C., Kelly, J.R., Martinson, J., Okum, S. and Pilgrim, E. (2017) Sensitivity and accuracy of high-throughput metabarcoding methods for early detection of invasive fish species. *Scientific Reports* 7: 46393. DOI: 10.1038/srep46393

Hebert, P.D., Cywinska, A., Ball, S.L. and deWaard, J.R. (2003) Biological identifications through DNA barcodes. *Proceedings of the Royal Society of London. Series B: Biological Sciences* 270, 313–321. DOI: 10.1098/rspb.2002.2218

Hebert, P.D., Penton, E.H., Burns, J.M., Janzen, D.H. and Hallwachs, W. (2004) Ten species in one: DNA barcoding reveals cryptic species in the neotropical skipper butterfly *Astraptes fulgerator*. *Proceedings of the National Academy of Sciences USA* 101, 14812–14817. DOI: 10.1073/pnas.0406166101

Hebert, P.D., Braukmann, T.W., Prosser, S.W., Ratnasingham, S., deWaard, J.R. *et al.* (2018) A sequel to Sanger: amplicon sequencing that scales. *BMC Genomics* 19: 219. DOI: 10.1186/s12864-018-4611-3

Hebert, P.D., Floyd, R., Jafarpour, S. and Prosser, S.W. (2025) Barcode 100k specimens: in a single nano-pore run. *Molecular Ecology Resources* 25: e14028. DOI: 10.1111/1755-0998.14028

Hilton, H.W. (1957) Herbicide tolerant strains of weeds. In: *Hawaiian Sugar Planters Association Annual Report*. Hawaiian Sugar Planters Association, Honolulu, Hawaii, pp. 69–72.

Hodgins, K.A., Battlay, P. and Bock, D.G. (2025) The genomic secrets of invasive plants. *New Phytologist* 245, 1846–1863. DOI: 10.1111/nph.20368

Holliday, R.J. and Putwain, P.D. (1980) Evolution of herbicide resistance in *Senecio vulgaris*: variation in susceptibility to simazine between and within populations. *Journal of Applied Ecology* 17, 779–791. DOI: 10.2307/2402655

Hollingsworth, P.M., Graham, S.W. and Little, D.P. (2011) Choosing and using a plant DNA barcode. *PLOS ONE* 6: e19254. DOI: 10.1371/journal.pone.0019254

Holman, L.E., de Bruyn, M., Creer, S., Carvalho, G., Robidart, J. and Rius, M. (2019) Detection of intro-duced and resident marine species using environmental DNA metabarcoding of sediment and water. *Scientific Reports* 9: 11559. DOI: 10.1038/s41598-019-47899-7

Holman, L.E., de Bruyn, M., Creer, S., Carvalho, G., Robidart, J. and Rius, M. (2021) Animals, protists and bacteria share marine biogeographic patterns. *Nature Ecology and Evolution* 5, 738–746. DOI: 10.1038/s41559-021-01439-7

Holman, L.E., Chng, Y. and Rius, M. (2022) How does eDNA decay affect metabarcoding experiments? *Environmental DNA*, 4, 108–116. DOI: 10.1002/edn3.201

Hudson, J., Bourne, S.D., Seebens, H., Chapman, M.A. and Rius, M. (2022) The reconstruction of invasion histories with genomic data in light of differing levels of anthropogenic transport. *Philosophical Trans-actions of the Royal Society B: Biological Sciences* 377: 20210023. DOI: 10.1098/rstb.2021.0023

Huskins, C.L. (1931) Origin of *Spartina townsendii*. *Nature* 127, 781. DOI: 10.1038/127781b0

Jousson, O., Pawlowski, J., Zaninetti, L., Meinesz, A. and Boudouresque, C.F. (1998) Molecular evidence for the aquarium origin of the green alga *Caulerpa taxifolia* introduced to the Mediterranean Sea. *Marine Ecology Progress Series* 172, 275–280. DOI: 10.3354/meps172275

Jousson, O., Pawlowski, J., Zaninetti, L., Zechman, F.W., Dini, F. *et al.* (2000) Invasive alga reaches Califor-nia. *Nature* 408, 157–158. DOI: 10.1038/35041623

Kaduskar, B., Kushwah, R.B., Auradkar, A., Guichard, A., Li, M. *et al.* (2022) Reversing insecticide resistance with allelic-drive in *Drosophila melanogaster*. *Nature Communications* 13: 291. DOI: 10.1038/s41467-021-27654-1

Kaiser, J., (2000) California algae may be feared European species. *Science* 289, 222–223. DOI: 10.1126/science.289.5477.222b

Karatayev, A.Y. and Burlakova, L.E. (2022) What we know and don't know about the invasive zebra (*Dreissena polymorpha*) and quagga (*Dreissena rostriformis bugensis*) mussels. *Hydrobiologia* 852, 1029–1102. DOI: 10.1007/s10750-022-04950-5

Keck, F., Blackman, R.C., Bossart, R., Brantschen, J., Couton, M. *et al.* (2022) Meta-analysis shows both congruence and complementarity of DNA and eDNA metabarcoding to traditional methods for biological community assessment. *Molecular Ecology* 31, 1820–1835. DOI: 10.1111/mec.16364

Keller, R.P., Lodge, D.M. and Finnoff, D.C. (2007) Risk assessment for invasive species produces net bioeconomic benefits. *Proceedings of the National Academy of Sciences USA* 104, 203–207. DOI: 10.1073/pnas.0605787104

Kerr, P.J. (2012) Myxomatosis in Australia and Europe: a model for emerging infectious diseases. *Antiviral Research* 93, 387–415. DOI: 10.1016/j.antiviral.2012.01.009

Kim, A.S., Kreiner, J.M., Hernández, F., Bock, D.G., Hodgins, K.A. and Rieseberg, L.H. (2023) Temporal collections to study invasion biology. *Molecular Ecology* 32, 6729–6742. DOI: 10.1111/mec.17176

Knipling, E.F. (1960) The eradication of the screw-worm fly. *Scientific American* 203, 54–61. DOI: 10.1038/scientificamerican1060-54

Kreiner, J.M., Stinchcombe, J.R. and Wright, S.I. (2018) Population genomics of herbicide resistance: adaptation via evolutionary rescue. *Annual Review of Plant Biology* 69, 611–635. DOI: 10.1146/annurev-arplant-042817-040038

Kreiner, J.M., Latorre, S.M., Burbano, H.A., Stinchcombe, J.R., Otto, S.P. *et al.* (2022) Rapid weed adaptation and range expansion in response to agriculture over the past two centuries. *Science* 378, 1079–1085. DOI: 10.1126/science.abo7293

Lacoursiere-Roussel, A., Bock, D.G., Cristescu, M.E., Guichard, F., Girard, P. *et al.* (2012) Disentangling invasion processes in a dynamic shipping–boating network. *Molecular Ecology* 21, 4227–4241. DOI: 10.1111/j.1365-294X.2012.05702.x

Lagator, M., Colegrave, N., Neve, P. (2014) Selection history and epistatic interactions impact dynamics of adaptation to novel environmental stresses. *Proceedings of the Royal Society B: Biological Sciences* 281: 20141679. DOI: 10.1098/rspb.2014.1679

Larson, E.R., Graham, B.M., Achury, R., Coon, J.J., Daniels, M.K. *et al.* (2020) From eDNA to citizen science: emerging tools for the early detection of invasive species. *Frontiers in Ecology and the Environment* 18, 194–202. DOI: 10.1002/fee.2162

Layton, K.K.S., Brieuc, M.S.O., Castilho, R., Diaz-Arce, N., Estévez-Barcia, D. *et al.* (2024) Predicting the future of our oceans – evaluating genomic forecasting approaches in marine species. *Global Change Biology* 30: e17236. DOI: 10.1111/gcb.17236

Le Roux, J. (2021) *The Evolutionary Ecology of Invasive Species*, 1st edn. Elsevier, Amsterdam.

Le Roux, J. and Wieczorek, A.M. (2009) Molecular systematics and population genetics of biological invasions: towards a better understanding of invasive species management. *Annals of Applied Biology* 154, 1–17. DOI: 10.1111/j.1744-7348.2008.00280.x

Lewontin, R.C. and Hubby, J.L. (1966) A molecular approach to the study of genic heterozygosity in natural populations. II. Amount of variation and degree of heterozygosity in natural populations of *Drosophila pseudoobscura*. *Genetics* 54, 595–609. DOI: 10.1093/genetics/54.2.595

Lodge, D.M. (1993) Biological invasions: lessons for ecology. *Trends in Ecology and Evolution*. 8, 133–137. DOI: 10.1016/0169-5347(93)90025-K

Lodge, D.M. (2024) Lessons learned from eDNA adoption in the management of bigheaded carps in Chicago IL USA area waterways. *Environmental DNA* 6: e528. DOI: 10.1002/edn3.528

Lucas, J.A., Hawkins, N.J. and Fraaije, B.A. (2015) The evolution of fungicide resistance. *Advances in Applied Microbiology* 90, 29–92. DOI: 10.1016/bs.aambs.2014.09.001

Mack, R.N., Simberloff, D., Lonsdale, W.M., Evans, H., Clout, M. and Bazzaz, F.A. (2000) Biotic invasions: causes, epidemiology, global consequences, and control. *Ecological Applications* 10, 689–710. DOI: 10.1890/1051-0761(2000)010[0689:BICEGC]2.0.CO;2

Manser, A., Cornell, S.J., Sutter, A., Blondel, D.V., Serr, M. *et al.* (2019) Controlling invasive rodents via synthetic gene drive and the role of polyandry. *Proceedings of the Royal Society B: Biological Sciences* 286: 20190852. DOI: 10.1098/rspb.2019.0852

Marshall, I.D. and Fenner, F. (1958) Studies in the epidemiology of infectious myxomatosis of rabbits. V. Changes in the innate resistance of wild rabbits between 1951 and 1959. *Journal of Hygiene* 56, 288–302. DOI: 10.1017/s0022172400037773

May, B. and Marsden, J.E. (1992) Genetic identification and implications of another invasive species of dreissenid mussel in the Great Lakes. *Canadian Journal of Fisheries and Aquatic Sciences* 49, 1501–1506. DOI: 10.1139/f92-166

McCouch, S., Baute, G., Bradeen, J., Bramel, P., Bretting, P.K. *et al.* (2013) Feeding the future. *Nature* 499, 23–24. DOI: 10.1038/499023a

McGaughran, A., Dhami, M.K., Parvizi, E., Vaughan, A.L., Gleeson, D.M. *et al.* (2024) Genomic tools in biological invasions: current state and future frontiers. *Genome Biology and Evolution* 16: evad230. DOI: 10.1093/gbe/evad230

McKinney, M.L. and Lockwood, J.L. (1999) Biotic homogenization: a few winners replacing many losers in the next mass extinction. *Trends in Ecology and Evolution* 14, 450–453. DOI: 10.1016/S0169-5347(99)01679-1

Meyerson, L.A. and Reaser, J.K. (2002) Biosecurity: moving toward a comprehensive approach: a comprehensive approach to biosecurity is necessary to minimize the risk of harm caused by non-native organisms to agriculture, the economy, the environment, and human health. *BioScience* 52, 593–600. DOI: 10.1641/0006-3568(2002)052[0593:BMTACA]2.0.CO;2

Nathan, L.M., Simmons, M., Wegleitner, B.J., Jerde, C.L. and Mahon, A.R. (2014) Quantifying environmental DNA signals for aquatic invasive species across multiple detection platforms. *Environmental Science and Technology* 48, 12800–12806. DOI: 10.1021/es5034052

Newman, J.F. (1957) Resistance to insecticides. *Outlook on Agriculture* 1, 235–239. DOI: 10.1177/003072705700100604

Noble, C., Min, J., Olejarz, J., Buchthal, J., Chavez, A. *et al.* (2019) Daisy-chain gene drives for the alteration of local populations. *Proceedings of the National Academy of Sciences USA* 116, 8275–8282. DOI: 10.1073/pnas.1716358116

North, H.L., McGaughran, A. and Jiggins, C.D. (2021) Insights into invasive species from whole-genome resequencing. *Molecular Ecology* 30, 6289–6308. DOI: 10.1111/mec.15999

North, H.L., Fu, Z., Metz, R., Stull, M.A., Johnson, C.D. *et al.* (2024) Rapid adaptation and interspecific introgression in the North American crop pest *Helicoverpa zea*. *Molecular Biology and Evolution* 41: msae129. DOI: 10.1093/molbev/msae129

Osmond, M. and Coop, G. (2024) Estimating dispersal rates and locating genetic ancestors with genome-wide genealogies. *eLife* 13: e72177. DOI: 10.7554/eLife.72177

Pheloung, P.C., Williams, P.A. and Halloy, S.R. (1999) A weed risk assessment model for use as a biosecurity tool evaluating plant introductions. *Journal of Environmental Management* 57, 239–251. DOI: 10.1006/jema.1999.0297

Pimentel, D., Lach, L., Zuniga, R. and Morrison, D. (2000) Environmental and economic costs of nonindigenous species in the United States. *BioScience* 50, 53–65. DOI: 10.1641/0006-3568(2000)050[0053:EAECON]2.3.CO;2

Powles, S.B. and Yu, Q. (2010) Evolution in action: plants resistant to herbicides. *Annual Review of Plant Biology* 61, 317–347. DOI: 10.1146/annurev-arplant-042809-112119

Pukk, L., Kanefsky, J., Heathman, A.L., Weise, E.M., Nathan, L.R. *et al.* (2021) eDNA metabarcoding in lakes to quantify influences of landscape features and human activity on aquatic invasive species prevalence and fish community diversity. *Diversity and Distributions* 27, 2016–2031. DOI: 10.1111/ddi.13370

Pyšek, P. and Richardson, D.M. (2010) Invasive species, environmental change and management, and health. *Annual Review of Environment and Resources* 35, 25–55. DOI: 10.1146/annurev-environ-033009-095548

Ravet, K., Patterson, E.L., Krähmer, H., Hamouzová, K., Fan, L. *et al.* (2018) The power and potential of genomics in weed biology and management. *Pest Management Science* 74, 2216–2225. DOI: 10.1002/ps.5048

Reaser, J.K., Burgiel, S.W., Kirkey, J., Brantley, K.A., Veatch, S.D. and Burgos-Rodríguez, J. (2020) The early detection of and rapid response (EDRR) to invasive species: a conceptual framework and federal capacities assessment. *Biological Invasions* 22, 1–9. DOI: 10.1007/s10530-019-02156-w

Rendón, P., McInnis, D., Lance, D. and Stewart, J. (2004) Medfly (Diptera: Tephritidae) genetic sexing: large-scale field comparison of males-only and bisexual sterile fly releases in Guatemala. *Journal of Economic Entomology* 97, 1547–1553. DOI: 10.1603/0022-0493-97.5.1547

Ricciardi, A. and MacIsaac, H.J. (2022) Vector control reduces the rate of species invasion in the world's largest freshwater ecosystem. *Conservation Letters* 15: e12866. DOI: 10.1111/conl.12866

Rius, M. and Darling, J.A. (2014) How important is intraspecific genetic admixture to the success of colonising populations? *Trends in Ecology and Evolution* 29, 233–242. DOI: 10.1016/j.tree.2014.02.003

Rius, M. and Teske, P.R. (2013) Cryptic diversity in coastal Australasia: a morphological and mitonuclear genetic analysis of habitat-forming sibling species. *Zoological Journal of the Linnean Society* 168, 597–611. DOI: 10.1111/zoj.12036

Rius, M., Turon, X., Ordóñez, V. and Pascual, M. (2012) Tracking invasion histories in the sea: facing complex scenarios using multilocus data. *PLOS ONE* 7: e35815. DOI: 10.1371/journal.pone.0035815 Rius, M., Bourne, S., Hornsby, H.G. and Chapman, M.A. (2015) Applications of next-generation sequencing to the study of biological invasions. *Current Zoology* 61, 488–504. DOI: 10.1093/czoolo/61.3.488

Robertson, B.C. and Gemmell, N.J. (2004) Defining eradication units to control invasive pests. *Journal of Applied Ecology* 41, 1042–1048. DOI: 10.1111/j.0021-8901.2004.00984.x

Rollins, L.A., Woolnough, A.P., Wilton, A.N., Sinclair, R.O. and Sherwin, W.B. (2009) Invasive species can't cover their tracks: using microsatellites to assist management of starling (*Sturnus vulgaris*) populations in Western Australia. *Molecular Ecology*. 18, 1560–1573. DOI: 10.1111/j.1365-294X.2009.04132.x

Rollins, L.A., Woolnough, A.P., Sinclair, R., Mooney, N.J. and Sherwin, W.B. (2011) Mitochondrial DNA offers unique insights into invasion history of the common starling. *Molecular Ecology* 20, 2307–2317. DOI: 10.1111/j.1365-294X.2011.05101.x

Sax, D.F., Stachowicz, J.J., Brown, J.H., Bruno, J.F., Dawson, M.N. *et al.* (2007) Ecological and evolutionary insights from species invasions. *Trends in Ecology and Evolution* 22, 465–71. DOI: 10.1016/j.tree.2007.06.009

Schierenbeck, K.A., Hamrick, J.L. and Mack, R.N. (1995) Comparison of allozyme variability in a native and an introduced species of *Lonicera*. *Heredity* 75, 1–9. DOI: 10.1038/hdy.1995.97

Seebens, H., Blackburn, T.M., Dyer, E.E., Genovesi, P., Hulme, P.E. *et al.* (2017) No saturation in the accumulation of alien species worldwide. *Nature Communications* 8: 14435. DOI: 10.1038/ncomms14435

Sheth, B.P. and Thaker, V.S. (2017) DNA barcoding and traditional taxonomy: an integrated approach for biodiversity conservation. *Genome* 60, 618–628. DOI: 10.1139/gen-2015-0167

Shine, R. and Baeckens, S. (2023) Rapidly evolved traits enable new conservation tools: perspectives from the cane toad invasion of Australia. *Evolution* 77, 1744–1755. DOI: 10.1093/evolut/qpad102

Simberloff, D., Martin, J.L., Genovesi, P., Maris, V., Wardle, D.A. *et al.* (2013) Impacts of biological invasions: what's what and the way forward. *Trends in Ecology and Evolution* 28, 58–66. DOI: 10.1016/j.tree.2012.07.013

Sjodin, B.M., Irvine, R.L., Ford, A.T., Howald, G.R. and Russello, M.A. (2020) *Rattus* population genomics across the Haida Gwaii archipelago provides a framework for guiding invasive species management. *Evolutionary Applications* 13, 889–904. DOI: 10.1111/eva.12907

Song, Y., Endepols, S., Klemann, N., Richter, D., Matuschka, F.R. *et al.* (2011) Adaptive introgression of anticoagulant rodent poison resistance by hybridization between old world mice. *Current Biology* 21, 1296–1301. DOI: 10.1016/j.cub.2011.06.043

Stuart, K.C., Woolnough, A.P. and Rollins, L.A. (2023) Invasive species detection and management using genomic methods. In: Berry, O.F., Holleley, C.E. and Jarman, S.N. (eds) *Applied Environmental Genomics*. CSIRO Publishing, Clayton, Australia, pp. 286–298.

Teem, J.L., Alphey, L., Descamps, S., Edgington, M.P., Edwards, O. *et al.* (2020) Genetic biocontrol for invasive species. *Frontiers in Bioengineering and Biotechnology* 8: 452. DOI: 10.3389/fbioe.2020.00452

Vasta, V., Ng, S.B., Turner, E.H., Shendure, J. and Hahn, S.H. (2009) Next generation sequence analysis for mitochondrial disorders. *Genome Medicine* 1: 100. DOI: 10.1186/gm100

Viard, F. and Comtet, T. (2015) Applications of DNA-based methods for the study of biological invasions. In: João Canning-Clode (ed.) *Biological Invasions in Changing Ecosystems*. De Gruyter Open, Warsaw/Berlin, pp. 411–435.

Vila-Aiub, M.M., Neve, P. and Powles, S.B. (2009) Fitness costs associated with evolved herbicide resistance alleles in plants. *New Phytologist* 184, 751–767. DOI: 10.1111/j.1469-8137.2009.03055.x

Waldvogel, A.M., Feldmeyer, B., Rolshausen, G., Exposito-Alonso, M., Rellstab, C. *et al.* (2020) Evolutionary genomics can improve prediction of species' responses to climate change. *Evolution Letters* 4, 4–18. DOI: 10.1002/evl3.154

Warwick, S.I. (1990) Allozyme and life history variation in five northwardly colonizing North American weed species. *Plant Systematics and Evolution* 169, 41–54. DOI: 10.1007/BF00935983

Wellenreuther, M. and Bernatchez, L. (2018) Eco-evolutionary genomics of chromosomal inversions. *Trends in Ecology and Evolution* 33, 427–440. DOI: 10.1016/j.tree.2018.04.002

Wood, S.A., Pochon, X., Ming, W., von Ammon, U., Woods, C. *et al.* (2019) Considerations for incorporating real-time PCR assays into routine marine biosecurity surveillance programmes: a case study targeting the Mediterranean fanworm (*Sabella spallanzanii*) and club tunicate (*Styela clava*). *Genome* 62, 137–146. DOI: 10.1139/gen-2018-0021

Wu, Y., Colborne, S.F., Charron, M.R. and Heath, D.D. (2023) Development and validation of targeted environmental DNA (eDNA) metabarcoding for early detection of 69 invasive fishes and aquatic invertebrates. *Environmental DNA* 5, 73–84. DOI: 10.1002/edn3.359

Zhan, A., Briski, E., Bock, D.G., Ghabooli, S. and MacIsaac, H.J. (2015) Ascidians as models for studying invasion success. *Marine Biology* 162, 2449–2470. DOI: 10.1007/s00227-015-2734-5

13 Invasion Genomics: A Transformative Lens for Understanding and Managing Biological Invasions

Marc Rius[1,2]* and Dan G. Bock[3]

[1]*Department of Marine Ecology, Centre for Advanced Studies of Blanes (CEAB), Spanish Research Council (CSIC), Blanes, Spain; [2]Department of Zoology, Centre for Ecological Genomics and Wildlife Conservation, University of Johannesburg, Auckland Park Johannesburg, South Africa; [3]School of Environment and Science, Griffith University, Nathan, Queensland, Australia*

Abstract

This chapter synthesizes key insights from the emerging research field of invasion genomics and reflects on its transformative impact for invasion science. As illustrated in the preceding chapters of this book, genomics has facilitated critical discoveries on the roles of genomic variation, hybridization and epigenetics during biological invasions. As the field continues to expand, fostering close collaborations between invasion ecologists and geneticists is critical both for achieving progress on understanding of the mechanisms shaping bioinvasions and for informing effective management actions. In this context, this chapter – and the book as a whole – highlights the importance of interdisciplinary collaboration, data sharing and methodological innovation to overcome persistent challenges in the field. Looking ahead, ongoing innovation in DNA sequencing and bioinformatics will continue to expand the scope and resolution of genomic studies in invasion science, while also enabling more proactive and effective responses to the accelerating global spread of invasive species.

This book offers a comprehensive overview of the state-of-the-art of invasion science through the lens of genomics and provides insights into the field's future development. Organized into four parts, the preceding chapters cover methodological approaches (Chapters 2 and 3), mechanisms shaping biological invasions (Chapters 4–7), insightful case studies (Chapters 8–10) and applied uses of genomics to understand bioinvasions (Chapters 11 and 12). The international nature of this book represents a concerted effort to provide comprehensive global coverage of this research field. Each chapter identifies key knowledge gaps and outlines priorities for future research. We hope this book inspires future studies in invasion genomics, enhances our understanding of the genomic processes driving biological invasions and supports the development of more effective management strategies.

The book begins with a reflection on the emergence and development of invasion science (Chapter 1), detailing how it has grown from a subdiscipline of ecology to a vibrant, interdisciplinary research field that integrates ecology, evolution, sociology, biodiversity conservation, molecular biology and genetics/genomics. The

*Corresponding author: mrius@ceab.csic.es

DOI: 10.1079/9781800626263.0013

burgeoning use of genomics in invasion science has largely been driven by the adoption of innovative methodologies. For example, the application of high-throughput sequencing and associated computational approaches over the past two decades has revolutionized our understanding of biological invasions. Researchers are now able to tackle unresolved questions that go beyond research on invasions per se, and extend to general ecological and evolutionary principles such as demographic history, range shifts and adaptive evolution. The expanding scope of the field also highlights a crucial interdisciplinary gap: many invasion ecologists would benefit from genomics training, just as genomics researchers would benefit from a deeper ecological knowledge. Thus, collaborative work is key for modern invasion science research. Indeed, a recurring theme throughout this book has been that invasion science researchers should strive to integrate genomic and ecological data to produce cutting-edge research that continues to advance the field. Chapters 2 and 3 discuss how different genomic technologies and analytical approaches are advancing research in invasion science. Population genomics and comparative genomics are identifying how genomic variation and genome architecture contribute to invasion success. While genome-wide surveys using single-nucleotide polymorphisms have become commonplace, recent attention has turned to other complex genomic features such as structural variants and transposable elements. Chapter 4 explains these genomic features in detail, including how they may significantly influence rapid evolution and adaptive responses by modifying gene regulation, introducing novel trait variation and affecting chromosome recombination. Additionally, machine learning and deep learning are increasingly powerful tools used to reveal patterns across genomic data sets of non-indigenous species (NIS). These methods offer innovative ways to understand how NIS adapt to new environments, although their utility is tempered by current limitations in data interoperability and experimental design (see Chapter 3).

Chapters 4–7 provide a detailed view on key (epi)genetic and evolutionary mechanisms that drive invasions and how these can be studied using the genomic technologies discussed in the preceding chapters. Adaptation is a cornerstone of successful invasions. It can occur via pre-adaptation, where source populations are already suited to new environments, or post-introduction adaptation, involving rapid evolutionary change after the establishment of NIS populations. Genomic studies increasingly reveal that selection on standing genetic variation, interspecific hybridization and genetic admixture are key mechanisms facilitating adaptation in NIS, as detailed in Chapter 5. The role of hybridization, one of the first evolutionary mechanisms to be convincingly linked to invasion success, is then explored in detail in Chapter 6. In addition, this chapter focuses on transgressive segregation in invaders, and its impacts on recipient ecosystems. Finally, recent research has highlighted the role of epigenetics, particularly mechanisms such as DNA methylation and non-coding RNAs, which introduces an additional layer of complexity to our understanding of invasion success. These modifications can occur more rapidly than genetic mutations, often in response to environmental changes. This enables phenotypic plasticity and potentially alleviates any issues associated with reduced genetic diversity in introduced populations (see Chapter 7 for an in-depth discussion on this topic). However, distinguishing adaptive epigenetic responses from maladaptive ones, as well as understanding the relative contribution of genetic versus epigenetic variation to invasion success, remain ongoing challenges.

Our understanding of evolutionary and (epi)genetic mechanisms of invasions has advanced, to a large extent due to the use of model organisms in invasion science that are often genome-enabled species. Chapters 8–10 of this book detail genomic studies in three such well-known study systems: the common ragweed (*Ambrosia artemisiifolia*; Chapter 8), Japanese knotweeds (*Reynoutria japonica*, *R. sachalinensis* and *R. × bohemica*; Chapter 9) and estuarine copepods (*Eurytemora affinis* species complex; Chapter 10). Chapter 8 explores how the genomic study of herbarium specimens can provide invasion scientists with a window into the past. Due to improvements in DNA extraction and sequencing techniques, herbarium genomics now allows researchers to extract genetic information from degraded DNA in historical specimens. This capability helps trace introductions, identify evolutionary changes, and track geographical and ecological shifts over time. Herbarium collections also facilitate the study of historical plant–microbe interactions, broadening our understanding of how invasion dynamics interact with recipient communities. In the near future, we anticipate that invasion genomics studies will increasingly make use of historical DNA retrieved from herbarium or museum specimens. Chapter 9 highlights the breadth of ecological and genetic research that is frequently needed to understand invasion success. This is critically important because genomic data can only reach their full potential in advancing the study of invasive species when detailed information is available on evolutionary biology, invasion history, trait heritability, and the links between species traits and invasion success. Finally, Chapter 10 illustrates how population genomics, comparative genomics and careful experimentation under controlled settings can help research progress in invasion genomics. In the *E. affinis* species complex, this approach has helped reveal shared and lineage-specific genetic mechanisms that underpin repeated saline-to-freshwater invasions.

Invasion genomics has partially emerged in response to the practical challenges of managing NIS. For example, population genomics of invasive species plays a crucial role in mapping their genetic connectivity, identifying adaptive variation and informing targeted management. When combined with ecological niche modelling (ENM), genomic data can more accurately predict the potential introduced range of NIS (see Chapter 11). This is an important, urgently needed development, given that genomics and ENM have advanced independently, and their integration is still in its infancy. Genomics-informed ENM can trace the origin of invaders, predict future range shifts and improve NIS surveillance. Chapter 12 highlights how genomic technologies enhance the management of NIS, from fine-scale genome sequencing to broad-scale biome-level monitoring. Genome sequencing, for example, improves traditional control strategies – including chemical, biological and genetic methods – by increasing their precision and effectiveness. In this context, revolutionary genome-enabled control strategies such as genome editing are poised to make significant contributions to invasion management in the following decades. However, their deployment will depend on appropriate risk management and meaningful consultation among researchers, policy makers and the broader public. At the

other end of the spectrum, whole-community biome sequencing, particularly through environmental DNA metabarcoding, is revolutionizing NIS detection, which is enabling faster and more accurate preventative and mitigation measures (Chapter 12). In addition, this approach provides key information on community changes as a result of NIS establishment and spread, significantly contributing to research progress on the ecological and evolutionary consequences of species invasions. Despite these advances, practical challenges remain. Financial constraints, limited taxonomic expertise and infrastructure gaps continue to hinder the widespread use of genomics in policy and management of invasive species. Overcoming these barriers is essential to fully harness the potential of genomics in managing biological invasions.

As illustrated throughout this book, genomic tools offer powerful ways to decipher complex biological processes across temporal and spatial scales. With ongoing innovation and collaboration, invasion genomics is poised to unravel the past and present of species invasions, while shaping proactive, informed responses to their future. Studies using whole-genome resequencing and/or metagenomics will increasingly provide new insights into the roles of range-core and invasion-front populations. This will shed light on how rapidly spatial genetic structuring arises during biological invasions. Furthermore, these approaches will identify new genomic regions under positive selection that may be associated with traits critical to invasion success (e.g. dispersal ability, metabolism, immune response and stress tolerance), significantly advancing our understanding of how selection shapes NIS during range expansions. More research is, however, needed to explain how selection influences population persistence at invasion fronts, where NIS encounter new environments, and where populations are additionally exposed to increased genetic drift. These topics align with key themes in invasion genomics, such as how rapid evolution can drive range expansion, or how contemporary and repeated adaptation shapes biological invasions, especially under sharp environmental gradients.

As global invasions accelerate under anthropogenic pressures, the integration of genomics into both basic science and applied management will become increasingly impactful. Despite remarkable research advances fuelled by the recent use of high-throughput sequencing, our understanding of the genomic mechanisms that govern invasion success remains incomplete. Future work must address how standardizing data generation and analysis can lead to meaningful synthesis across studies. This is particularly important for bioinformatic pipelines used for biome-level monitoring (Chapter 12). Such standardization will improve data sharing and the refinement of analytical methods to fully realize the potential of genomic data. In addition, future research will broaden our understanding of genome architecture – such as chromosome number, inversions and gene linkage – that are known to influence evolutionary outcomes. These genomic features may influence whether adaptation is more likely to arise from standing genetic variation, from *de novo* mutations or by interactions among genes (epistasis). Understanding these mechanisms will be key for predicting future invasions and the adaptive potential of NIS.

Today, genomics is an essential component of invasion science, helping to shed light on invasion routes and ecological mechanisms shaping the success of invasive species. In addition, genomic studies are advancing our understanding of how contemporary evolution can drive invasion success. Studies increasingly show how rapid evolution can also be an important determinant of whether invasive populations establish and spread, or fail and become extinct. This stands in sharp contrast to the beginnings of the field, when studies focused almost exclusively on species interactions and ecological barriers to invasions. Thanks to decades of genomic and evolutionary biology research, biological invasions are no longer viewed exclusively as ecological events that disrupt biological communities but also as unique evolutionary experiments unfolding in real time. Although challenges remain, including limited data on failed invasions, these experiments are uniquely positioned to provide a long-term and global-scale perspective on how populations respond to rapid environmental change and overcome the demographic and genetic constraints of small population sizes. Enabled by improvements in DNA extraction and sequencing techniques, researchers can now analyse degraded DNA to trace introductions, identify evolutionary changes, and assess geographical and ecological shifts over time. Such research also offers a window into historical species interactions (including NIS), broadening our understanding of the role of biotic resistance or community dynamics. Future research will draw on other sources such as field samples (e.g. ancient sedimentary DNA) to unravel new research questions on invasion science related to the temporal dimension of biological invasions (e.g. how different invasion stages, failed invasions and lag phases shape invasion success). All in all, invasion genomics is enhancing our understanding of the natural world by building on the legacy of early naturalists, while enhancing our capacity to manage these complex and rapidly accelerating global-change events.

Index

Note: Page numbers in italics refers to figures those in bold refers to tables.